Introduction to control theory

O. L. R. JACOBS

Introduction to control theory

CLARENDON PRESS · OXFORD

1974

Oxford University Press, Ely House, London W.1

GLASGOW NEW YORK TORONTO MELBOURNE WELLINGTON
CAPE TOWN IBADAN NAIROBI DAR ES SALAAM LUSAKA ADDIS ABABA
DELHI BOMBAY CALCUTTA MADRAS KARACHI LAHORE DACCA
KUALA LUMPUR SINGAPORE HONG KONG TOKYO

ISBN 0 19 856108 3

© OXFORD UNIVERSITY PRESS 1974

PRINTED IN GREAT BRITAIN BY
WILLIAM CLOWES & SONS, LIMITED
LONDON, BECCLES AND COLCHESTER

Preface

CONTROL theory is a branch of applied mathematics devoted to analysis and design of control systems. Control systems are systems in which a controller interacts with a real process in order to influence its behaviour. A primary objective for most control systems is to make some real variable take a desired value, for example to regulate the temperature of an oven or to make the direction of a receiving aerial track a moving target. The objective is usually to be achieved by adjusting some other variable, such as heat input to the oven or force applied to the aerial, although the response to such adjustments in most real controlled processes is neither instantaneous nor certain. The non-instantaneous response is accounted for by regarding the controlling and controlled variables as input and output of a *dynamic* system described by differential or difference equations. The effect of uncertainties is reduced by using *feedback* to provide the controller with continuous indication of what adjustment is needed; for example, if the oven is too cold more heat must be supplied and if the aerial points to the left of its target it must be forced to turn to the right. Feedback is a characteristic feature of control systems: in addition to reducing the effects of uncertainties it modifies the dynamic behaviour of controlled processes and can cause instability. The main preoccupations of control theory are with analysis of system dynamics and with applications of probability theory to describe the behaviour of dynamic systems in the presence of uncertainty. Control theory is similar to other branches of applied mathematics in that the majority of solved theoretical problems are linear and the majority of real control systems non-linear. Techniques for applying the theoretical results to specific practical problems are beyond the scope of the theory.

Control theory originally developed as a branch of engineering science, but subsequently found applications elsewhere; for example in economics, in studies of social systems, and in biology.† Its development has passed through three stages.

† Most of the books about control in the Bibliography have an introductory chapter on engineering applications of control theory. Applications to economic systems are described in Tustin (1953), Allen (1968), IFAC/IFORS (1973); to social systems in Forrester (1961, 1968, 1971) and to biological systems in Bayliss (1966), Milsum (1966), Milhorn (1966), Stark (1968), McFarland (1971).

(i) A classical stage originating with a study by Maxwell (1868) of speed-control systems and culminating during the second world war in designs of gunnery, radar, and other military control systems. In this stage classical analysis of ordinary linear differential equations was interpreted for and applied to control systems.

(ii) A more modern stage during the 1950s and 1960s when the attention of applied mathematicians was directed to aerospace and to complex industrial problems. In this stage multi-variable optimization methods were developed and applied under the influence of the contemporary development of digital computers.

(iii) The most recent stage which emphasizes the importance of uncertainty, almost completely neglected at earlier stages. In this stage control systems are regarded as stochastic systems and probability theory is applied.

The history of control theory is reflected in the existing range of books on the subject. There are many books about the classical theory of stage (i), often written by engineers for engineers and with emphasis on engineering applications. There are books about the more modern theory of stage (ii), often written with emphasis on mathematical rigour and on computational feasibility. There are a few books about the recent stochastic control theory of stage (iii), mainly restricted to linear systems. Books covering all three stages are still rare.

The purpose of this book is to provide an introduction to the main results from all three stages of control theory and to unify them in a single volume at a mathematical level suitable for final year undergraduates in engineering and for graduates.† The material is presented without much reference to specific applications and the book is not intended only for engineers. It is suitable for anyone whose mathematical background includes the classical solution of ordinary linear differential or difference equations and the elements of probability theory, which are reviewed in Chapters 1 and 12 respectively, complex numbers and matrix algebra, which are used without any special explanation, and a certain confidence in manipulating mathematical expressions.

The book is divided into three parts on a mathematical, rather than historical, basis.

Part I Deterministic linear systems described by equations having analytic solutions.

Part II Deterministic non-linear systems described by equations without analytic solutions.

---

† The book is in some of these respects similar to certain other recent books: Takahashi, Rabins, and Auslander (1970) covers a similar wide range of topics from all three stages but with less emphasis on theory; Anderson and Moore (1971) is written at a similar mathematical level and discusses modern control theory from stage (ii); Kwakernaak and Sivan (1972) unifies the theory of linear systems from all three stages.

Part III Systems with uncertainty, described with the help of probability theory.

This framework leads to several unconventional features in presentation.

(i) The uniform treatment of both continuous-time and discrete-time systems in the chapters of Part I about classical control theory. For example the z-transform precedes the Laplace transform in the presentation of Section 1.4 because the z-transform is mathematically simpler.

(ii) Chapter 9 on optimal control theory precedes the material on phase-plane analysis in Chapter 10 in order that it shall immediately follow the presentation of optimal linear control theory in Chapter 8 at the end of Part I.

(iii) Chapter 12 is an introduction to probability theory rather than to control theory. It is included as necessary background for Part III in the same way that Chapter I provides necessary background on ordinary dynamic equations for Part I.

Other features of the book are as follows.

(iv) Dynamic programming and Bayes's rule are emphasized as foundations for optimal and stochastic control theory.

(v) Chapter 13 on stochastic control theory precedes and motivates the introduction to estimation theory of Chapter 14. With this order of presentation the Kalman filter equations are used in Chapter 13 in advance of their derivation in Chapter 14.

(vi) The discussion in Chapter 15 of the structure of stochastic controllers has not previously appeared outside the research literature.

The book derives largely from lectures given to undergraduates in engineering science at the University of Edinburgh, the University of Oxford, and the University of California, San Diego, and should be suitable for similar course work elsewhere. Problems and their answers are provided for every chapter except Chapter 15 which presents research material. Selected chapters might be used for various courses, for example:

(i) *Classical control theory*
(Chapter 1 contains necessary background material.)
Chapters 2, 3, 4, 5.
(Chapter 10 and Section 11.3 optional.)
(ii) *Optimal control theory*
(Chapter 6 provides a link with classical control theory.)
Chapters 7, 8, 9; or Chapters 7, 8; or Chapters 8, 9; or Chapter 8.
(iii) *Non-linear control theory*
(Classical control theory may be a necessary prerequisite.)
Chapters 10, 11.

(iv) *Stochastic control theory*
(Chapter 12 contains necessary background material, Chapter 8 is a necessary prerequisite.)
Chapters 13, 14.
(Chapter 15 optional.)

Each chapter ends with a bibliography indicating further or alternative reading; the references are mainly to other books rather than to primary sources and are inevitably incomplete.

The book also is inevitably incomplete. The theory of observers mentioned at the end of Section 7.4, Popov's method mentioned in Section 11.2, and recent developments in the theory of linear multi-variable systems† might well have been included. Power spectral density functions were not used in Section 12.6, in spite of their intuitive appeal, because they are not needed for the derivation of any results. Practical applications, including computer realizations of control algorithms, were deliberately excluded in the belief that insight into the main results of control theory can be achieved by concentrating on the theory.

It is a pleasure to acknowledge the debts I have incurred in writing this book. The debt to other authors is evident from the Bibliography. The students who have attended my various courses and questioned what I told them may not know how much I learned from them. David Clarke, David Hughes, and David Witt read drafts of the book and made many helpful suggestions. Most of all I am grateful to my wife Sheila who encouraged me to persevere and who, together with our children, had to live with me while I did so.

† See MacFarlane (1973).

Contents

PART III
SYSTEMS WITH UNCERTAINTY

xii   *Contents*

Part I
Deterministic linear systems

1. Ordinary linear systems

1.1. Dynamic equations

MOST control systems are dynamic systems characterized by variables that are functions of time. Such variables may be continuous-time variables, like the temperature at some particular place, or discrete-time variables, like a series of mid-day observations of temperature at the same place; Fig. 1.1 shows the difference between the two sorts of variable.

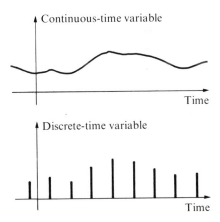

FIG. 1.1. Two sorts of function of time.

Dynamic systems are described for purposes of mathematical analysis by dynamic equations; continuous-time systems by differential equations and discrete-time systems by difference equations. A linear system is one that can be described by a linear dynamic equation. If a linear system is single-variabled, like that shown in Fig. 1.2, its dynamic equation would be either

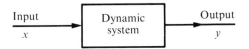

FIG. 1.2. Single-variabled dynamic system.

the continuous-time linear differential equation

$$a_n \frac{d^n y}{dt^n} + a_{n-1} \frac{d^{n-1} y}{dt^{n-1}} + \cdots + a_1 \frac{dy}{dt} + a_0 y = b_m \frac{d^m x}{dt^m} + \cdots + b_0 x, \qquad (1.1)$$

where t represents time; or the discrete-time linear difference equation

$$a_n y(i + n) + \cdots + a_1 y(i + 1) + a_0 y(i) = b_m x(i + m) + \cdots + b_0 x(i), \qquad (1.2)$$

where i is an integer counting the discrete-time instants. When these equations describe real physical systems the coefficients $a_n, \ldots, a_0, b_m, \ldots, b_0$ are all real and n, the order of the equation, is greater than or equal to m, the order of the forcing function. This condition $n \geqslant m$ reflects the fact that the system cannot respond to an input before the input has been applied. In addition the coefficients $a_n, \ldots, a_0, b_m, \ldots, b_0$ are often constant although the equations remain linear even when the coefficients vary with time.

Linear equations satisfy the principle of superposition, which states that

if input x_1 causes output y_1 and input x_2 causes output y_2,

then input $x_1 + x_2$ causes output $y_1 + y_2$.

Figure 1.3 illustrates the principle for a continuous-time equation.

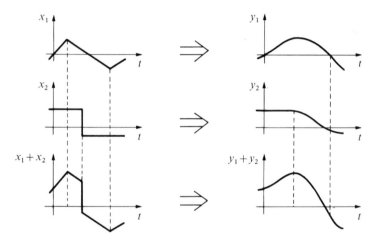

FIG. 1.3. Superposition.

It follows from the principle of superposition that the complete solution to a linear dynamic equation is the sum of two parts,

complete solution = particular integral + complementary function.

The particular integral, or 'driven response', is due to the input x, and the complementary function, or 'natural response', follows from the initial condition of the equation. In order to find the complete solution it is necessary to know the form of the input function x and to know n initial conditions.

The complementary function is found by setting the right-hand side of the equation to zero and seeking solutions to the resulting homogeneous equation in the general form

$$y_{CF} = \sum_{k=1}^{n} A_k \, e^{s_k t}$$

for continuous-time equations, or

$$y_{CF} = \sum_{k=1}^{n} A_k s_k{}^{i}$$

for discrete-time equations; in both cases the A_k are constants determined by the initial conditions and the s_k, which may be complex numbers, are the roots of the characteristic equation

$$a_n s^n + a_{n-1} s^{n-1} + \cdots + a_1 s + a_0 = 0. \tag{1.3}$$

These roots of the characteristic equation determine the dynamic behaviour of the equation.

The most important aspect of the dynamic behaviour is the stability of the equation. A linear dynamic system is said to be 'stable' if its response to any input tends to a finite steady value after the input is removed; this implies that the complementary function must remain finite as time goes to infinity. It follows that the condition for stability of a continuous-time system is that the roots of the characteristic equation must all have negative real parts and that the condition for stability of a discrete-time system is that the roots of the characteristic equation must all have magnitude less than unity. Figure 1.4

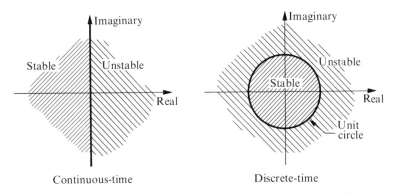

Continuous-time   Discrete-time

FIG. 1.4. Stability conditions on roots of the characteristic equation.

shows how these conditions can be interpreted on an Argand diagram, sometimes described as the 's-plane', as regions where the roots must, or must not, be for stability, or instability.

Stability is discussed more fully in Chapter 3. Chapter 1 is devoted to standard descriptions of the responses of ordinary linear dynamic equations.

Problem 1.1

1.2. Continuous-time responses

The typical, first-order differential equation

$$T\frac{dy}{dt} + y = x \tag{1.4}$$

has characteristic equation

$$Ts + 1 = 0$$

and complementary function

$$y_{CF} = A\,e^{-t/T}.$$

If the input x has the constant value $x = X$ the particular integral is

$$y_{PI} = X$$

and if the initial condition is $y(t = 0) = 0$, the general solution is

$$y = X(1 - e^{-t/T}). \tag{1.5}$$

This solution is sketched in Fig. 1.5 which shows how the constant T determines the time scale of the response. T is called the 'time constant' of the first-order equation. The time constant of an unknown first-order system can

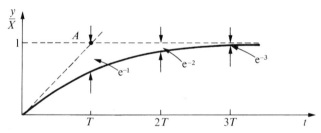

FIG. 1.5. Step response of first-order system.

easily be found from an experimental step response by noticing that the tangent to the initial response intersects the final value (point A in Fig. 1.5) after time T.

The typical, second-order differential equation is written

$$\frac{d^2y}{dt^2} + 2\zeta\omega_0\frac{dy}{dt} + \omega_0^2 y = \omega_0^2 x. \tag{1.6}$$

It has complementary function

$$y_{CF} = A_1\,e^{-\omega_0(\zeta + \sqrt{(\zeta^2-1)})t} + A_2\,e^{-\omega_0(\zeta - \sqrt{(\zeta^2-1)})t},$$

and when the input has the constant value $x = X$ and the initial conditions are $y(t = 0) = \mathrm{d}y/\mathrm{d}t\,(t = 0) = 0$ the general solution is as sketched in Fig. 1.6, which shows how the constant ω_0 determines the time scale and the

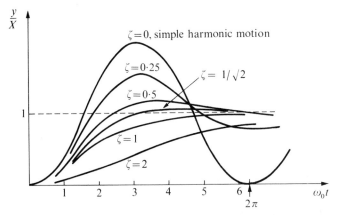

FIG. 1.6. Step responses of second-order systems.

constant ζ determines the shape of the response. ω_0 is called the 'natural frequency' and is expressed in radians per unit time; ζ is called the 'damping ratio' of the equation and is dimensionless. When ζ is less than unity the response is oscillatory and the equation is said to be 'under-damped', when ζ is greater than unity there is no overshoot and the equation is said to be 'over-damped', and when ζ is equal to unity the equation is said to be 'critically damped'.

The second-order equation can describe a position-control system where the angular position y of a rotating load of inertia J is controlled by a motor that applies torque u to the load. Figure 1.7 shows such a system in which the

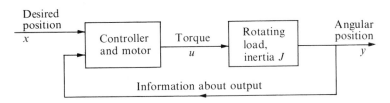

FIG. 1.7. Position-control system.

torque u is controlled by comparing information about the actual output y with information about the desired output x. The differential equation of the rotating load is given by Newton's equation of motion

$$J \frac{\mathrm{d}^2 y}{\mathrm{d}t^2} = u$$

and the controller must be designed so that the system has acceptable performance. One possibility would be to make the torque proportional to the position error $x - y$:

$$u = K(x - y),$$

but this gives the system the differential equation

$$J\frac{d^2y}{dt^2} + Ky = Kx,$$

which has no damping and shows that the response to any disturbance would be an undamped simple harmonic motion, which is not acceptable. The usual way to stabilize this sort of system is to measure the velocity of the output and to make the controller generate an additional decelerating torque proportional to velocity so that

$$u = K(x - y) - k\frac{dy}{dt}.$$

The differential equation of the system then becomes

$$J\frac{d^2y}{dt^2} + k\frac{dy}{dt} + Ky = Kx$$

with natural frequency $\omega_0 = \sqrt{(K/J)}$ and damping ratio $\zeta = \tfrac{1}{2}k/\sqrt{(KJ)}$ which can be chosen to give acceptable responses.

All roots of the general characteristic equation (1.3) are either real or complex conjugate. The real roots give rise to exponential complementary functions like that of the first-order equation (1.4) and the complex conjugate roots give rise to oscillatory complementary functions like that of an under-damped second-order equation (1.6). The transient response of any high-order linear system is therefore a sum of typical first- and second-order responses with time constants, natural frequencies, and damping ratios determined by the characteristic equation, and with magnitudes determined by initial conditions.

When the input to a linear differential equation is a sine wave the particular integral, sometimes called the 'steady-state' output, is a sine wave of the same frequency as the input but with different amplitude and phase. This steady-state output can be found by writing the sinusoidal input in the general form

$$x = X\,e^{j\omega t}$$

and the output

$$y_{PI} = MX\,e^{j(\omega t + \phi)},$$

where M is the amplitude ratio and ϕ the phase difference between input and output. Substituting these into the general equation (1.1) and performing the differentiations

$$\frac{d}{dt}(e^{j\omega t}) = j\omega\,e^{j\omega t}$$

gives

$$(a_n(j\omega)^n + \cdots + a_1 j\omega + a_0)MX\, e^{j(\omega t + \phi)} = (b_m(j\omega)^m + \cdots + b_0)X\, e^{j\omega t}$$

so that

$$Me^{j\phi} = \frac{b_m(j\omega)^m + \cdots + b_0}{a_n(j\omega)^n + \cdots + a_0} = G(j\omega). \qquad (1.7)$$

The symbol $G(j\omega)$ is used here to represent the ratio of the polynomials

$$(b_m(j\omega)^m + \cdots + b_0)/(a_n(j\omega)^n + \cdots + a_0),$$

which is a complex number with frequency-dependent modulus M and argument ϕ. The function $G(j\omega)$ is called the frequency-response function of the equation because it specifies the steady-state sinusoidal output at all frequencies.

The frequency response of the typical first-order equation (1.4) is

$$G(j\omega) = \frac{1}{1 + j\omega T}$$

and this has

$$M = (1 + \omega^2 T^2)^{-1/2},$$

$$\phi = -\tan^{-1}\omega T.$$

Figure 1.8 shows how M and ϕ vary with frequency ω and Fig. 1.9 shows the same information on a logarithmic scale. It is often convenient to represent

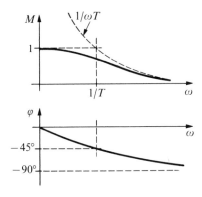

Fig. 1.8. Frequency response of first-order equation.

frequency responses thus, with graphs of $20\log_{10} M$ and of ϕ against $\log \omega$; this representation is called a Bode diagram. Another convenient representation shows the locus traced out on an Argand, or polar, diagram by the com-

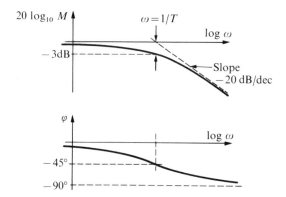

FIG. 1.9. Bode diagram of first-order frequency response.

plex number $G(j\omega)$ as ω varies from zero to infinity; Fig. 1.10 shows the semicircular locus of the first-order frequency-response function.

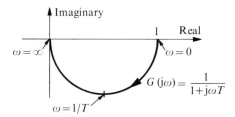

FIG. 1.10. Polar diagram of first-order frequency response.

Frequency response is a particularly suitable form for experimental determination of the dynamics of unknown linear systems.† Figure 1.11 shows an arrangement for measuring the frequency response of a continuous-time

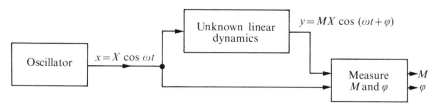

FIG. 1.11. Measuring frequency response.

linear system with unknown dynamics. It is easy to measure the amplitude ratio M and phase difference ϕ at different frequencies ω; these results can be presented graphically as in Fig. 1.8, 1.9, or 1.10.

Problems 1.2–1.8

† See Bruns and Saunders (1955) Chapter 14.

1.3. Discrete-time responses

The typical first-order difference equation

$$a_1 y(i + 1) + a_0 y(i) = x(i) \qquad (1.8)$$

has characteristic equation

$$a_1 s + a_0 = 0$$

and complementary function

$$y_{\text{CF}} = A(-a_0/a_1)^i.$$

If the input x has the constant value $x = X$ the particular integral is

$$y_{\text{PI}} = X/(a_1 + a_0)$$

and if the initial condition is $y(i = 0) = 0$, the general solution is

$$y = \frac{X}{a_1 + a_0}\left(1 - (-a_0/a_1)^i\right). \qquad (1.9)$$

This solution is sketched in Fig. 1.12 which shows how the ratio a_0/a_1 determines the shape of the response; when the ratio is positive the response is oscillatory, when the ratio is negative there is no overshoot, and when the

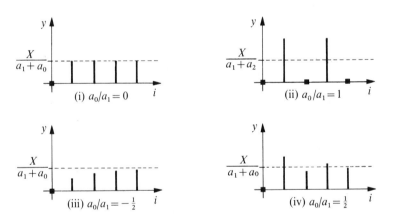

FIG. 1.12. Step responses of first-order equation.

ratio is zero the output follows the input exactly but with a delay of one time interval. The ratio a_0/a_1 of the coefficients of the first-order difference equation (1.8) describes the shape of the transient responses in the same way as does the damping ratio ζ of the second-order differential equation (1.6). The time scale of the response described by equation (1.9) is determined by the timing of the instants where the discrete-time variables are defined; this timing does not appear explicity in the difference equation (1.8) but it affects the horizontal scales illustrated in Fig. 1.12.

The first-order difference equation thus has a family of step responses like those of the second-order differential equation. A unique discrete-time response, like that of a first-order differential equation, would be produced by the difference equation of a time delay

$$y(i + k) = x(i).\tag{1.10}$$

One example of the step response of this equation is shown in Fig. 1.12(i) for the case where k, the number of instants by which the output lags behind the input, is unity. Step responses for other values of k would be similar in shape but would have different time scales; so the constant k affects these step responses in the same way that the time constant T affects the continuous-time responses illustrated by Fig. 1.5.

Problems 1.9–1.13

1.4. Transformation of dynamic equations to algebraic equations

Linear difference equations can be converted to algebraic equations, which are much easier to manipulate, by using a special notation known as the z-transformation. The z-transform of a discrete-time variable $x(i)$ is defined by constructing a polynomial in a general complex variable z having coefficients equal to values of the discrete-time variable. The polynomial is expressed in powers of z^{-1}:

$$z\text{-transform } \{x(i)\} \equiv x(0) + x(1)z^{-1} + x(2)z^{-2} + \cdots$$

$$= \sum_{i=0}^{\infty} x(i)z^{-i};$$

it is written $X(z)$ and defined only for positive values of time i. Some typical z-transforms are shown in the table of Fig. 1.13.

$x(i)\ (i \geqslant 0)$	1	i	i^2	a^i	$f(i - j)$
$X(z) \equiv \displaystyle\sum_{i=0}^{\infty} x(i)z^{-i}$	$\dfrac{z}{z-1}$	$\dfrac{z}{(z-1)^2}$	$\dfrac{z(z+1)}{(z-1)^3}$	$\dfrac{z}{z-a}$	$z^{-j}F(z)$

FIG. 1.13. Some z-transforms.

It follows from the definition that

$$z\text{-transform } \{x(i + 1)\} = -zx(i = 0) + zX(z)$$

so that transforming the general equation (1.2) term by term gives

$$(a_n z^n + \cdots + a_0)\, Y(z)$$
$$- \big((a_n z^n + \cdots + a_1 z)y(i = 0) + \cdots + a_n z y(i = n - 1)\big)$$
$$= (b_m z^m + \cdots + b_0)X(z)$$
$$- \big((b_m z^m + \cdots + b_1 z)x(i = 0) + \cdots + b_m z x(i = m - 1)\big),$$

and when all the initial values of x and y prior to $x(i = m)$ and $y(i = n)$ are zero this becomes

$$(a_n z^n + \cdots + a_0)\, Y(z) = (b_m z^m + \cdots + b_0)\, X(z).$$

The ratio

$$\frac{Y(z)}{X(z)} = \frac{b_m z^m + \cdots + b_0}{a_n z^n + \cdots + a_0} = G(z) \tag{1.11}$$

is called the z-transfer function of the equation.

The response of a linear equation can thus be specified by an algebraic equation in the complex variable z; for zero initial conditions the z-transform of the output is

$$Y(z) = G(z) X(z). \tag{1.12}$$

For example the z-transfer function of the typical first-order equation (1.8) is

$$G(z) = \frac{1}{a_0 + a_1 z}$$

and the z-transform of a constant input $x = X$ is

$$X(z) = \frac{Xz}{z - 1},$$

so the z-transform of the step response of the first-order system in Section 1.3 is

$$Y(z) = \frac{Xz}{(a_0 + a_1 z)(z - 1)}.$$

Conversion from z-transforms back to functions of the integer variable i can be done by expressing the z-transform, say $Y(z)$, as a polynomial in z^{-1}:

$$Y(z) = y_0 + y_1 z^{-1} + y_2 z^{-2} + \cdots.$$

Then it follows from the definition of the z-transformation,

$$z\text{-transform } \{y(i)\} \equiv y(0) + y(1) z^{-1} + y(2) z^{-2} + \cdots,$$

that the coefficients y_i in the expansion of $Y(z)$ are the values $y(i)$ of the discrete-time variable. The z-transform of the step response in the above example can thus be expressed

$$Y(z) = \frac{X}{a_1} \left(z^{-1} + \left\{ 1 + \left(-\frac{a_0}{a_1} \right) \right\} z^{-2} \right.$$

$$\left. + \left\{ 1 + \left(-\frac{a_0}{a_1} \right) + \left(-\frac{a_0}{a_1} \right)^2 \right\} z^{-3} + \cdots \right)$$

and it follows that values of $y(i)$ are

$$y(0) = 0$$

$$y(1) = \frac{X}{a_1}$$

$$y(2) = X\left(\frac{a_1 - a_0}{a_1}\right)$$

and so on, as illustrated in Fig. 1.12.

It is often more convenient to convert back from z-transforms through a table like that in Fig. 1.13. When the z-transform is a ratio of polynomials like the output of the general linear equation,

$$Y(z) = G(z)X(z) = \frac{b_m z^m + \cdots + b_0 z}{a_n z^n + \cdots + a_0 z} X(z),$$

it must first be expanded in partial fractions.† The expansion is based on the factors of the denominator polynomial $a_n z^n + \cdots + a_0 z$ and these are identical with the roots of the characteristic equation (1.3), which must therefore be solved before $y(i)$ can be found this way. In the above example, the partial fraction expansion is

$$Y(z) = \frac{X}{a_1 + a_0}\left(\frac{1}{z - 1} + \frac{a_0}{a_0 + a_1 z}\right)$$

and comparison with the z-transforms in the table gives

$$y(i) = \frac{X}{a_1 + a_0}\left(1 - (-a_0/a_1)^i\right),$$

which is the same equation (1.9) as before.

Linear differential equations can be similarly converted to algebraic equations by the Laplace transformation. This is the continuous-time equivalent of the z-transformation and is defined by

$$X(s) = \text{Laplace transform } \{x(t)\} \equiv \int_0^\infty x(t)\, e^{-st}\, dt,$$

where the notation $X(s)$ indicates that the Laplace transform is a function of the general complex variable s. Some typical Laplace transforms are shown in the table of Fig. 1.14.

† See Truxal (1955) Section 1.2 or Kreysig (1972) Section 4.4.

$x(t) \, (t \geqslant 0)$	1	t	t^n	e^{-at}	$\sin at$	$\cos at$	$f(t - T)$
$X(s) \equiv \displaystyle\int_0^\infty x(t)\, e^{-st}\, dt$	s^{-1}	s^{-2}	$\dfrac{n!}{s^{n+1}}$	$\dfrac{1}{s+a}$	$\dfrac{a}{s^2 + a^2}$	$\dfrac{s}{s^2 + a^2}$	$e^{-sT} F(s)$

FIG. 1.14. Some Laplace transforms.

It follows from the definition that

$$\text{Laplace transform} \left\{ \frac{dx}{dt} \right\} = -x(t = 0) + sX(s)$$

and that when all initial conditions are zero the transform of the general equation (1.1) is

$$(a_n s^n + \cdots + a_0)\, Y(s) = (b_m s^m + \cdots + b_0)X(s).$$

The ratio

$$\frac{Y(s)}{X(s)} = \frac{b_m s^m + \cdots + b_0}{a_n s^n + \cdots + a_0} = G(s) \tag{1.13}$$

is called the transfer function of the equation. This function $G(s)$ is the same function as the frequency response $G(j\omega)$ of equation (1.7) but with the general complex argument s in place of the purely imaginary number $j\omega$ associated with frequency response. The algebraic equation specifying response for zero initial conditions is

$$Y(s) = G(s)X(s). \tag{1.14}$$

Conversion from Laplace transforms back to functions of the real variable t can be done by contour integration in the complex plane of s, but it is usually simpler to go back through a table of Laplace transforms using partial fraction expansions, as described for z-transforms.

The transfer function of the typical first-order equation (1.4) is

$$G(s) = \frac{1}{1 + sT}$$

and the Laplace transform of a steady input $x = X$ is

$$X(s) = \frac{X}{s},$$

so the Laplace transform of the step response of the first-order system in Section 1.2 is

$$Y(s) = \frac{X}{s(1 + sT)}$$

and the partial fraction expansion is

$$Y(s) = X\left(\frac{1}{s} - \frac{T}{1 + sT}\right).$$

Comparison with the Laplace transforms in the table gives

$$y(t) = X(1 - e^{-t/T}),$$

which is the same equation (1.5) as before.

Any discrete-time variable can be regarded as a sampled version of a continuous-time variable. The most usual case is that of a continuous-time variable sampled periodically at time intervals T to generate the discrete-time variable $x(iT)$. The table in Fig. 1.15 gives z-transforms of such sampled variables together with the Laplace transforms of the original continuous-time variables.

Continuous-time $x(t)$ ($t \geqslant 0$)		Laplace transform $X(s) \equiv \int_0^\infty e^{-st} x(t)\, dt$	z-transform $X(z) \equiv \sum_{i=0}^\infty z^{-i} x(iT)$
Impulse	$\delta(t)$	1	1
Step	1	s^{-1}	$\dfrac{z}{z - 1}$
Ramp	t	s^{-2}	$\dfrac{Tz}{(z - 1)^2}$
Acc'n	t^2	$2s^{-3}$	$T^2\dfrac{z(z + 1)}{(z - 1)^3}$
Exp	e^{-at}	$\dfrac{1}{s + a}$	$\dfrac{z}{z - e^{-aT}}$
Sine	$\sin at$	$\dfrac{a}{s^2 + a^2}$	$\dfrac{z \sin aT}{z^2 - 2z \cos aT + 1}$
Cos	$\cos at$	$\dfrac{s}{s^2 + a^2}$	$\dfrac{z^2 - z \cos aT}{z^2 - 2z \cos aT + 1}$
Sum	$f_1(t) + f_2(t)$	$F_1(s) + F_2(s)$	$F_1(z) + F_2(z)$
Const factor	$af_1(t)$	$aF(s)$	$aF(z)$
Exp factor	$e^{-at}f(t)$	$F(s + a)$	$F(e^{aT}z)$
Time delay	$f(t - jT)$	$e^{-jTs}F(s)$	$z^{-j}F(z)$
Pulse train	$\sum_{j=0}^\infty \delta(t - jT)$	$\dfrac{1}{1 - e^{-sT}}$	$\dfrac{z}{z - 1}$

FIG. 1.15. Laplace transforms and z-transforms of periodically sampled variables.

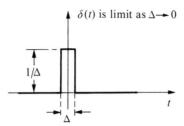

FIG. 1.16. Unit impulse function.

The unit impulse function $\delta(t)$ included in the table is the limiting case of the pulse shown in Fig. 1.16 and is defined by

$$\delta(t) = 0 \quad \text{for } t \neq 0,$$

$$\int_{-\infty}^{\infty} \delta(t)\, dt = 1.$$

This function provides a mathematical way of defining a continuous-time variable $x^*(t)$

$$x^*(t) \equiv x(t) \sum_{i=-\infty}^{\infty} \delta(t - iT)$$

which is equivalent to the discrete-time variable $x(iT)$ as shown in Fig. 1.17.

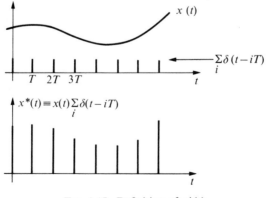

FIG. 1.17. Definition of $x^*(t)$.

The Laplace transform of $x^*(t)$ is

$$X^*(s) \equiv \int_0^\infty \left(x(t) \sum_i \delta(t - iT) \right) e^{-st} \, dt$$

$$= x(0) + x(T) e^{-sT} + x(2T) e^{-2sT} + \cdots$$

$$= \sum_{i=0}^\infty x(iT)(e^{sT})^{-i},$$

which is the same as the z-transform of $x(iT)$ if the complex variables z and s are related by

$$z = e^{sT} \tag{1.15}$$

Problems 1.14–1.23

1.5. Impulse responses

The unit impulse function of Fig. 1.16 can provide a basis for describing continous-time systems even though it is never exactly realized in practice. Its Laplace transform is unity and so if it were the input x to a linear system having transfer function $G(s)$ the output y would have Laplace transform

$$Y(s) = G(s) \times 1$$

and could be written

$$y(t) = g(t),$$

where $g(t)$, the inverse Laplace transform of the transfer function $G(s)$, is known as the *impulse response* of the system. Figure 1.18 shows typical impulse responses of first-order and second-order systems. The idea of a function $g(t)$ which describes the response to a unit impulse need not be derived, as here, from the Laplace transform; it could be regarded as the fundamental starting point for describing system dynamics. As such it would have the

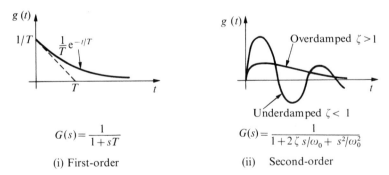

(i) First-order

$$G(s) = \frac{1}{1+sT}$$

(ii) Second-order

$$G(s) = \frac{1}{1 + 2\zeta s/\omega_0 + s^2/\omega_0^2}$$

FIG. 1.18. Typical impulse responses.

advantage of corresponding to the simple, if idealized, experiment of applying an impulse and observing the resulting response.

Any input $x(t)$ can be described as a sequence of closely-spaced impulses of width Δ, as illustrated in Fig. 1.19, where the area of the pulse at time t is

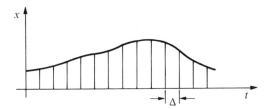

FIG. 1.19. A variable $x(t)$ regarded as a sequence of pulses.

approximately $x(t)\Delta$. In the limit as Δ goes to zero the impulses tend, as in Fig. 1.16, to idealized δ-function impulses of strength $x(t)\Delta$. The output $y(t)$ caused by an input impulse $x(u)\Delta$ at some time u previous to t is approximated by the impulse response $g(t - u)x(u)\Delta$ as shown in Fig. 1.20. The output $y(t)$

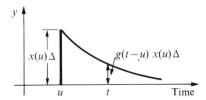

FIG. 1.20. Response at time t due to impulse $x(u)\Delta$ at time u.

caused by all input impulses previous to t is approximated, using the principle of superposition, by a summation in which each impulse $x(u)\Delta$ is weighted by the appropriate value of impulse response

$$y(t) \simeq \sum_{u \leqslant t} g(t - u)x(u)\Delta.$$

In the limit as Δ goes to zero this becomes,

$$y(t) = \int_{-\infty}^{t} g(t - u)x(u)\,\mathrm{d}u \tag{1.16a}$$

and is sometimes written with a change in the variable of integration

$$y(t) = \int_{0}^{\infty} g(\tau)x(t - \tau)\,\mathrm{d}\tau. \tag{1.16b}$$

The impulse function $g(t)$ is sometimes called a weighting function because it weights values of $x(u)$ in the convolution integrals of equation (1.16).

The disadvantage of using impulse response as a general-purpose description of dynamic systems is that the convolution integrals of equations (1.16) are less convenient to manipulate and evaluate than are the Laplace transforms of equation (1.14).

Impulse responses also exist for discrete-time systems. The z-transform of a unit impulse is unity and the impulse response of a system having z-transfer function $G(z)$ is given by expressing $G(z)$ as a polynomial in z^{-1}

$$G(z) = g(z^{-1})$$
$$= g_0 + g_1 z^{-1} + g_2 z^{-2} + \cdots .$$

When the input to the system is a unit impulse, the output has z-transform

$$Y(z) = G(z) \times 1$$
$$= g(z^{-1})$$

and it follows from the definition of the z-transformation that the coefficients g_i in the expansion $g(z^{-1})$ are the values of the response $y(i)$ to the unit impulse.

Problems 1.24, 1.25

1.6. Transfer functions and block diagrams

It is often convenient to represent continuous-time linear systems by a block diagram like Fig. 1.21 which shows the names of the input and output and the

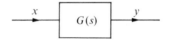

FIG. 1.21. Block diagram of a linear system.

transfer function of the equation relating them. Such block diagrams provide the basis for a pictorial algebra that can be used to build up descriptions of complex linear systems as follows.

(i) *Non-interacting cascaded systems*

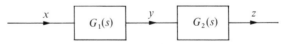

FIG.1.22. Cascaded systems.

Provided that the systems in Fig. 1.22 are non-interacting, in the sense that G_1 is unaffected by whether or not its output is connected as input to G_2, the overall transfer function is

$$\frac{Z(s)}{X(s)} = \frac{Y(s)}{X(s)} \cdot \frac{Z(s)}{Y(s)} = G_1(s)G_2(s).$$

(ii) *Unity feedback systems*

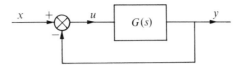

FIG. 1.23. Unity feedback system.

In Fig. 1.23 $\qquad\qquad\qquad u = x - y$

and so $\qquad\qquad\qquad U(s) = X(s) - Y(s),$

but $\qquad\qquad\qquad\qquad Y(s) = G(s)U(s)$

so the overall transfer function is

$$\frac{Y(s)}{X(s)} = \frac{G(s)}{1 + G(s)}.$$

(iii) *Non-unity feedback systems*

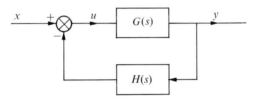

FIG. 1.24. Non-unity feedback system.

In Fig.1.24 $\qquad\qquad\qquad u \neq x - y$

but $\qquad\qquad\qquad U(s) = X(s) - H(s)\,Y(s),$

and, as before, $\qquad\qquad Y(s) = G(s)U(s)$

so the overall transfer function is

$$\frac{Y(s)}{X(s)} = \frac{G(s)}{1 + G(s)H(s)}$$

For example, if

$$G(s) = \frac{K}{1 + sT} \quad \text{and} \quad H(s) = B,$$

Fig. 1.24 could represent an amplifier of gain K and time constant T with negative feedback By from its output. The overall transfer function for such a feedback amplifier, from (iii) above, is

$$\frac{Y(s)}{X(s)} = \frac{K/(1 + sT)}{1 + BK/(1 + sT)} = \frac{K}{1 + BK + sT}$$

$$= \frac{K'}{1 + sT'},$$

which has gain

$$K' = \frac{K}{1 + KB}$$

and time constant

$$T' = \frac{T}{1 + KB}.$$

This analysis shows that the effect of feedback is to increase the speed of response of the amplifier

$$T' < T$$

at the expense of a reduction in gain

$$K' < K.$$

It also shows that for large values of K the overall gain with feedback depends only on the feedback constant B,

$$K' = \frac{1}{B + 1/K} \simeq \frac{1}{B} \text{ for } K \gg 1.$$

Discrete-time linear systems can be represented similarly by block diagrams using z-transfer functions.

In sampled-data systems with discrete-time variables which are sampled versions of continuous-time variables, as illustrated for example in Fig. 1.25,

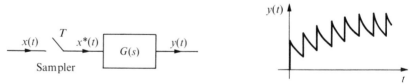

FIG. 1.25. Sampled-data system and typical step response.

the equivalence of equation (1.15) relating z-transforms to Laplace transforms is used. The output $y(t)$ is the sum of a series of impulse responses and has Laplace transform

$$Y(s) = G(s)X^*(s),$$

which is likely to be complicated; however values of $y(t)$ at the sampling instants are given by considering the discrete-time equivalent $y(iT)$ of the sampled version $y^*(t)$, which has z-transform

$$Y(z) = G(z)X(z).$$

This z-transform analysis gives no information about values of variables in sampled-data systems at times other than the sampling instants. Any number of different continuous-time functions could have the same values at the sampling instants and give rise to the same z-transform. The step and the pulse train in the table of Fig. 1.15 are examples of a pair of such functions.

The z-transfer function in sampled-data system analysis specifies the relationship between sampled versions of variables that are separated by continuous-time dynamics. If the dynamics are described by cascaded transfer functions $G_1(s)$ and $G_2(s)$, as in Fig. 1.22, the overall z-transfer function would be that associated with the product $G_1G_2(s)$; this is written $G_1G_2(z)$ and it is not the same as the product of the z-transforms associated with $G_1(s)$ and $G_2(s)$:

$$G_1G_2(z) \neq G_1(z)G_2(z).$$

The product $G_1(z)G_2(z)$ is the overall z-transform for a system where there is a sampler between the cascaded dynamic elements so that the input to the transfer function $G_2(s)$ is a sampled version of the output from $G_1(s)$. Figure 1.26 shows the difference between the two cases.

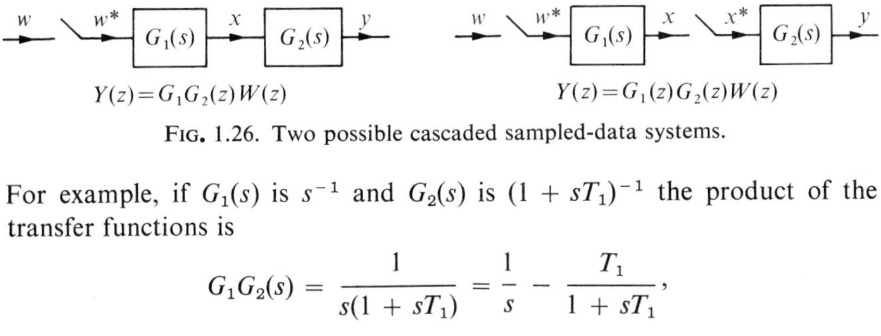

$$Y(z)=G_1G_2(z)W(z) \qquad\qquad\qquad Y(z)=G_1(z)G_2(z)W(z)$$

FIG. 1.26. Two possible cascaded sampled-data systems.

For example, if $G_1(s)$ is s^{-1} and $G_2(s)$ is $(1 + sT_1)^{-1}$ the product of the transfer functions is

$$G_1G_2(s) = \frac{1}{s(1 + sT_1)} = \frac{1}{s} - \frac{T_1}{1 + sT_1},$$

and the table of Fig. 1.15 gives the corresponding z-transform

$$G_1G_2(z) = \frac{z}{z - 1} - \frac{z}{z - e^{-T/T_1}}$$

$$= \frac{z(1 - e^{-T/T_1})}{(z - 1)(z - e^{-T/T_1})},$$

which is not the same as the product

$$G_1(z)G_2(z) = \frac{z^2}{(z - 1)(z - e^{-T/T_1})}.$$

If the input $w(t)$ were a unit step, the two possible inputs to $G_2(s)$ are $x(t)$ and $x^*(t)$ as shown in Fig. 1.27, and these two will not cause the same output $y(t)$.

FIG. 1.27. Two possibilities for the input to $G_2(s)$.

The block diagram algebra for sampled-data systems has thus to take account of the positioning of samplers, as in the following examples.

(i) *'Error'-sampled feedback systems*

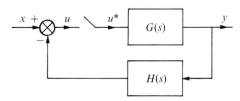

FIG. 1.28. 'Error'-sampled feedback system.

In Fig. 1.28 $U(z) = X(z) - GH(z)U(z)$

and $Y(z) = G(z)U(z),$

so the overall z-transfer function is

$$\frac{Y(z)}{X(z)} = \frac{G(z)}{1 + GH(z)}.$$

(ii) *Output-sampled feedback systems*

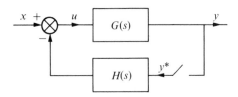

FIG. 1.29. Output-sampled feedback system.

In Fig. 1.29 it is convenient to consider the Laplace transform

$$Y(s) = G(s)X(s) - G(s)H(s)Y^*(s),$$

and the z-transform equivalent to this is

$$Y(z) = GX(z) - GH(z)Y(z),$$

so the z-transform of the output is

$$Y(z) = \frac{GX(z)}{1 + GH(z)}$$

and there is not an easily derived overall z-transfer function.

(iii) *Completely sampled feedback systems*

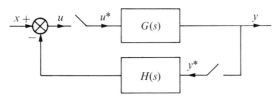

FIG. 1.30. Completely sampled feedback system.

In Fig. 1.30 $\qquad\qquad U(z) = X(z) - H(z)Y(z)$

and $\qquad\qquad\qquad Y(z) = G(z)U(z),$

so the overall z-transfer function is

$$\frac{Y(z)}{X(z)} = \frac{G(z)}{1 + G(z)H(z)}.$$

Problems 1.26–1.28

1.7. Minimum-phase equations

There is a special class of linear differential equations called 'minimum-phase' which includes the equations describing most ordinary control systems. The class consists of equations with transfer functions that are ratios of finite polynomials having no factors with positive real parts; the examples in Section 1.2 were minimum-phase. The name arises in connection with the frequency response function $G(j\omega)$ and refers to theorems by Bode† which show that:

(i) For minimum-phase equations there is a unique relationship between the phase angle $\underline{/G(j\omega)}$ and the amplitude $|G(j\omega)|$ of the frequency response function. The phase angle at frequency ω_1 is expressed in terms of new variables,

$$A \equiv \log_e|G(j\omega)| \quad \text{and} \quad U \equiv \log_e\left(\frac{\omega}{\omega_1}\right),$$

† See Bode (1945) Chapter 14.

and is

$$\underline{/G(j\omega_1)} = \frac{1}{\pi} \int_{-\infty}^{\infty} \frac{dA}{dU} \log_e \coth \left| \frac{U}{2} \right| dU \text{ radians.}$$

(ii) For a non-minimum-phase equation the phase shift at any frequency is greater than it would be for a minimum-phase equation having the same frequency response amplitude function $|G(j\omega)|$.

A common non-minimum-phase equation is that describing a time delay T,

$$y(t) = x(t - T).$$

Its transfer function e^{-sT} cannot be expressed as a ratio of finite polynomials. The corresponding frequency response function $e^{-j\omega T}$ has amplitude and phase

$$|G(j\omega)| = 1, \qquad \underline{/G(j\omega)} = -\omega T,$$

and the minimum-phase frequency response with the same amplitude function is

$$G_1(j\omega) = 1,$$

which has

$$G_1(j\omega) = 1, \qquad \underline{/G_1(j\omega)} = 0.$$

The phase lag ωT of the non-minimum-phase equation is thus greater than the phase lag, zero, of the minimum-phase equation.

Non-minimum-phase equations also arise when the polynomials of a transfer function have unstable factors; that is continuous-time factors in the right-half plane, or discrete-time factors outside the unit circle of Fig. 1.4. Unstable systems, where factors of the denominator polynomial are unstable, are thus non-minimum-phase. Stable non-minimum-phase systems, where

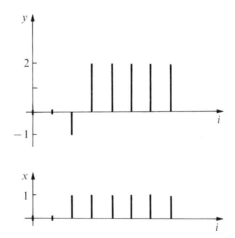

FIG. 1.31. Step response of a non-minimum-phase system.

factors of the numerator polynomial are unstable, are sometimes characterized by step responses in which the initial response is in the opposite direction to the final steady-state response. Figure 1.31 illustrates this effect in a
discrete-time system having difference equation

$$y(i) = -x(i) + 3x(i - 1)$$

and z-transfer function

$$\frac{Y(z)}{X(z)} = -(1 - 3z^{-1})$$

with numerator factor of value $+3$.

1.8. Bibliography

Ordinary linear differential equations are discussed in many books about mathematics, for example:
    PIAGGIO (1920) Chapter 3,
    KREYSIG (1972) Chapter 2.
Corresponding discussions of difference equations are in
    BOOLE (1872) Chapters 9, 10, and 11,
    MILNE-THOMPSON (1933) Chapter 13,
    GOLDBERG (1961) Chapter 3.
Continuous-time responses, the Laplace transform, and transfer functions are
discussed in most books on classical control engineering, for example:
    JAMES, NICHOLS, and PHILLIPS (1947) Chapter 2,
    BROWN and CAMPBELL (1948) Chapters 3 and 5,
    TSIEN (1954) Chapters 2 and 3,
    MURPHY (1957) Chapters 4 and 5,
    CHESTNUT and MAYER (1959) Chapters 3, 4, and 5,
    GILLE, PELEGRIN, and DECAULNE (1959) Chapters 3–8,
    DEL TORO and PARKER (1960) Chapters 2–7,
    TAYLOR (1960) Chapters 14, 16, and 17,
    THALER and BROWN (1960) Chapters 2–6,
    WILTS (1960) Chapter 2,
    RAVEN (1961) Chapters 5 and 6,
    BARBE (1963) Chapter 4,
    DOUCE (1963) Chapters 2 and 3,
    SHINNERS (1964) Chapter 2,
    D'AZZO and HOUPIS (1966) Chapters 3, 4, and 9,
    DORF (1967) Chapters 2 and 7,
    CHEN and HAAS (1968) Chapter 2,
    HARRISON and BOLLINGER (1969) Chapters 5–7,
    GUPTA and HASDORFF (1970) Chapter 2,
    HEALEY (1970) Chapter 4.
More extensive discussion of the Laplace transform is in
    GARDNER and BARNES (1942),

JAEGER (1949),
ASELTINE (1958) Chapters 1–8.
Continuous-time and discrete-time responses and transforms are presented together, as here, in
  KUO (1967) Chapters 1 and 2,
  DE RUSSO, ROY, and CLOSE (1965) Chapters 2 and 3.
Discrete-time sampled-data responses and the *z*-transform are discussed in
  TRUXAL (1955) Chapter 9,
  ASELTINE (1958) Chapter 16,
  JURY (1958) Chapter 1,
  RAGAZZINI and FRANKLIN (1958) Chapters 4 and 5,
  GILLE, PELEGRIN, and DECAULNE (1959) Chapter 20,
  TOU (1959) Chapter 5,
  WILTS (1960) Chapter 11,
  BROWN (1961) Chapter 13,
  KUO (1963) Chapters 4 and 5; (1970) Chapters 2 and 3,
  SHINNERS (1964) Chapter 9,
  LINDORFF (1965) Sections 2.1–2.9,
  SCHWARZ and FRIEDLAND (1965) Chapter 8,
  ELGERD (1967) Chapter 9,
  SAUCEDO and SCHIRING (1968) Chapter 4,
  VIDAL (1969) Chapter 2,
  WATKINS (1969) Chapter 11,
  CADZOW and MARTENS (1970) Chapters 2, 3, and 4,
  GUPTA and HASDORFF (1970) Chapter 5,
  OGATA (1970) Chapter 13,
  SMYTH (1972) Chapters 6 and 7.
Time delays in continuous-time systems are discussed in
  TSIEN (1954) Chapter 8,
  TRUXAL (1955) Chapter 9,
  MURPHY (1959) Chapter 7,
  CHOKSY (1962).
Non-minimum-phase equations are discussed in
  MELSA and SCHULTZ (1969) Section 5.4.

1.9. Problems

1.1. Which of the following equations represent linear systems?

(i) $\dfrac{d^2 y}{dt^2} + 6y = x$

(ii) $\dfrac{d^2 y}{dt^2} + 3\left(\dfrac{dy}{dt}\right)^2 + 6y = x$

(iii) $y(i + 2) + 5y(i + 1) = x(i)$

(iv) $\dfrac{d^2y}{dt^2} + 9\dfrac{dy}{dt} + 6y = x(1 - x^2)$

(v) $y(i + 4) + 3y(i + 3)y(i + 2) + 7y(i) = 10x(i + 2) + x(i)$

(vi) $y(i + 9) + 4y(i + 2) - 7 = \log\{\cos x(i)\}$

(vii) $\dfrac{d^3y}{dt^3} + 4\left(1 - \dfrac{y^2}{100}\right)\dfrac{dy}{dt} + 60y = 0.$

1.2. A vehicle of mass m is subject to an accelerating force F that is proportional to a control signal x, $(F = k_1 x)$. Assuming that the only other force acting on the vehicle is a resistance R proportional to vehicle velocity y, $(R = k_2 y)$, write down the differential equation relating velocity to control signal.

What is the time constant of this equation? Check that the expression for the time constant has the dimensions of time (N.B. dimensions of force are MLT^{-2}).

It is often more accurate to represent the resistance R as proportional to the square of vehicle velocity. Rewrite the differential equation using this assumption. What can be said about the equation now?

1.3 Figure 1.32 shows a low-pass filter in a feedback circuit; sketch on a single diagram step responses of

(i) the low-pass filter without feedback;
(ii) the closed-loop system with feedback gain $K = 1$;
(iii) the closed-loop system with feedback gain $K = 3$;
(iv) the closed-loop system with feedback gain $K = 10$.

How does the ratio of closed-loop time constant to open-loop time constant vary with gain K?

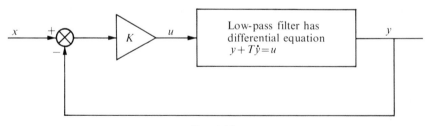

FIG. 1.32.

1.4. Suppose that in the position-control system of Fig. 1.7 the rotating load has differential equation

$$0\cdot1\dfrac{d^2y}{dt^2} + 0\cdot5\dfrac{dy}{dt} = u$$

and the torque u is given by

$$u = K(x - y).$$

(i) What value of K gives the system a damping ratio of unity? What is the natural frequency of the system?

(ii) If the value of K is increased by a factor of 4 what will be the values of natural frequency and damping ratio?

(iii) If the original value of K is decreased by a factor of 4, what will be the values of natural frequency and damping ratio?

Sketch step responses for the system, with the three different values of K, on a single graph. (Copy the step responses, with appropriate changes of time scaling, from Fig. 1.6.)

1.5. An instrument to record the e.m.f. of a thermocouple is based on the principle of a motor-driven self-balancing potentiometer. The e.m.f. of the thermocouple is compared with that picked up on the brush of a slide-wire potentiometer. Any difference between the two voltages is amplified and causes a proportionate current to flow in the armature of a small permanent-magnet motor. The motor is connected through a gear box and lead-screw to the brush of the slide-wire and moves it in such a direction as to restore the balance of voltages. A tachometer generator is attached direct to the motor, and its output voltage is reduced by an attenuating resistance network and added to the error voltage at the amplifier input in such a sense as to provide damping; there is no appreciable mechanical damping.

The slide-wire is 0·05 m long and the p.d. across it is 10 mV. The motor 'sees' a moment of inertia of 2×10^{-5} kg m^2 and produces a torque of 7×10^{-3} Nm for 100 mA of armature current. The motor rotates 10 000 times per linear metre of travel of the brush. The tacho-generator produces 10 V output per 100 rad s^{-1}.

If the system is to have damping ratio 0·85 and natural frequency 1·25 rad s^{-2}, find the gain required of the amplifier, in mA V^{-1}, and the transmission factor through the attenuating network.

1.6. Prove that the polar diagram of the first-order frequency response shown in Fig. 1.10 is semicircular.

1.7. The differential equation of a second-order system is

$$\frac{d^2y}{dt^2} + 2\zeta\omega_0\frac{dy}{dt} + \omega_0^2 y = \omega_0^2 x.$$

Sketch its frequency-response characteristics on a Bode diagram and on a polar diagram for systems having $\zeta = \frac{1}{2}, 1, 2$.

1.8. A certain control system has differential equation

$$\frac{d^2y}{dt^2} + k\frac{dy}{dt} + 5Ky = 5Kx,$$

where the constant k is such that the damping ratio is one-half. If the system is to follow a sinusoidal input signal x of frequency 0·5 Hz with not less than 5° of phase lag, find its natural frequency and the value of the constant K.

1.9. The difference equation of a discrete-time system is

$$y(i + 1) + ay(i) = x(i).$$

If $x(i)$ has the constant value 4, and the initial value of y is $y(i = 0) = 10$, find and sketch the response of the system when

(i) $a = 0{\cdot}2$

(ii) $a = -0{\cdot}2$.

1.10. The difference equation of a discrete-time system is
$$y(i + 2) + a_1 y(i + 1) + a_0 y(i) = x(i + 1) - x(i).$$
Prove that if x has the constant value $x = X$ it is not possible for the particular integral solution to be $y_{PI} = X$ if the transient response of the system is to be stable.

1.11. A discrete-time process having difference equation
$$y(i + 1) = y(i) + au(i)$$
is controlled by making its input u proportional to a previous error signal according to the difference equation
$$u(i) = K\big(x(i - 1) - y(i - 1)\big).$$
What value of the gain K will make both roots of the characteristic equation of the resulting control system have the value $\tfrac{1}{2}$?

1.12. In a discrete-time position-control system the position y satisfies a first-order difference equation
$$y(i + 1) = y(i) + av(i),$$
where v is the velocity and satisfies another first-order difference equation
$$v(i + 1) = v(i) + bu(i).$$
Show that the difference equation relating y to u is therefore
$$y(i + 2) - 2y(i + 1) + y(i) = abu(i).$$
The system is controlled by making u proportional to the error and is stabilized by velocity feedback so that
$$u(i) = K\big(x(i) - y(i)\big) - kv(i).$$

(i) What values of the constants K and k will make both roots of the characteristic equation of the resulting control system have the value $\tfrac{1}{2}$?

(ii) If x has the constant value X, what is the particular integral (steady-state value) of y?

(iii) Does the result in (ii) above depend on the values of K and k?

1.13. If $x(i)$ is a discrete-time function which is zero for all negative values of i and has z-transform $X(z)$, prove that
$$z\text{-transform } \{x(i - 1)\} = z^{-1}X(z).$$

1.14. Derive the z-transforms in the table of Fig. 1.13.

1.15. Express the function
$$X(z) = \frac{z^2}{(z - a)^2}$$
as a polynomial in z^{-1} and hence find the function $x(i)$ for which $X(z)$ is the z-transform.

1.16. Show that the z-transfer function of the position-control system in Problem 1.12 is

$$\frac{Y(z)}{X(z)} = \frac{1}{(1 - 2z)^2}.$$

Find the step-response function $y(i)$ of the system:

(i) by expressing $Y(z)$ as a polynomial in z^{-1};

(ii) by expressing $Y(z)$ as a sum of partial fractions and using the table of z-transforms in Fig. 1.13 and the result of Problem 1.15.

1.17. If $x(t)$ is a continuous-time function which is zero for all negative values of t and has Laplace transform $X(s)$, prove that

$$\text{Laplace transform} \left\{ \int_0^t x(t) \, dt \right\} = \frac{1}{s} X(s).$$

1.18. Derive the Laplace transforms in the table of Fig. 1.14.

1.19. Prove the final-value theorems:

(i) $\lim_{t \to \infty} x(t) = \lim_{s \to 0} s X(s)$ for continuous-time functions;

(ii) $\lim_{i \to \infty} x(i) = \lim_{z \to 1} (1 - z^{-1}) X(z)$ for discrete-time functions.

1.20. A continuous-time system has transfer function

$$\frac{Y(s)}{X(s)} = \frac{1 + sT_1}{1 + sT_2}.$$

Use Laplace transforms to find the output of the system if $x = at$ and $y(t = 0) = 0$.

1.21. Sketch Bode diagram asymptotes, and also polar diagrams, for systems having the following transfer functions.

(i) s

(ii) $\dfrac{1}{(1 + sT)(1 + s2T)}$

(iii) $s(1 + sT)$

(iv) $\dfrac{1}{(1 + sT)^4}$

(v) $\dfrac{1 + 0\cdot05s}{s^2(1 + 0\cdot01s)(1 + 0\cdot005s)}$

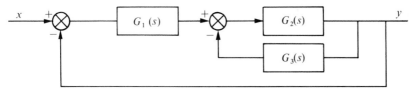

FIG. 1.33.

1.22. The block diagram of a certain control system is as shown in Fig. 1.33.

The transfer functions of the separate blocks are investigated by frequency-response tests, and the resulting Bode-diagram amplitude-ratio asymptotes are as shown in Fig. 1.34.

(i) Write down the transfer function of each block.

(ii) Determine the damping ratio of system response.

(iii) Why might this system be considered unsatisfactory?

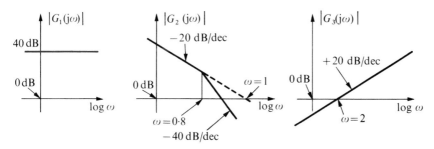

FIG. 1.34.

1.23. A continuous-time variable $x = Ae^{-t/T}$ is sampled aperiodically at times defined by a difference equation

$$t(i + 1) = t(i) + i\Delta,$$

where i is the integer counting sampling instants and $t(i = 0) = 0$.

Prove that the z-transform of the resulting discrete-time sampled variable is

$$X(z) = \sum_{i=0}^{\infty} z^{-1} \exp\left(-\frac{i(i + 1)\Delta}{2}\right).$$

1.24. Find and sketch the impulse response of the discrete-time systems

(i) of Problem 1.9;

(ii) of Problem 1.12.

1.25. Find the transfer function of a linear system having the rectangular impulse response shown in the Fig. 1.35. Such a system is known as a 'zero-order hold'; how might it be used in a sampled-data control system?

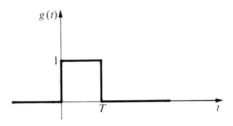

$g(t)$

1

T

t

FIG. 1.35.

1.26. Prove that the sampled-data control system in Fig. 1.36 will be unstable if the time T between sampling instants is greater than $2/K$:

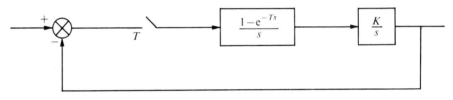

$+$

$-$

T

$\dfrac{1-e^{-Ts}}{s}$

$\dfrac{K}{s}$

FIG. 1.36.

1.27. If the transfer functions in Figs 1.28, 1.29, and 1.30 are

$$G(s) = K \frac{1 - e^{-sT}}{s^2}, \qquad H(s) = \frac{1}{1 + sT_1},$$

derive expressions for the step response of each of the three sampled-data systems when the input x has Laplace transform

$$X(s) = \frac{X}{s}.$$

1.28. How is the overall z-transfer function for the system of Fig. 1.29 specified?

2. Feedback control of steady-state errors

2.1. Steady-state errors

A PRIMARY objective for most control systems is to make some physical variable take a desired value; for example to regulate the temperature of an oven or to make the direction of a receiving aerial track a moving target. This is to be achieved by adjusting some other variable, like heat input to the oven or force applied to the aerial, although the response of most physical systems to such adjustments is neither instantaneous nor certain. The non-instantaneous nature of the response is accounted for by regarding the physical controlling and controlled variables as the input and output of a dynamic system called the 'controlled process'. The effect of uncertainties is reduced by using feedback, as shown in Fig. 2.1, where a controller adjusts the controlling variable u on the basis of information about the desired value x and of information fed back about the actual value y of the controlled variable. Feedback

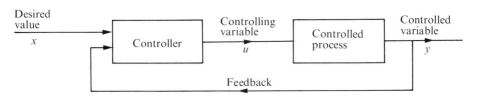

FIG. 2.1. Feedback control system.

provides the controller with continuous indication of what adjustment is needed to make the controlled variable approach its desired value; for example if the oven is too cold more heat must be supplied and if the aerial points to the left of its target it must be forced to turn to the right. In this way feedback reduces the effect of uncertainties about the exact relationship between u and y; it also reduces the effect of other uncertainties, for example about the values of unpredictable variables describing the weather or future target motions.

Ideally the error e, defined by

$$e \equiv x - y,$$

should be zero at all times and for all desired values x, but this ideal is not usually achievable in real control systems which tend to have errors that vary with time as shown in Fig. 2.2. One source of errors is the dynamic nature of real controlled processes which makes instantaneous changes of y impossible

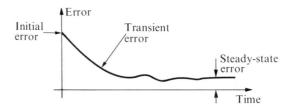

FIG. 2.2. Typical error-response function.

and so causes transient errors. In a linear system the transient error is associated with the complementary function solution of the dynamic equation; in a stable system the complementary function always decays to zero and so the transient error also decays to zero. Any error that remains after the transients have decayed is a steady-state error; in a linear system the steady-state error is associated with the particular integral solution of the dynamic equation and therefore depends both on the dynamics of the system and on how the desired value x varies with time.

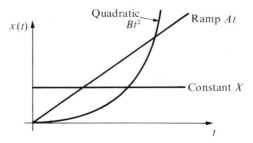

FIG. 2.3. Simple desired-value functions.

Steady-state errors can be zero, finite, or unbounded, and these three types correspond to control systems where the controlled variable y eventually takes its desired value x exactly, approximately, or not at all. In most systems different types of steady-state error result from different forms of variation of the desired value x in time. The response to a few simple forms, shown in Fig. 2.3, gives a useful indication of what a control system can achieve; a good system is usually one that can follow rapidly changing desired values with very little error.

Chapter 2 discusses the steady-state response of continuous-time linear system to simple desired-value functions. Transient response is discussed in the following chapters.

Problem 2.1

2.2. Unity feedback

In a system with unity feedback, as shown in Fig. 2.4, the input to the control-

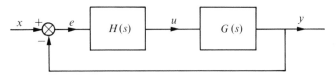

FIG. 2.4. Unity-feedback system.

ler is the error e. The simplest possible linear feedback controller makes the controlling variable u proportional to e:

$$u = K_0 e$$

and has transfer function

$$H_0(s) = K_0.$$

The overall transfer function is then

$$\frac{Y(s)}{X(s)} = \frac{K_0 G(s)}{1 + K_0 G(s)},$$

and the Laplace transform of the error is

$$E(s) = \frac{Y(s)}{K_0 G(s)} = \frac{X(s)}{1 + K_0 G(s)}.$$

If the controlled process transfer function $G(s)$ has the general form

$$G(s) = \frac{b_0 + b_1 s + \cdots + b_m s^m}{a_0 + a_1 s + \cdots + a_n s^n},$$

the Laplace transform of the error becomes

$$E(s) = \frac{(a_0 + a_1 s + \cdots) X(s)}{a_0 + a_1 s + \cdots + K_0 (b_0 + b_1 s + \cdots)},$$

which can be converted into a differential equation

$$a_0 e + a_1 \frac{de}{dt} + \cdots + K_0 \left(b_0 e + b_1 \frac{de}{dt} + \cdots \right) = a_0 x + a_1 \frac{dx}{dt} + \cdots.$$

If the desired value is constant,

$$x(t) = X,$$

the particular integral solution for the steady-state error is

$$e_{\text{PI}} = \frac{a_0}{a_0 + K_0 b_0} X = \frac{X}{1 + K}, \quad \text{where } K = \frac{K_0 b_0}{a_0}.$$

The steady-state error therefore depends on K, the amount of amplification, or 'gain', in the system. When K is large compared to unity the steady-state error is small compared to X and the controlled variable approximates to its constant desired value regardless of the exact value of X or of coefficients in the transfer function $G(s)$ relating u and y. Uncertainty about these values thus has a negligible effect on the eventual value of the controlled variable.

If the desired value is a ramp function At, or any other function that increases with time, there is no finite particular integral solution to the differential equation. The steady-state error is therefore unbounded and the control system eventually fails in its objective of making the controlled variable take the desired value.

Steady-state performance can be improved by designing a controller which prevents the system from settling with non-zero errors. This can be done by making the rate of change of the controlling variable u proportional to the error e,

$$\frac{du}{dt} = K_1 e$$

with controller transfer function,

$$H_1(s) = \frac{K_1}{s},$$

which is sometimes called 'integral control' because the controller generates an output u proportional to the time-integral of its input e. Combining this control law with the same controlled process $G(s)$ as before the Laplace transform of the error is

$$E(s) = \frac{(a_0 s + a_1 s^2 + \cdots)X(s)}{a_0 s + a_1 s^2 + \cdots + K_1(b_0 + b_1 s + \cdots)},$$

which can be converted into a differential equation

$$a_0 \frac{de}{dt} + a_1 \frac{d^2 e}{dt^2} + \cdots + K_1 \left(b_0 e + b_1 \frac{de}{dt} + \cdots \right) = a_0 \frac{dx}{dt} + a_1 \frac{d^2 x}{dt^2} + \cdots.$$

If the desired value x is now constant the particular integral solution is zero and the steady-state error is therefore zero. If the desired value is a ramp function

$$x(t) = At,$$

the particular integral

$$e_{\text{PI}} = \frac{a_0 A}{K_1 b_0} = \frac{A}{K}$$

again depends on the gain K in such a way that when the gain is made large the steady-state error becomes small and the controlled variable approximates to its desired value. Integral control thus improves steady-state performance by eliminating steady-state errors when the desired value is constant, and by making the controlled variable approximate to its desired value when it is a ramp function. If the desired value is a quadratic function Bt^2, or any other function increasing more rapidly than a ramp, the steady-state error is infinite and the system fails in its objective.

Steady-state performance can be further improved by using further integrations. If the same controlled process $G(s)$ as before were combined with a controller having two integrations, giving transfer function

$$H_2(s) = \frac{K_2}{s^2},$$

the error differential equation would be

$$a_0\frac{d^2e}{dt^2} + a_1\frac{d^3e}{dt^3} + \cdots + K_2\left(b_0e + b_1\frac{de}{dt} + \cdots\right) = a_0\frac{d^2x}{dt^2} + a_1\frac{d^3x}{dt^3} + \cdots.$$

This has zero particular integral, and no steady-state error, if the desired value $x(t)$ is constant or a ramp function; and when the desired value is a quadratic function

$$x(t) = \tfrac{1}{2}Bt^2,$$

the particular integral

$$e_{\text{PI}} = \frac{a_0 B}{K_2 b_0} = \frac{B}{K}$$

is inversely proportional to gain K, as before. The second integration thus eliminates steady-state errors when the desired value is a ramp function, and makes the controlled variable approximate to its desired value when this is a quadratic function. If the desired value increases more rapidly than a quadratic function the system again fails.

Integrations have a good effect on steady-state performance, but they usually have a bad effect on transient performance. So, for reasons that are discussed further in Chapter 3, it is unusual to find more than two integrations in the combined transfer function of controller and controlled process.

The above results about steady-state errors in the single-input unity-feedback system of Fig. 2.4 depend on the combined transfer function $H(s)G(s)$; integrations and gain have the same effect whether they arise in the controller $H(s)$ or in the controlled process $G(s)$. The results are summarized in Fig. 2.5 by writing the combined transfer function

$$H(s)G(s) = \frac{K(1 + \beta_1 s + \cdots)}{s^N(1 + \alpha_1 s + \cdots)}.$$

where K is the gain and N is the number of integrations, sometimes called the 'type-number' of the system. The type of steady-state error depends on the type-number of the system; when the steady-state error is finite its value

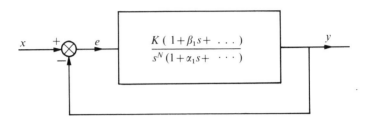

		Desired value $x(t)$		
		X	At	$\frac{1}{2}Bt^2$
	0	$\dfrac{X}{1+K}$	∞	∞
Type-number N	1	0	$\dfrac{A}{K}$	∞
	2	0	0	$\dfrac{B}{K}$

FIG. 2.5. Steady-state errors in unity feedback systems.

depends only on the magnitude of the input function and on the gain K, and it can be made small by making the gain large. Steady-state performance of the feedback control system is thus insensitive to exact values of the system's parameters.

Problems 2.2–2.6

2.3. Non-unity feedback

In a system with transfer function $G_2(s)$ other than unity in the feedback, as shown in Fig. 2.6, the input to the forward transfer function $KG_1(s)$ is not the error e and the results summarized in Fig. 2.5 do not apply. However the steady-state error in any particular system can be found by analysis similar to that of Section 2.2: the overall transfer function is

$$\frac{Y(s)}{X(s)} = \frac{KG_1(s)}{1 + KG_1(s)G_2(s)},$$

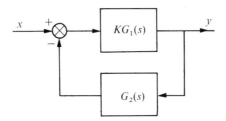

FIG. 2.6. Non-unity feedback system.

and the Laplace transform of the error is

$$E(s) = \frac{1 + KG_1(s)G_2(s) - KG_1(s)}{1 + KG_1(s)G_2(s)} X(s),$$

which can be converted into a differential equation for any given transfer functions $KG_1(s)$ and $G_2(s)$.

For example the position-control system with velocity feedback of Section 1.2 is a non-unity feedback system having the block diagram of Fig. 2.7 (the

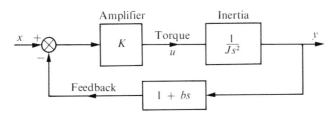

FIG. 2.7. Position-control system.

product Kb of amplifier gain K and velocity-feedback coefficient b corresponds to the constant of proportionality k used previously). The Laplace transform of the error

$$E(s) = \frac{Js^2 + Kbs}{Js^2 + Kbs + K} X(s)$$

converts into the differential equation

$$J\frac{d^2e}{dt^2} + Kb\frac{de}{dt} + Ke = J\frac{d^2x}{dt^2} + Kb\frac{dx}{dt},$$

which shows that when $x(t)$ is constant the particular integral is zero, when $x(t)$ is a ramp function At the particular integral is

$$e_{PI} = bA,$$

and when $x(t)$ increases more rapidly than a ramp function there is no finite particular integral. The steady-state performance is thus similar to that of a type-one unity-feedback system, although there are two integrations in the position-control system, and the finite error with a ramp function is proportional to the feedback coefficient b rather than inversely proportional to the gain in the system.

Problem 2.7

2.4. Disturbances

Most controlled processes are subject to uncontrollable and unpredictable disturbances affecting their performance. A typical source of such disturbances is the weather, which could cause unexpected heat transfers to or from a temperature-controlled oven and which could apply unexpected windage forces to a direction-controlled aerial. Disturbances are one of the most common reasons why a controlled variable might depart from its desired value and why feedback control is necessary. Steady-state analysis, like that in Sections 2.2 and 2.3, shows how feedback makes the controlled variable insensitive to disturbances.

For example the position-control system of Fig. 2.7 might be subject to a disturbance torque z, as shown in Fig. 2.8, due to wind resistance or due to

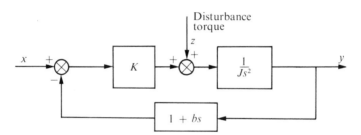

FIG. 2.8. Position control with disturbance torque.

fluctuating loads. Expressions for the controlled variable y are derived using the principle of superposition to write

$$y = y_x + y_z,$$

where y_x is the output due to x when z is zero, and y_z is the output due to z when x is zero. The Laplace transform of the error is

$$E(s) = X(s) - Y_x(s) - Y_z(s)$$

$$= \frac{(Js^2 + Kbs)X(s) - Z(s)}{Js^2 + Kbs + K},$$

which can be converted into a differential equation to show that the steady-state error is finite for ramp desired-value functions $x(t)$, and for constant disturbance torques. The steady-state error due to the disturbance when

$$z(t) = Z$$

is

$$(x - y_z)_{\text{PI}} = -\frac{Z}{K},$$

which becomes small when K is made large.

Steady-state errors due to disturbances z can be reduced by gain and eliminated by integrations, in the same way as other steady-state errors, provided that the integrations are associated with the transfer function $G_1(s)$ shown in

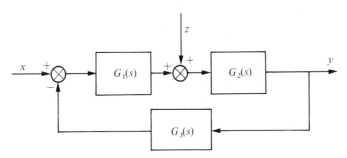

FIG. 2.9. Feedback system with disturbance torque.

the general feedback system of Fig. 2.9. The integration must be associated with $G_1(s)$ so that a signal can build up at the output of $G_1(s)$ to back off the unwanted disturbance z.

Problem 2.8

2.5. Bibliography

Most books on classical control engineering discuss steady-state errors in continuous-time systems. The material is grouped together, as here, in

JAMES, NICHOLS, and PHILLIPS (1947), Chapter 4,
BROWN and CAMPBELL (1948) Chapter 9,
SAVANT (1958) Chapter 3,
CHESTNUT and MAYER (1959) Chapter 8,
GILLE, PELEGRIN, and DECAULNE (1958) Chapter 15,
WILTS (1960) Chapter 3,
BARBE (1963) Chapter 9,
D'AZZO and HOUPIS (1966) Chapter 6,
DORF (1967) Chapter 3,
VERNON (1967) Chapter 6,

Steady-state errors in discrete-time systems are discussed in
  Tou (1959) Sections 6.5 and 6.6,
  Freeman (1965),
  Lindorff (1965) Sections 4.3 and 4.4,
  Gupta and Hasdorff (1970) Section 6.3.

2.6. Problems

2.1. A water turbine is fitted with a governor to control its speed in the face of changing loads. The speed y of the turbine is given by the equation

$$y = K_1 Q - K_2 L \quad \text{rad s}^{-1},$$

where Q is the water flow to the turbine in $\text{m}^3 \text{ s}^{-1}$, L is the load torque in Nm and K_1, K_2 are constants.

The governor controls water flow according to the equation

$$Q = K_3(x - y) \quad \text{m}^3 \text{ s}^{-1},$$

where x is the desired speed and K_3 is a constant.

Draw a block diagram of the system, and express y in terms of x and L. What is the loop gain, and how does the speed change from no-load to full-load L_{max} depend on its magnitude?

If $K_1 = 0.042$, $K_2 = 7 \times 10^{-5}$, $L_{\text{max}} = 3 \times 10^5$, and speed is to be kept within $\pm 2\%$ of a nominal 21 rad s^{-1} with a constant setting of x, suggest a value for K_3. Should x be set to 21 for this?

Sketch a graph showing how speed y varies with load L for constant x. Show on the same graph how speed would vary with L if the water flow Q were held constant (no feedback).

2.2. Figure 2.10 shows how the temperature of an oven may be regulated by automatic control. The rate at which heat is supplied to the oven (u Watts) is adjusted by a controller having input equal to the error between desired temperature x and

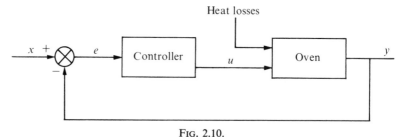

FIG. 2.10.

actual temperature y. Heat losses from the oven to its surroundings are at a rate proportional (K_2 W K^{-1}) to the difference between oven temperature and ambient temperature z. The thermal capacity of the oven is Q J K^{-1}.

(i) If there is proportional control

$$u = K_1 e,$$

write down the differential equation relating oven temperature to desired temperature and ambient temperature.

What is the steady-state value of oven temperature when both x and z are constant?

Under what conditions is the steady-state error e zero?

What is the time constant of the control system?

(ii) If there is integral control

$$du/dt = K_1 e,$$

write down the differential equation for oven temperature. Under what conditions is the steady-state error zero? What value of gain K_1 gives the control system a damping ratio $1/\sqrt{2}$?

What then is the natural frequency of the system?

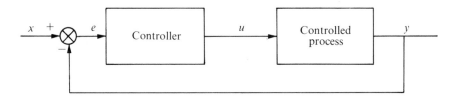

FIG. 2.11.

2.3. Figure 2.11 shows a feedback control system. The controlled process can be described by a second-order differential equation

$$\ddot{y} + 2\zeta\omega_0\dot{y} + \omega_0^2 y = \omega_0^2 u.$$

(i) For proportional control ($u = Ke$) write expressions for: the steady-state error when x is constant; the natural frequency of the control system; the damping ratio of the control system.

What is the effect on system performance of an increase in the value of gain K?

(ii) For integral control $du/dt = Ke$, write down the differential equation of the system.

What is the steady-state error when x is constant?

2.4. Figure 2.12 shows a speed-control system. The rotating inertia is $J =$

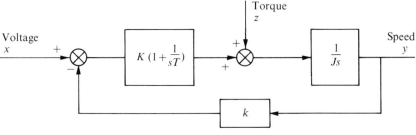

FIG. 2.12.

2 kg m² and the speed is measured by a tachometer with a proportionality constant $k = 0.25$ V rad^{-1} s. It is known that the load torque z cannot change more
rapidly than a ramp function with slope ±0.1 Nm s^{-1}.

It is required that the speed y be 50 rev s^{-1} with error $\pm1\%$ and that the system
have damping ratio 0·6. Determine the values of the controller reference voltage x,
the gain K, and the time constant T.

2.5. Derive an expression for the sensitivity $\partial y/\partial K_1$ of speed to the value of the
parameter K_1 in the control system of Problem 2.1. By what percentage can the
parameter K_1 (or K_3) depart from its nominal value without causing the speed y
to change by more than 2% of its nominal value?

(i) when the load L is zero?

(ii) when the load has its nominal maximum value?

2.6. What is the relationship between the steady-state speed y and the value of the
inertia J in the control system of Problem 2.4?

2.7. A position-control system can be stabilized either by velocity feedback or by
differentiating the error signal, as shown in Fig. 2.13. Which is preferable from the
point of view of the steady-state response of y to x?

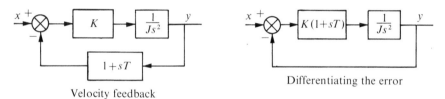

Velocity feedback   Differentiating the error

FIG. 2.13.

2.8. A closed-loop servomechanism is arranged so that a torque proportional to
the error (constant of proportionality K) is brought to bear on an output member
of inertia J_1. The system is mechanically damped by another inertia J_2 that is
coupled to the output shaft by a viscous friction coupling such that the torque
transmitted is proportional to the difference of the angular velocities of J_1 and J_2
(constant of proportionality f). The second inertia is elastically coupled to a fixed
reference by a spring k. The output position that is fed back to form the error signal is the position of J_1. (For this particular problem, a sketch can be more helpful than a block diagram.)

(i) Derive the differential equation relating output position (of J_1) to input
signal x.

(ii) For a ramp input $x = at$, what is the steady-state error?

2.9. Summarize the effects of integrations on the steady-state performance of
control systems.

3. Stability in linear systems

3.1. Oscillations, stability, and transient response

MOST control systems are feedback systems, as discussed in Chapter 2, and the outstanding feature of dynamic feedback systems is that they can sometimes sustain self-excited oscillations. An example is the howl that appears in a speech amplification system when the microphone receives too much of the loudspeaker output; the system is useless for speech amplification while it is howling, and a control system is similarly unable to make its controlled variable take a desired value while it is oscillating. Self-excited oscillations are usually of large amplitude and in a control system they can be dangerous; large fluctuations of heat input in temperature-control processes could cause explosions, and large fluctuations in economic systems could help to cause bankruptcies, famines, or wars. One of the first objectives of control theory is to predict conditions under which such oscillations will develop, so that they can be avoided.

In linear dynamic systems self-excited oscillations are associated with the complementary function solution of the dynamic equations. A familiar example is simple harmonic motion, which has differential equation

$$\frac{d^2y}{dt^2} + \omega_0^2 y = 0$$

with roots of the characteristic equation

$$s_1, s_2 = \pm j\omega_0$$

and oscillating complementary function

$$y_{CF} = A_1 \sin \omega_0 t + A_2 \cos \omega_0 t.$$

Simple harmonic oscillations mark the boundary between stable and unstable response and the conditions for there to be no oscillations are the same as the conditions for stability summarized in the s-plane in Fig. 1.4; the roots of a continuous-time characteristic equation must all lie to the left of the imaginary axis, and the roots of a discrete-time characteristic equation must all lie within the unit circle.

In most control systems it is necessary not only to avoid self-excited oscillations but also to ensure that transient response is not too oscillatory; Fig. 3.1

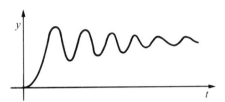

FIG. 3.1. Stable but oscillatory response.

shows a transient response which is technically stable but which is too oscillatory to be accepted in a control system. In linear systems stability, in this wider sense of whether or not the transient response is acceptable, depends on the roots $s_1, \ldots, s_n$ of the characteristic equation (1.3). These roots may be real or complex, but in ordinary systems any complex roots appear in conjugate pairs.

For continuous-time systems it is convenient to write the real roots in the form

$$s_k = -1/T_k$$

and the complex conjugate roots in the form

$$s_l, s_{l+1} = -1/T_l \pm j\omega_l,$$

so that the complementary function, which determines transient response, can be written

$$y_{\mathrm{CF}} = \sum_{i=1}^{n} A_i \, e^{s_i t}$$

$$= \sum_k A_k \, e^{-t/T_k} + \sum_l e^{-t/T_l}(A_l \cos \omega_l t + B_l \sin \omega_l t)$$

to show that the real roots give rise to non-oscillatory exponentially decaying transients with time constants T_k, as in Fig. 3.2(i), and the complex conjugate

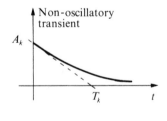

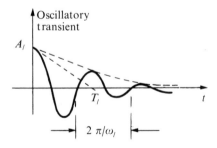

(i) Term from real root   (ii) Term from complex conjugate roots.

FIG. 3.2. Typical components of complementary function.

roots give rise to oscillatory transients with frequencies ω_l and amplitudes decaying exponentially with time constants T_l, as in Fig. 3.2(ii).

Each oscillatory component of transient response is caused by a pair of complex conjugate roots which can be regarded as the solution of a typical second-order characteristic equation:

$$s^2 + 2\zeta\omega_0 s + \omega_0^2 = 0$$

with natural frequency ω_0 and damping ratio ζ. The roots can therefore be written

$$s_1, s_2 = -\omega_0\zeta \pm j\omega_0\sqrt{(1 - \zeta^2)},$$

and then Fig. 3.3 shows how the position of the roots in the s-plane is related to the natural frequency and the damping ratio. It is a matter of simple

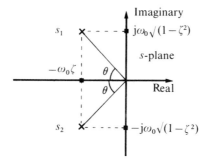

FIG. 3.3. Complex conjugate roots.

trigonometry to prove that each root is distant ω_0 from the origin and lies on a line at angle

$$\theta = \cos^{-1}\zeta$$

to the real axis. Oscillatory response is associated with low values of damping ratio, and so the requirement that transient response is not too oscillatory can be interpreted as a requirement that no roots of the characteristic equation have damping ratio less than some stated value ζ_1; this can be interpreted as a requirement that no roots lie to the right of the boundary shown in Fig. 3.4. A typical value for ζ_1 is $1/\sqrt{2}$ and then the angle $\cos^{-1}\zeta_1$ is 45°.

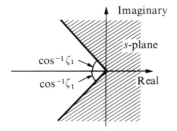

FIG. 3.4. Requirement that transient response be not too oscillatory.

Another common requirement that can be interpreted as a boundary in the s-plane is the requirement that transient response should be rapid. Rapid response is associated with small values of the time constants shown in Fig. 3.2 and these are inversely proportional to the real part of the roots of the characteristic equation. The requirement for rapid response can therefore be interpreted as a requirement that no roots have time constant greater than some stated value T_1 and this can be interpreted as a requirement that no roots lie to the right of the boundary shown in Fig. 3.5. Values of T_1 vary from problem to problem: in position control it might be one second; in temperature control one minute; in an economic system one year.

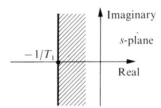

FIG. 3.5. Requirement for rapid response.

The requirements for non-oscillatory, rapid response can be combined into a requirement that no roots of the characteristic equation lie to the right of the boundary shown in Fig. 3.6. This includes the original condition for stability, that no roots lie to the right of the imaginary axis.

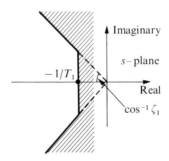

FIG. 3.6. Requirement for not too oscillatory, rapid response.

For discrete-time systems the transient response depends principally on the magnitude of roots of the characteristic equation. The exact relationship with the position of complex roots in the s-plane is not so clear as that illustrated in Fig. 3.3 for continuous-time systems. However the analysis in Section 1.3

suggests that good transient response requires that the roots should be as close to the origin of the s-plane as possible; Fig. 1.12 showed how the response becomes oscillatory as a real root tends to the value minus one and sluggish as a real root tends to the value plus one. These considerations are summarized in Fig. 3.7 by a circular boundary, corresponding to the boundary of Fig 3.6 for continuous-time systems, defining a region of the s-plane which

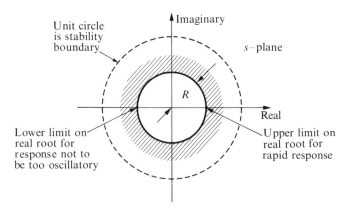

FIG. 3.7. Requirements for good transient response of discrete-time systems.

must contain all roots of the characteristic equation if transient reponse is to be acceptable. The radius R of this boundary must be less than unity for stability; it determines how oscillatory the response can be and also, for given intervals between the discrete-time instants, how slow.

The conditions about stability and transient response are all expressed with reference to the position in the s-plane of the roots of a characteristic equation. To use the known conditions, it would appear to be necessary first to solve the equation, but most characteristic equations of control systems have polynomials of higher order than two and cannot be solved analytically. Numerical solution, using a computer, is always possible but several more helpful special techniques are available to overcome this difficulty. Stability can be investigated by means of mathematical criteria which determine from the coefficients of a polynomial whether or not its roots are in the stable part of the s-plane; the Routh–Hurwitz criterion for continuous-time systems and the Schur–Cohn criterion for discrete-time systems are presented, without proof, in Sections 3.2 and 3.3. Transient response is often investigated by graphical techniques without solving the characteristic equation; the Nyquist stability criterion for continuous-time systems is the subject of Section 3.4, and phase margin and gain margin are introduced in Section 3.5. A more general technique, the root-locus diagram, is presented in Chapter 4.

3.2. Mathematical stability criteria for continuous-time systems

A continuous-time, linear, dynamic system is stable provided that all the roots $s_1, \ldots, s_n$ of its characteristic equation

$$a_n s^n + \cdots + a_1 s + a_0 \equiv a_n(s - s_1)(s - s_2)\ldots(s - s_n) = 0$$

have negative real parts.

A necessary condition for stability is that all the coefficients $a_0, \ldots, a_n$ should have the same sign and be non-zero. If the condition is not satisfied it must be because some of the factors $s_1, \ldots, s_n$ have positive real parts which account for the negative and cancelled coefficients in the polynomial, so the system will be unstable. This condition can sometimes be used to recognize unstable systems: for example, a second-order equation with negative damping,

$$\frac{d^2 y}{dt^2} - a^2 \frac{dy}{dt} + \omega^2 y = 0$$

and a high-order equation with some terms missing

$$\frac{d^9 y}{dt^9} + y = 0$$

can both be recognised as unstable without further analysis. When all the coefficients have the same sign and none are missing, as for example

$$a_3 \frac{d^3 y}{dt^3} + a_2 \frac{d^2 y}{dt^2} + a_1 \frac{dy}{dt} + a_0 y = 0, \tag{3.1}$$

further analysis is required to determine stability.

A complete test for stability is provided by the Routh–Hurwitz criterion which gives a condition that is both necessary and sufficient. The criterion can be expressed with reference to determinants formed from the coefficients of the characteristic equation. The first determinant is

$$\Delta_n \equiv \begin{vmatrix} a_{n-1} & a_n & 0 & 0 & 0 & . & . & . & 0 \\ a_{n-3} & a_{n-2} & a_{n-1} & a_n & 0 & . & . & & . \\ a_{n-5} & . & & . & & . & & . & . \\ . & . & . & . & . & . & . & . & a_4 \\ . & . & . & . & . & 0 & a_0 & a_1 & a_2 \\ 0 & . & . & . & . & 0 & 0 & 0 & a_0 \end{vmatrix},$$

and other determinants are derived from the first by removing the last row and the last column to give

$$\Delta_{n-1} \equiv \begin{vmatrix} a_{n-1} & a_n & 0 & . & . & 0 \\ a_{n-3} & a_{n-2} & a_{n-1} & . & . & . \\ a_{n-5} & . & . & . & . & . \\ . & . & . & . & a_2 & a_3 \\ . & . & . & 0 & a_0 & a_1 \end{vmatrix},$$

and so on down to

$$\Delta_3 \equiv \begin{vmatrix} a_{n-1} & a_n & 0 \\ a_{n-3} & a_{n-2} & a_{n-1} \\ a_{n-1} & a_{n-4} & a_{n-3} \end{vmatrix}$$

$$\Delta_2 \equiv \begin{vmatrix} a_{n-1} & a_n \\ a_{n-3} & a_{n-2} \end{vmatrix}, \qquad \Delta_1 \equiv a_{n-1}.$$

The criterion then states that the roots of the characteristic equation all have negative real parts if and only if all the determinants $\Delta_1, \ldots, \Delta_n$ are greater than zero. This statement assumes that at least the leading coefficient a_n is positive and uses the notational convention that any element with negative suffix in a determinant is to be interpreted as zero.

Applying the Routh–Hurwitz criterion, for example, to the general third-order equation (3.1), the determinants are

$$\Delta_3 = \begin{vmatrix} a_2 & a_3 & 0 \\ a_0 & a_1 & a_2 \\ 0 & 0 & a_0 \end{vmatrix} = a_0(a_2a_1 - a_3a_0)$$

$$\Delta_2 = \begin{vmatrix} a_2 & a_3 \\ a_0 & a_1 \end{vmatrix} = a_2a_1 - a_3a_0, \qquad \Delta_1 = a_2,$$

and requiring these to be positive gives the necessary conditions about coefficients being positive and also an additional condition for stability:

$$a_2a_1 - a_3a_0 > 0.$$

The third-order equation describes the position-control system, discussed previously in Sections 1.2 and 2.3, when account is taken of the dynamic nature of the amplifier which cannot produce instantaneous adjustments of torque. Figure 3.8 is a block diagram of the system with amplifier dynamics

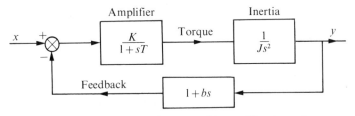

FIG. 3.8. Position-control system with amplifier dynamics.

described by a first-order differential equation with time constant T. The overall transfer function is

$$\frac{Y(s)}{X(s)} = \frac{K}{Js^2(1 + sT) + K(1 + bs)},$$

so the characteristic equation is

$$JTs^3 + Js^2 + Kbs + K = 0$$

and the condition for stability becomes

$$JKb - JTK > 0,$$

or

$$b > T,$$

which shows that large values of the amplifier time constant T can make the system unstable.

Problems 3.1, 3.2

3.3. Mathematical stability criteria for discrete-time systems

A discrete-time, linear, dynamic system is stable provided that all the roots of its characteristic equation lie within the unit circle in the s-plane.

The Routh–Hurwitz criterion is not immediately applicable to discrete-time equations but it can nevertheless be used in conjunction with the bilinear transformation

$$s = \frac{w + 1}{w - 1} \quad \text{or} \quad w = \frac{s + 1}{s - 1}.$$

This conformal transformation maps the region outside the unit circle in the s-plane to the positive real half of a w-plane, and the region inside the unit circle in the s-plane to the negative real half of the w-plane, as shown in Fig. 3.9. If the bilinear transformation is used to convert a discrete-time

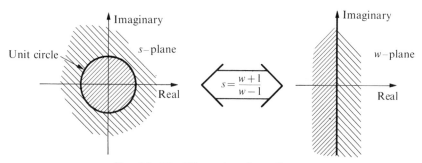

Fig. 3.9. The bilinear transformation.

characteristic equation in s to an equation in w the Routh–Hurwitz criterion can then be applied to the resulting equation in w to determine whether any roots lie in the unstable region.

For example if the general third-order discrete-time characteristic equation were written

$$a_3 s^3 + a_2 s^2 + a_1 s + a_0 = 0,$$

the bilinear transformation would give

$$a_3\left(\frac{w+1}{w-1}\right)^3 + a_2\left(\frac{w+1}{w-1}\right)^2 + a_1\frac{w+1}{w-1} + a_0 = 0,$$

which simplifies to

$$b_3 w^3 + b_2 w^2 + b_1 w + b_0 = 0,$$

where

$$b_0 = a_3 - a_2 + a_1 - a_0$$
$$b_1 = 3a_3 - a_2 - a_1 + 3a_0$$
$$b_2 = 3a_3 + a_2 - a_1 - 3a_0$$
$$b_3 = a_3 + a_2 + a_1 + a_0.$$

The roots of the transformed equation in w all lie in the negative real part of the w-plane provided that the Routh–Hurwitz criterion

$$b_1 b_2 - b_0 b_3 > 0$$

is satisfied. The bilinear transformation maps roots from inside the unit circle in the s-plane to the negative real part of the w-plane, so the above w-plane Routh–Hurwitz criterion is also the stability criterion that all roots of the discrete-time characteristic equation lie within the unit circle in the s-plane.

    An alternative, more direct, test for stability of discrete-time systems is provided by the Schur–Cohn criterion. This, like the Routh–Hurwitz criterion, is expressed in terms of determinants formed from coefficients of the characteristic equation. There are n determinants for an equation of order n; the typical one, Δ_j, is of order $2j$:

$$\Delta_j \equiv \left|
\begin{array}{cccccccccc}
a_0 & 0 & & . & . & 0 & a_n & a_{n-1} & . & . & a_{n-j+1} \\
a_1 & a_0 & & 0 & . & . & 0 & a_n & & . & a_{n-j+2} \\
. & . & & . & . & . & . & . & & . & . \\
. & . & & . & . & 0 & . & 0 & & . & . \\
a_{j-1} & a_{j-2} & & . & . & a_0 & 0 & . & & 0 & a_n \\
\hline
a_n & 0 & & . & . & 0 & a_0 & a_1 & . & . & a_{j-1} \\
a_{n-1} & a_n & & 0 & . & . & 0 & a_0 & . & . & a_{j-2} \\
. & . & & . & . & . & . & 0 & & . & . \\
. & . & & . & . & 0 & . & . & & . & . \\
a_{n-j+1} & a_{n-j+2} & & . & . & a_n & 0 & . & & 0 & a_0
\end{array}
\right|,$$

and $j = 1, \ldots, n$. The criterion states that the roots of the characteristic equation all have magnitude less than unity if and only if

$$\Delta_j < 0 \text{ for } j \text{ odd},$$
$$\Delta_j > 0 \text{ for } j \text{ even}.$$

This statement assumes that the leading coefficient a_n is positive. There is no simple discrete-time equivalent of the continuous-time necessary condition that all the coefficients of the characteristic equation must be non-zero and of the same sign.

Applying the Schur–Cohn criterion, for example, to a second-order discrete-time characteristic equation:

$$a_2 s^2 + a_1 s + a_0 = 0,$$

the determinants are

$$\Delta_1 = \begin{vmatrix} a_0 & a_2 \\ a_2 & a_0 \end{vmatrix},$$

$$\Delta_2 = \begin{vmatrix} a_0 & 0 & a_2 & a_1 \\ a_1 & a_0 & 0 & a_2 \\ a_2 & 0 & a_0 & a_1 \\ a_1 & a_2 & 0 & a_0 \end{vmatrix},$$

which simplify to

$$\Delta_1 = a_0^2 - a_2^2,$$

$$\Delta_2 = (a_2 - a_0)^2 \big((a_2 + a_0)^2 - a_1^2 \big).$$

The conditions on the sign of the determinants then give the stability criteria, for positive a_2,

$$a_0^2 < a_2^2,$$
$$a_1^2 < (a_0 + a_2)^2.$$

An example of a discrete-time control system is the problem of controlling the number of items, say door-mats, held in stock, say on the shelf in a shop, to meet a fluctuating demand, say customers wanting door-mats, where control is exercised at discrete intervals, say once a fortnight when a commercial traveller calls, by deciding what number of new items should be acquired to meet future demand. If $y(i)$ is the number of items in stock at time i, if $u(i)$ is the number of new items acquired, and if $w(i)$ is the demand in the coming time interval, the number of items in stock at time $i + 1$ will be given by the difference equation:

$$y(i + 1) = y(i) + u(i) - w(i).$$

A simple feedback control law would be to make the stock up to some desired level x by choosing

$$u(i) = x - y(i),$$

which combines with the difference equation for $y(i + 1)$ to give the stable difference equation

$$y(i + 1) = x - w(i).$$

The above control law requires that the number $y(i)$ of items currently in stock be known before the order $u(i)$ is placed, but such up-to-date information is not always available. When there is delay in the availability of information about y a potential alternative control law might be based on the previous value $y(i - 1)$,

$$u(i) = x - y(i - 1).$$

This combines with the difference equation $y(i + 1)$ to give

$$y(i + 1) - y(i) + y(i - 1) = x - w(i),$$

a second-order difference equation in y with characteristic equation

$$s^2 - s + 1 = 0,$$

having roots

$$s_1, s_2 = \tfrac{1}{2} \pm j \frac{\sqrt{3}}{2}$$

on the unit circle. The coefficients

$$a_0 = 1, \qquad a_1 = -1, \qquad a_2 = 1$$

here do not satisfy the Schur–Cohn stability criterion. The control law using delayed information would therefore lead to oscillations and should not be used without further modification, as discussed in Section 4.5.

Problems 3.3, 3.4

3.4. The Nyquist diagram

Mathematical stability criteria, like those in Sections 3.2 and 3.3, are not always entirely suitable for the analysis of control systems. The chief disadvantages are as follows.

(i) The dynamics of both the controlled process and the controller must be known in sufficient detail that it is possible to write down the overall dynamic equation and its characteristic equation.

(ii) Evaluation of the determinants used to express the criteria can be awkward for high-order systems.

(iii) The effect of change in any part of the system can be investigated only by completely recalculating the criteria.

(iv) When a system is stable the criteria give no further information about the roots of the characteristic equation and the transient response.

These disadvantages are avoided by the Nyquist diagram which provides a graphical criterion for continuous-time systems.

The Nyquist stability criterion is expressed with reference to the general, continuous-time, negative feedback system of Fig. 3.10(i). The closed-loop feedback system is stable if the open-loop frequency response locus $G(j\omega)$ on

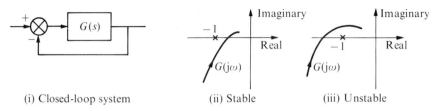

(i) Closed-loop system (ii) Stable (iii) Unstable

FIG. 3.10. The Nyquist stability criterion.

a polar diagram leaves the -1 point to the left as frequency increases.† In this context the polar diagram of open-loop frequency response is sometimes called a Nyquist diagram; its advantages are as follows.

(i) The dynamics of the system are represented by the open-loop frequency response which can sometimes be found experimentally even if exact differential equations are not known.

(ii) Stability is determined by straightforward inspection, as in Fig. 3.10(ii) and (iii).

(iii) Relationships between the shape of the frequency-response locus and the components of systems are well known, so the effect of changes is easily investigated.

(iv) The Nyquist diagram carries information about the values of roots of the closed-loop characteristic equation, and thus about transient response, as discussed below and in Section 3.5.

The open-loop transfer function $G(s)$ in Fig. 3.10(i) is the product of all transfer functions in the feedback loop. From the point of view of stability and the characteristic equation of the closed-loop system, the position of individual elements in the loop is irrelevant and systems like those in Fig. 3.11 are indistinguishable.

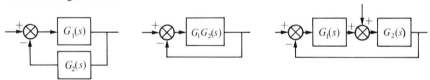

FIG. 3.11. Systems having the same characteristic equation $1 + G_1G_2(s) = 0$.

In the position-control system example of Fig. 3.8 the open-loop transfer function

$$G(s) = \frac{K(1 + bs)}{Js^2(1 + sT)}$$

† This statement of the Nyquist criterion is not completely precise because, for certain frequency response functions (e.g. $G(j\omega) = 1/\omega$), there could be ambiguity about what was 'left' and what 'right'. The precise statement requires reference to a closed contour around the positive half of the s-plane, see Nyquist (1932).

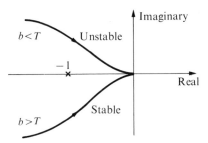

FIG. 3.12. Open-loop frequency response of position-control system.

has the two possible frequency-response loci shown in Fig. 3.12, and the
stability criterion $b > T$ follows directly from the Nyquist criterion.

The plausibility of the Nyquist criterion can be demonstrated by regarding
$G(s)$, which is a function of the complex variable s, as a mapping from the
s-plane to a corresponding $G(s)$-plane. The frequency response $G(j\omega)$ is then
the map of the imaginary axis from the s-plane, and other lines from the
s-plane would be mapped as shown in Fig. 3.13. For ordinary transfer functions the mapping is conformal; lines which intersect orthogonally in the

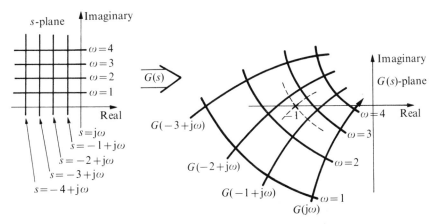

FIG. 3.13. $G(s)$ regarded as a conformal mapping.

s-plane map into lines which intersect orthogonally in the $G(s)$-plane, and
squares in the s-plane map into pseudo-squares in the $G(s)$-plane which can
be used for graphical solution of the closed-loop characteristic equation

$$1 + G(s) = 0$$

or

$$G(s) = -1.$$

The critical -1 point is marked on the $G(s)$-plane and then interpolation between the pseudo-squares identifies a value of s satisfying the equation; in the example of Fig. 3.13 the resulting root is

$$s_1 \simeq -1{\cdot}6 + 2{\cdot}7\mathrm{j}.$$

Other roots of the characteristic equation are not so easily found by this graphical method because the grid of straight lines in the s-plane maps into curved lines which can overlap each other in the $G(s)$-plane, so the mapping is not one-to-one and overlapping parts of the curved grid at the critical point are not easily sketched. However if the critical point is to the right of the $G(\mathrm{j}\omega)$ locus there must be at least one value of s which is to the right of the imaginary axis and which is a root of the characteristic equation: it follows that if there are to be no roots with positive real parts the critical point must be to the left of the $G(\mathrm{j}\omega)$ locus, which confirms the Nyquist criterion.

Many open-loop transfer functions can be expressed in the form used in Section 2.2:

$$G(s) = \frac{K(1 + \beta_1 s + \cdots)}{s^N(1 + \alpha_1 s + \cdots)} = KG_1(s),$$

where K is the gain and N the number of integrations in the closed loop. The Nyquist criterion can then be expressed with reference to the system of Fig. 3.14(i); the closed-loop system is stable if the open-loop $G_1(\mathrm{j}\omega)$ locus on the

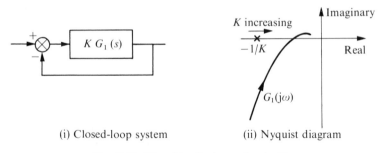

(i) Closed-loop system   (ii) Nyquist diagram

Fig. 3.14. Nyquist criterion and gain K.

Nyquist diagram leaves the critical $-1/K$ point to the left for increasing frequencies. Figure 3.14(ii) shows how the critical point moves to the right as the gain K increases and thus demonstrates the well-known tendency of closed-loop systems to become unstable as loop gain is increased: in the speech amplification system mentioned in Section 3.1 howling arises when there is too much gain, and can usually be eliminated by reducing the amount of feedback from loudspeaker to microphone. The Nyquist diagram also shows that integrations in the loop tend to make the closed-loop system unstable. Writing

$$G_1(s) = \frac{1}{s} G_2(s)$$

and drawing the frequency responses in Fig. 3.15, it can be seen that integration, which has frequency response $1/j\omega$, rotates the frequency-response locus clockwise so that it is more likely to pass the wrong side of the critical point.

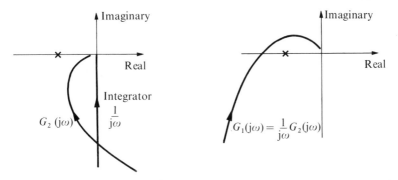

Fɪɢ. 3.15. Nyquist diagram showing the effect of integration.

High values of gain and many integrations tend to cause instability, so there is conflict between the requirements for acceptable transient response and the requirements for good steady-state performance, which were shown in Chapter 2 to be precisely high values of gain and many integrations. Much of control theory is devoted to ways of compromising between these conflicting requirements.

Problems 3.5–3.10

3.5. Phase margin and gain margin

The requirement that the transient response of a continuous-time system be not too oscillatory was interpreted in Section 3.1 in terms of damping ratio and the s-plane boundary of Fig. 3.4. The same requirement is interpreted on the Nyquist diagram in terms of *phase margin* and *gain margin*, both of which are a measure of the distance between the critical point and the open-loop frequency-response locus. The conformal mapping of Fig. 3.13 shows that when the critical point is close to the frequency-response locus there will be a root of the characteristic equation with small real part compared to its imaginary part, and it was shown in Section 3.1 that such a root would be one of a complex conjugate pair with small damping ratio. The distance between the critical point and the open-loop frequency-response locus is thus an indication of how well damped the closed-loop system is; the greater the distance the greater the damping.

Phase margin could be defined with reference to either of Figs 3.10(i) and 3.14(i). With reference to Fig. 3.14(i), phase margin is defined as the difference between 180° and the phase lag − $\underline{/G_1(j\omega_1)}$ at the frequency ω_1 for which the amplitude ratio $|G_1(j\omega_1)|$ is equal to $1/K$; Fig. 3.16 illustrates the definition and shows that phase margin increases with the distance between the critical $-1/K$ point and the $G_1(j\omega)$ locus.

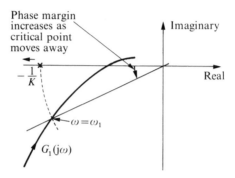

FIG 3.16. Phase margin on Nyquist diagram.

The relationship between phase margin and damping ratio is demonstrated by calculating the phase margin of a second-order system in which the open-loop transfer function is

$$G_1(s) = \frac{1}{s(1 + sT)}$$

and the gain K is chosen to make the damping ratio of the closed-loop system take the value $1/\sqrt{2}$. The closed-loop characteristic equation

$$1 + KG_1(s) = 0$$

here is

$$Ts^2 + s + K = 0$$

with natural frequency

$$\omega_0 = \sqrt{(K/T)}$$

and damping ratio

$$\zeta = \frac{1}{2}\frac{1}{\sqrt{(KT)}},$$

so the value of gain for the chosen damping ratio is

$$K = \frac{1}{2T}.$$

To calculate the resulting phase margin it is necessary first to find the frequency ω_1 at which

$$|G_1(j\omega_1)| = 1/K.$$

Substituting the given transfer function $G_1(s)$ and the calculated value of K gives

$$\frac{1}{\omega_1\sqrt{(1 + \omega_1^2 T^2)}} = 2T$$

or

$$4(\omega_1 T)^4 + 4(\omega_1 T)^2 - 1 = 0,$$

so that

$$(\omega_1 T)^2 = \frac{-1 \pm \sqrt{2}}{2}$$

and taking the positive value for the positive number $(\omega_1 T)^2$ gives

$$\omega_1 T = \sqrt{\left(\frac{\sqrt{2} - 1}{2}\right)}.$$

The phase margin can now be calculated from the definition

$$180° - (-\underline{/G_1(j\omega_1)}) = 180° - (90 + \tan^{-1}\omega_1 T)$$

$$= 90 - \tan^{-1}\omega_1 T = \tan^{-1}(1/\omega_1 T)$$

$$= \tan^{-1}(1\cdot 414/\sqrt{0\cdot 414}) \simeq 65\cdot 5°.$$

The exact relationship between phase margin and damping ratio is different for different transfer functions $G_1(s)$, but phase margins of about 50° usually give acceptable transient response.

The alternative measure of distance between the critical point and the frequency-response locus is gain margin which could also be defined with reference to either of Figs 3.10(i) and 3.14(i). With reference to Fig. 3.14(i), gain margin is defined as the attentuation, measured in decibels, of the open-loop transfer function $KG_1(j\omega)$ at the frequency ω_2 for which the phase angle $\underline{/G_1(j\omega_2)}$ is $-180°$; Fig. 3.17 illustrates the definition; the gain margin is

$$-20\log_{10}\frac{OA}{OP}\text{ dB,}$$

and this is a positive number which increases with the distance between the $-1/K$ point and the $G_1(j\omega)$ locus. Gain margins of about 15 dB usually give acceptable transient response.

Phase margin and gain margin are sometimes expressed with reference to Fig. 3.10(i). Writing

$$G(j\omega) = KG_1(j\omega) = M\,e^{j\phi},$$

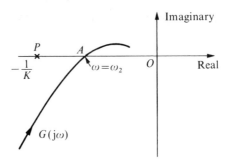

FIG. 3.17. Gain margin on Nyquist diagram.

the two margins can be conveniently shown on a Bode diagram of the open-loop frequency response, with the graphs for gain M and phase ϕ superposed by lining up 0 dB of M with $-180°$ of ϕ, as shown in Fig. 3.18. Negative values of either margin would be associated with instability and so the stability criterion appears on the Bode diagram as a requirement that the M curve must cross the horizontal axis at a lower frequency than does the ϕ curve, as in Fig. 3.18.

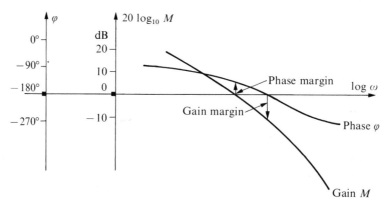

FIG. 3.18. Phase margin and gain margin on Bode diagram.

   The Bode diagram can be especially useful for rough calculations based on sketches of the asymptotes of the frequency-response graphs. Suppose for example that the transfer function in Fig. 3.14(i) were

$$KG_1(s) = \frac{K}{s(1 + 0{\cdot}1s)(1 + 0{\cdot}5s)},$$

and it was required to find a value of gain K to make the gain margin 15 dB. The frequency-response asymptotes can be sketched as in Fig. 3.19 which shows that the amplitude $|G_1(j\omega_2)|$ gives attenuation of about 20 dB at the

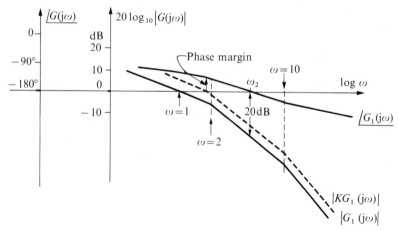

FIG. 3.19. Calculation of gain on sketched Bode diagram.

frequency ω_2 where the phase angle $\underline{/G_1(j\omega_2)}$ is $-180°$. If the gain margin is to be 15 dB, this attenuation can be reduced by 5 dB, so the gain K is estimated from

$$20 \log_{10} K = 5$$

to be

$$K \simeq 1·8,$$

and the dotted line for $|KG_1(j\omega)|$ shows that the resulting phase margin would be about 45°. In this particular example the results can be checked by exact calculation which gives

> actual gain margin with gain 1·8 = 16·4 dB,
> actual phase margin with gain 1·8 = 46°,
> correct gain K for 15 dB gain margin = 2·1,
> phase margin with correct gain 2·1 = 42°.

This comparison shows the sort of accuracy that can be achieved with calculations based on Bode diagram sketches: in many control systems the specification of a desired value for gain margin or phase margin is so arbitrary that the rough calculation is sufficient for practical purposes.

Problems 3.11–3.15

3.6. Bibliography

Most books on classical control engineering discuss continuous-time stability criteria, phase margin, and gain margin. These topics are presented together, as here, in

CHESTNUT and MAYER (1959) Chapter 6,
GILLE, PELEGRIN, and DECAULNE (1959) Chapters 9 and 16,

TAYLOR (1960) Chapter 19,
DOUCE (1963) Chapter 4,
SHINNERS (1964) Chapter 5,
DORF (1967) Chapter 5,
SAUCEDO and SCHIRING (1968) Chapter 10,
MELSA and SCHULTZ (1969) Chapter 6.

Discrete-time stability criteria are discussed in
RAGAZZINI and FRANKLIN (1958) Chapter 5,
JURY (1958) Section 1.12,
TOU (1959) Section 6.2,
KUO (1963) Chapter 6; (1967) Section 7.9; (1970) Chapter 5,
SHINNERS (1964) Chapter 9,
FREEMAN (1965) Section 7.13,
LINDORFF (1965) Sections 2.10–2.12.

3.7. Problems

3.1. Use the Routh–Hurwitz criterion to find which of the following polynomials
have factors with positive real parts:

(i) $s^3 + 10s^2 + 100s + 200$
(ii) $s^{10} + 9{\cdot}76s^9 + 7{\cdot}32s^8 + 85{\cdot}6s^7 + 101{\cdot}4s^6 - 23{\cdot}7s^5 + 435s^4 + 99{\cdot}6s^3$
$+ 258s^2 + 10{\cdot}5s + 3{\cdot}62$
(iii) $s^4 + 3s^3 + 2s^2 + 4s + 1$.

3.2. Explain why it is a necessary condition for stability of continuous-time
systems that all the coefficients in the characteristic equation must have the same
sign and be non-zero.

3.3. Use the Schur–Cohn criterion to investigate whether the polynomial
$$s^3 + s^2 + 1 = 0$$
has any factors outside the unit circle.

3.4. Use the bilinear transformation and the Routh–Hurwitz criterion to derive
the stability conditions for second-order discrete-time systems.

3.5. A continuous-time negative feedback system has open-loop transfer function
$$KG(s) = \frac{K}{(1 + sT_1)(1 + sT_2)(1 + sT_3)}.$$
Show that if the three time constants are all equal the value of gain that will just
give instability is 8, but that if only two of the time constants are equal and the
third is ten times as large then the limiting value of K is more than three times as
great.

3.6. The open-loop transfer function of a closed-loop system is
$$KG(s) = \frac{K}{s(1 + a_1s + a_2s^2)}.$$
Sketch the Nyquist diagram of open-loop frequency response and hence find the
range of values of K for stable performance if $a_1 = 2$ s and $a_2 = 1$ s^2.

3.7. Sketch the map in the $G(s)$-plane of a rectangular grid in the s-plane when $G(s) = 1/s$. Explain why the Nyquist stability criterion (that the $G(j\omega)$ locus must leave the $-1/K$ point to port) is plausible.

The open-loop transfer function of a closed-loop system is

$$KG(s) = K/s.$$

If $K = 2$ mark the $-1/K$ point on the map in the $G(s)$-plane and hence solve graphically the characteristic equation of the system and write down the time constant of closed-loop system response.

Consider the relation between the above map and one that would have been drawn for

$$G(s) = \frac{1}{1 + s},$$

and hence write down the time constant of the response of a closed-loop system with open-loop transfer function

$$KG(s) = \frac{2}{1 + s}.$$

3.8. Prove that the continuous-time system in Fig. 3.20 can only be stable if $a_1 > (a_2 + a_3)$. What then is the maximum value of K for which the system remains stable?

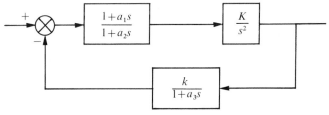

FIG. 3.20.

3.9. Figure 3.21(a) is the block diagram of a control system in which the transfer function $G(s)$ has the frequency-response asymptotes shown in Fig. 3.21(b). Over what range of values of K will the system be stable?

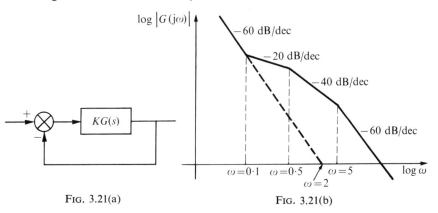

FIG. 3.21(a)   FIG. 3.21(b)

3.10. When the input x to the control system in Fig. 3.22 is a ramp function At, the steady-state error signal e has magnitude $10A$. The system becomes unstable if the gain K is increased by a factor of three.

What is the value of the time constant T?

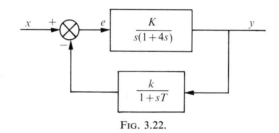

FIG. 3.22.

3.11. What value of gain K will give the system in Fig. 3.23.
   (i) phase margin $60°$?
   (ii) gain margin 12 dB?

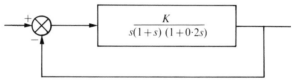

FIG. 3.23.

(This problem should be solved both by analysis and by estimation based on a *sketch* of the Bode diagram of open-loop frequency response.)

3.12. Prove that the phase margin ϕ and damping ratio ζ of the position-control system in Fig. 2.7 are related by the equation

$$\tan \phi = 2\zeta\sqrt{(2\zeta^2 + \sqrt{(4\zeta^4 + 1)})}.$$

3.13. Information about the transfer function $G(s)$ in the system shown in the Fig. 3.24 is derived from frequency-response tests as follows:

ω	0·5	1	2	3
$\|G(j\omega)\|$	0·72	0·28	0·08	0·02
$\angle G(j\omega)$	$-137°$	$-168°$	$-213°$	$-272°$

What should be the value of K if the system gain margin is to be 12 dB?

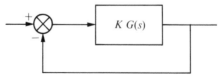

FIG. 3.24.

3.14. If the output y of the system in Fig. 3.25 is to follow an accelerating input $x = t^2$ with steady-state error $e = 0.02$ and to have phase margin $45°$, what should be the value of T?

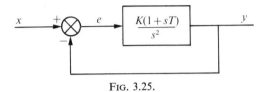

FIG. 3.25.

3.15. What is meant by the stability of a control system?

4. Root-locus diagrams

4.1. Introduction

THE root locus is a graphical technique for investigating roots of the characteristic equation

$$1 + G(s) = 0 \qquad\qquad (4.1)$$

of feedback systems like that in Fig. 4.1, without solving the equation analytically. It shows how the position of the roots in the s-plane changes as some

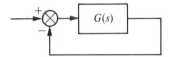

FIG. 4.1. Feedback system.

parameter, usually the gain, of the transfer function $G(s)$ takes different values, usually over the range from zero to infinity. The position of the roots determines the transient response of the system, as discussed in Section 3.1.

An introductory example is provided by considering a continuous-time system where

$$G(s) = \frac{K}{1 + s}$$

and the closed-loop characteristic equation

$$1 + s + K = 0$$

has a single root given by

$$s_1 = -(1 + K).$$

The complementary function of the differential equation of the system therefore has the exponentially decaying form $A\,e^{-t/T}$ with time constant

$$T = 1/(1 + K).$$

The effect of increasing the gain K is to decrease this time constant T, that is to make the transient response more rapid. The locus of the root s_1, as K varies from zero to infinity, is drawn in Fig. 4.2 which shows the root moving to the

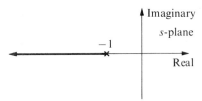

FIG. 4.2. Root locus when $G(s) = K/(1 + s)$.

left along the negative real axis as gain increases; distance to the left of the origin is proportional to rapidity of transient response.

A second introductory, continuous-time example is where

$$G(s) = \frac{K}{(1 + sT_1)(1 + sT_2)}$$

and the closed-loop characteristic equation

$$1 + s(T_1 + T_2) + s^2 T_1 T_2 + K = 0$$

has two roots given by

$$s_1, s_2 = -\frac{1}{2}\left(\frac{1}{T_1} + \frac{1}{T_2}\right) \pm \frac{1}{2}\sqrt{\left\{\left(\frac{1}{T_1} - \frac{1}{T_2}\right)^2 - \frac{4K}{T_1 T_2}\right\}}.$$

These roots are real or complex conjugate according to whether

$$\frac{4K}{T_1 T_2} \lessgtr \left(\frac{1}{T_1} - \frac{1}{T_2}\right)^2,$$

and their locus as K varies from zero to infinity is drawn in Fig. 4.3. The

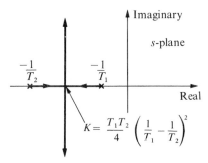

FIG. 4.3. Root locus when $G(s) = K/(1 + sT_1)(1 + sT_2)$.

closed-loop system has natural frequency ω_0 and damping ratio ζ given by

$$\omega_0 = \sqrt{\left(\frac{1 + K}{T_1 T_2}\right)}, \qquad \zeta = \frac{1}{2}\frac{T_1 + T_2}{\sqrt{(T_1 T_2(1 + K))}},$$

and the effect of increasing the gain K is to increase ω_0 and to decrease ζ. The root-locus diagram shows the roots moving away from the origin and subtending an increasing angle at the origin as gain increases; Fig. 3.3 showed how distance from the origin corresponds to natural frequency and how damping ratio decreases as the angle subtended at the origin increases.

4.2. Angle and magnitude conditions

The point of the root-locus technique is that it can be used when the characteristic equation cannot be solved analytically as it was in the examples of Section 4.1. To use the technique the open-loop transfer function $G(s)$ is written in the form

$$G(s) = \frac{k(s - z_1)(s - z_2) \times \cdots \times (s - z_m)}{(s - p_1)(s - p_2) \times \cdots \times (s - p_n)},$$

where k is proportional to the gain, $z_1, \ldots, z_m$ are the zeros and $p_1, \ldots, p_n$ the poles of the transfer function. The transfer function is a function of the complex variable s, and it is written in the above form so that it can itself be regarded as a complex number, in a $G(s)$-plane, having magnitude

$$|G(s)| = \frac{k|s - z_1| \, |s - z_2| \times \cdots \times |s - z_m|}{|s - p_1| \, |s - p_2| \times \cdots \times |s - p_n|}$$

and angle

$$\underline{/G(s)} = \underline{/s - z_1} + \underline{/s - z_2} + \cdots + \underline{/s - z_m} -$$
$$- \left(\underline{/s - p_1} + \underline{/s - p_2} + \cdots + \underline{/s - p_n} \right)$$

The basis of the root-locus technique is the observation that the characteristic equation

$$G(s) = -1$$

can only be satisfied by values of s such that the angle $\underline{/G(s)}$ is an odd integer multiple of π; Fig. 4.4 shows this is the angle associated with -1 when it is

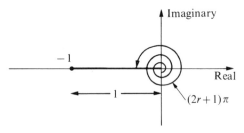

Fig. 4.4. -1 as a complex number (r is an integer).

regarded as a complex number in the $G(s)$-plane. The root-locus consists of all points in the s-plane satisfying the angle condition

$$\underline{/G(s)} = (2r + 1)\pi, \quad r \text{ an integer.}$$

The angle $\underline{/G(s)}$ is the sum of contributions from all the poles and zeros, and these contributions can be evaluated graphically for any value of s. A construction to find the contribution $\underline{/s - f}$ from a typical factor f is shown in the s-plane of Fig. 4.5 where the complex number $s - f$ is represented by the resultant of vectors representing s and $-f$; it follows from the figure that $\underline{/s - f}$ is the angle associated with the vector joining f to s.

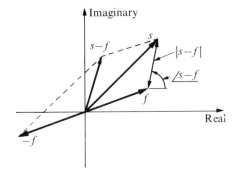

FIG. 4.5. Construction for $\underline{/s - f}$ and $|s - f|$.

An example of how the construction of Fig. 4.5 is used to evaluate $\underline{/G(s)}$ and find the root-locus is provided by a system with two real open-loop poles $-a_1$ and $-a_2$, and one real open-loop zero $-b$,

$$G(s) = \frac{k(s + b)}{(s + a_1)(s + a_2)}$$

The poles and the zero are marked on the s-plane of Fig. 4.6, and the angle

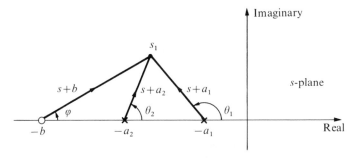

FIG. 4.6. Components of $\underline{/G(s)}$ when $G(s) = k(s + b)/(s + a_1)(s + a_2)$.

$\underline{/G(s_1)}$ associated with the complex number $G(s)$ when s takes the value s_1 is the sum of three contributions

$$\underline{/G(s_1)} = \phi - (\theta_1 + \theta_2).$$

The angle condition here can be written

$$\theta_1 + \theta_2 - \phi = \pm \pi$$

and the root-locus of s-plane points satisfying this condition is sketched in Fig. 4.7. The parts of the locus on the real axis are precisely determined from

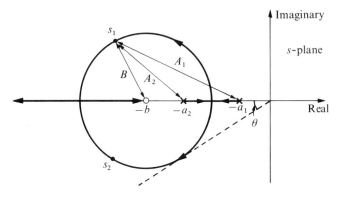

Fig. 4.7. Root locus when $G(s) = k(s + b)/(s + a_1)(s + a_2)$.

inspection of the angle condition: they are segments of the axis having an odd number of singularities to their right. The complex part of the locus for second-order systems such as this can be shown to be a circle centred on the zero at $-b$. The locus is symmetrical about the real axis because complex roots of characteristic equations of ordinary control systems arise in conjugate pairs. This diagram shows that the roots of the closed-loop characteristic equation are complex, giving oscillatory transient response, over some intermediate range of values of gain; that the damping ratio of the closed-loop system is never less than some value cos θ; and that for large values of gain the transient response consists of two exponentially decaying terms, the slower of which has time constant about $1/b$.

The angle condition is used to draw the root-locus diagram; correlation between points on the locus and values of the gain k is established by a magnitude condition

$$|G(s)| = 1$$

which must also be satisfied by values of s that are roots of the characteristic equation. The magnitude $|G(s)|$ is the product of contributions from all the poles and zeros, and these contributions can be evaluated graphically using the construction of Fig. 4.5 again. The figure shows that the contribution

$|s - f|$ from a typical factor f is the length of the vector joining f to s. Using the magnitude condition for the example of Fig. 4.7, the value of gain which gives the roots the complex conjugate values s_1, s_2 would be

$$k_1 = \frac{A_1 A_2}{B},$$

where A_1, A_2, B are the lengths of the vectors in the figure.

4.3. Rules for drawing root-locus diagrams

Root-locus diagrams are usually drawn with the help of the rule mentioned above.

(i) Segments of the real axis having an odd number of singularities to their right are parts of the root locus;

and of rules which follow from writing the characteristic equation in the factored form

$$(s - p_1)(s - p_2) \times \cdots \times (s - p_n) + k(s - z_1)(s - z_2) \times \cdots \times (s - z_m) = 0.$$

(ii) There are n branches to the root locus.

The number of branches is the number of roots of the characteristic equation and is the same as the order of the equation, which is n. The order is n, the number of open-loop poles, rather than m, the number of open-loop zeros, because of the property of real systems that $n \geqslant m$.

(iii) The branches of the locus run from open-loop poles to open-loop zeros.

At one end of the locus k goes to zero and the factored characteristic equation goes to

$$(s - p_1)(s - p_2) \times \cdots \times (s - p_n) = 0,$$

an equation with roots at the open-loop poles $p_1, \ldots, p_n$.

At the other end of the locus k goes to infinity and the factored characteristic equation goes to

$$k(s - z_1)(s - z_2) \times \cdots \times (s - z_m) = 0,$$

an equation with roots at the open-loop zeros $z_1, \ldots, z_m$.

(iv) When the number of open-loop poles exceeds the number of open-loop zeros $(n > m)$, there are $n - m$ branches of the locus going to infinity. These branches are asymptotic to straight lines having slopes

$$\frac{2r + 1}{n - m} \pi, \quad r \text{ an integer},$$

and intersect, on the real axis, at a point s_1 given by

$$s_1 = \big(p_1 + \cdots + p_n - (z_1 + \cdots + z_m)\big)/(n - m).$$

As s goes to infinity the open-loop transfer function $G(s)$ can be approximated by the limiting form

$$G(s) \to \frac{k}{s^{n-m}} \quad \text{as } s \to \infty,$$

which has $n - m$ zeros at infinity, where $n - m$ excess branches of the locus can terminate. Applying the angle condition to this limiting form gives

$$\underline{/G(s)} \to \left\lfloor \frac{k}{s^{n-m}} \right. = (2r + 1)\pi,$$

and so roots of the characteristic equation at infinity would lie on lines with angles

$$\underline{\lfloor s} = \frac{2r + 1}{n - m}\,\pi.$$

The point of intersection s_I is found by using a slightly closer approximation to $G(s)$:

$$G(s) \to \frac{k}{(s - s_I)^{n-m}} \quad \text{as } s \to \infty,$$

which would give rise to a root-locus diagram of $n - m$ straight lines intersecting at s_I as shown in Fig. 4.8. Neglecting all but the higher-order terms in

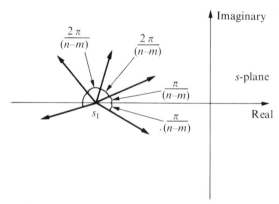

FIG. 4.8. Root locus when $G(s) = k/(s - s_I)^{n-m}$.

s, this approximation to $G(s)$ can be written

$$\frac{k}{(s - s_I)^{n-m}} \simeq \frac{k}{s^{n-m} - (n - m)s_I s^{n-m-1}},$$

and the coefficient of s^{n-m-1} can be related to the poles and zeros of the true transfer function $G(s)$ by dividing the denominator $s^n - (p_1 + \cdots + p_n)s^{n-1} + \cdots$ by the numerator $s^m - (z_1 + \cdots + z_m)s^{m-1} + \cdots$ to give

$$(n - m)s_I = (p_1 + \cdots + p_n) - (z_1 + \cdots + z_m).$$

The foregoing rules are sufficient for sketching root-locus diagrams in simple cases. An example is the third-order system discussed in connection with the Bode diagram in Section 3.5:

$$G(s) = \frac{K}{s(1 + 0{\cdot}1s)(1 + 0{\cdot}5s)}$$

$$= \frac{k}{s(s + 2)(s + 10)}, \quad \text{where } k = 20K.$$

Applying the rules:

(i) two segments of the real axis, between the origin and -2, and between -10 and $-\infty$, are parts of the root locus;

(ii) there are three open-loop poles and therefore three branches;

(iii) the branches start at the open-loop poles 0, -2, -10;

(iv) there are no open-loop zeros so all the branches go to infinity with asymptotes having slopes $\pi/3$, π, $5\pi/3$ and point of intersection s_I at -4;

results in the root-locus diagram of Fig. 4.9. This shows that when the gain is small there is a real root of the closed-loop characteristic equation near the

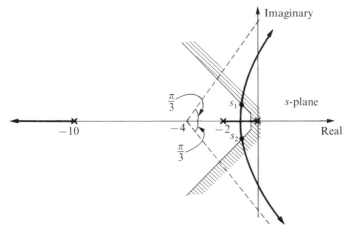

Fig. 4.9. Root locus when $G(s) = k/s(s + 2)(s + 10)$.

origin, which gives rise to a slowly decaying component of transient response, and when the gain is large there are a pair of complex conjugate roots crossing into the positive half of the s-plane, which give rise to instability. The figure

also shows a boundary, like that in Fig. 3.6, expressing the combined requirements for non-oscillatory, rapid response; the largest value of gain that could
be used while satisfying such requirements is the value that puts the complex
conjugate roots s_1, s_2 on the boundary. This value of gain can be roughly
calculated from the figure, using the magnitude condition. From the sketch
in the figure the lengths of vectors joining s_1 to the open-loop poles 0, -2,
-10 are estimated to be $1\frac{1}{2}$, $1\frac{1}{2}$, $9\frac{1}{2}$; then the magnitude condition gives

$$k_{max} \simeq 1\tfrac{1}{2} \times 1\tfrac{1}{2} \times 9\tfrac{1}{2} \simeq 21$$

and so

$$K_{max} = \frac{k_{max}}{20} \simeq 1.$$

This value for K is of the same order of magnitude as the value 2.1, quoted in
Section 3.5, to make the gain margin 15 dB.

Root-locus diagrams are like Bode diagrams in that they can sometimes
provide rough but useful numerical estimates of closed-loop transient response, from simple sketches based on knowledge of the open-loop transfer
function.

An example of the root-locus diagram for a system with an open-loop zero
is provided by the position-control system of Fig. 3.8 which has open-loop
transfer function

$$G(s) = \frac{K(1 + bs)}{Js^2(1 + sT)}$$

$$= \frac{k(s + 1/b)}{s^2(s + 1/T)}, \quad \text{where } k = \frac{Kb}{JT}.$$

Applying the rules:

(i) the segment of real axis between $-1/b$ and $-1/T$ is part of the root
locus;

(ii) there are three branches;

(iii) the branches start at the open-loop poles 0, 0, $-1/T$ and one branch
terminates at the open-loop zero $-1/b$;

(iv) two branches go to infinity with asymptotes having slopes $\pi/2$, $3\pi/2$
and point of intersection s_I at $\frac{1}{2}(-1/T + 1/b)$.

The position of s_I determines the stability of this particular system; Fig. 4.10
shows the two possible root-locus diagrams depending on whether s_I is in the
negative or the positive half of the s-plane. The condition for stability here

$$b > T$$

was also illustrated on the Nyquist diagram in Fig. 3.12. In a real system there

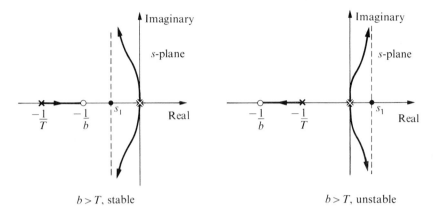

FIG. 4.10. Root locus when $G(s) = k(s + 1/b)/s^2(s + 1/T)$.

would be additional small lags, perhaps in instruments measuring the output, so that the open-loop transfer function might be

$$G(s) = \frac{k_1(s + 1/b)}{s^2(s + 1/T)(s + 1/T_1)}, \quad \text{where } k_1 = \frac{Kb}{JTT_1}$$

with the root-locus diagram shown in Fig. 4.11. This shows little change at low values of gain from the stable diagram of Fig. 4.10, but at high values of

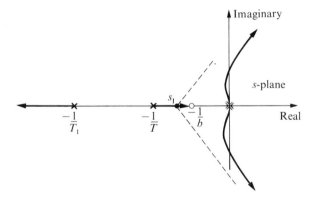

FIG. 4.11. Root locus when $G(s) = k_1(s + 1/b)/(s^2(s + 1/T)(s + 1/T_1))$.

gain the complex conjugate roots cross into the positive half of the *s*-plane and the system would become unstable even though the condition $b > T$ was satisfied.

Problems 4.1–4.3

4.4. Additional features of root-locus diagrams

Further guidance about the shape of root-locus diagrams at certain special points in the s-plane can be derived from the angle condition.

(i) At complex conjugate open-loop singularities, like the poles p_1, p_2 in Fig. 4.12, the direction of the locus can be found.

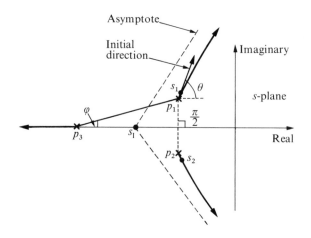

FIG. 4.12. Root locus when $G(s) = k/(s + p_1)(s + p_2)(s + p_3)$, p_1 and p_2 complex conjugates.

The direction of the locus at the complex pole p_1, specified by the angle θ in the figure, is found by considering a point s_1 on the locus very close to p_1. At such a point the contributions to $/G(s_1)$ due to the factors $s_1 - p_2$ and $s_1 - p_3$ can be approximated by the angles of lines joining p_1, rather than s_1, to p_2 and to p_3, that is by $\pi/2$ and ϕ. The angle condition then gives

$$\theta = \lfloor s_1 - p_1 \doteq \pi - (\pi/2 + \phi) = \pi/2 - \phi,$$

which determines the direction of the locus close to p_1.

Similar calculations can be used to find the direction of the locus at other points. Such calculations are usually done either at complex conjugate open-loop singularities, as here, or for points where the locus branches away from the real axis, as below.

(ii) Where the locus branches away from the real axis, as in Figs 4.3, 4.7, 4.8, 4.9, 4.10, 4.11, the direction of the locus depends only on the number N of branches at the point in question and is given by angles

$$\theta = \frac{(2r + 1)\pi}{N}, \quad r \text{ an integer.}$$

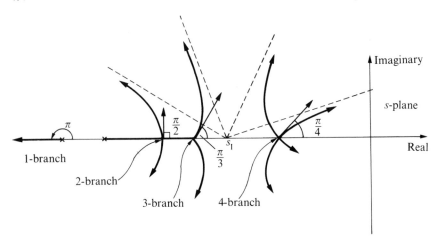

FIG. 4.13. Root locus with many branchings from real axis.

Figure 4.13 shows some typical branchings from points where two, three, and four branches leave the real axis.

(iii) Points, other than open-loop singularities, where a pair of branches meet the real axis, as in Figs 4.7, 4.9, and 4.13, can sometimes be calculated.

The calculation is based on the result from (ii) above, that a pair of branches must meet the real axis at angles $\pm \pi/2$. The method is demonstrated by considering an example with open-loop transfer function like that in Fig. 4.7 but with numerical values assigned to the singularities to give

$$G(s) = \frac{k(s + 3)}{(s + 1)(s + 2)}.$$

Figure 4.14 is part of the root-locus diagram for this example with the

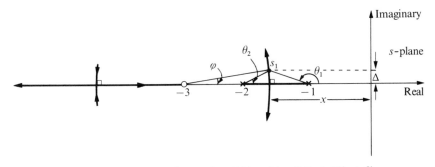

FIG. 4.14. Part of root locus when $G(s) = k(s+3)/(s + 1)(s + 2)$.

'imaginary' scale magnified to show a point s_1 on the locus close to a branching point distant x from the origin. The angle condition here can be written

$$\theta_2 - \phi = \pi - \theta_1,$$

and when s_1 is close to the axis these angles are small and can be approximated by their tangents

$$\frac{\Delta}{2 - x} - \frac{\Delta}{3 - x} = \frac{\Delta}{x - 1}.$$

This gives a quadratic equation with roots

$$x = 3 \pm \sqrt{2};$$

the root $3 - \sqrt{2}$ gives the point considered in detail in Fig. 4.14, the other root gives the point where the two branches rejoin the real axis.

Another way to calculate the same points is to notice that in any segment of locus on the real axis the branching point is where the gain k takes an extreme (maximum or minimum) value. In Fig. 4.14, for example, the branching point between -1 and -2 is where the gain has its maximum value in that segment of real axis, and the branching point between -3 and $-\infty$ is where the gain has its minimum value in that segment. The points can be calculated using the magnitude condition to write the gain corresponding to a point s_2 on the real axis in Fig. 4.14,

$$k = (2 - x)(x - 1)/(3 - x);$$

this takes extreme values when

$$\frac{\mathrm{d}k}{\mathrm{d}x} = 0,$$

which gives the same quadratic equation with the same roots as before.

Branching points are not so easily calculated when the number of open-loop singularities is high, because then the equations to be solved become high-order polynomials; nor when there are complex open-loop singularities, because then the equations are less simple.

Another source of guidance about the shape of root-locus diagrams is provided by an analogy with two-dimensional flow patterns satisfying Laplace's equation.† This analogy arises because the open-loop transfer function $G(s)$ is analytic in the s-plane, except at its singularities, so that its angle $\underline{/G(s)}$ is also analytic. It follows that the locus of points in the s-plane where $\underline{/G(s)}$ has the constant value π, that is the root locus, will correspond to loci where other, similar, analytic functions of s are constant. A suitable function, which can be proved to be similar to $\underline{/G(s)}$, is the stream function of two-dimensional flow around a number of point sources and sinks. The analogy is therefore expressed by regarding the s-plane as a uniform conducting sheet having unit

† See Tsien (1954) pp. 46–47 or Truxal (1955) Section 1.3.

sources at each open-loop pole, unit sinks at each open-loop zero, and $n - m$ sinks distributed around its edge at infinity so that the total flow out is the same as the flow in: then the root locus corresponds to a stream line of the flow. This statement of the analogy does not specify which streamline corresponds to the root locus, but it nevertheless gives a useful feel for the way in which the root locus tends to be drawn towards open-loop zeros.

Problems 4.4–4.9

4.5. Discrete-time systems

The root-locus technique can just as well be used to investigate the roots of characteristic equations for discrete-time systems when these have the form of equation (4.1). The resulting root-locus diagrams must be interpreted in terms of the discrete-time s-plane boundary of Fig. 3.7 instead of the continuous-time boundary of Fig. 3.6.

An example is provided by considering a simple feedback control law for the stock-control problem introduced in Section 3.3. Figure 4.15 is a block diagram of this example where it is assumed that the number of acquisitions

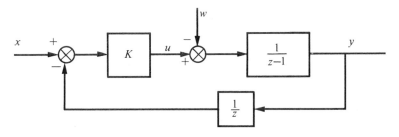

FIG. 4.15. Block diagram of stock-control system.

is made proportional to the difference between a desired value x and the actual stock y, but that there is a delay of one time interval in measuring the stock y. In the figure the z-transfer function $1/(z - 1)$ corresponds to the difference equation

$$y(i + 1) = y(i) + u(i) - w(i)$$

relating stock to acquisitions u and to demand w; the z-transfer function $1/z$ corresponds to the delay in measuring y. This discrete-time system corresponds to the completely-sampled feedback system in Fig. 1.27 and has characteristic equation

$$1 + \frac{K}{s(s - 1)} = 0.$$

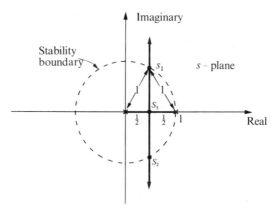

FIG. 4.16. Root locus when $G(s) = K/s(s-1)$.

Figure 4.16 is the corresponding root-locus diagram; it shows that

(i) the system becomes unstable when the roots are at s_1, s_2, and then the magnitude condition gives

$$K_{\text{crit}} = 1;$$

(ii) both roots are nearest the origin when they coincide at s_3, and then the magnitude condition gives

$$K = \tfrac{1}{4}.$$

For values of K less than $\tfrac{1}{4}$, one of the real roots is closer to unity and the transient response becomes sluggish; for values greater than $\tfrac{1}{4}$, both roots move away from s_3, their magnitude increases and the response becomes oscillatory.

Problems 4.10, 4.11

4.6. Bibliography

Many books on classical control engineering discuss root-locus diagrams for continuous-time systems. Some having a special chapter on this topic are

EVANS (1954) Chapter 7,
TRUXAL (1955) Chapter 4,
SAVANT (1958) Chapter 4,
CHESTNUT and MAYER (1959) Chapter 13,
GILLE, PELEGRIN, and DECAULNE (1959) Chapter 14,
TAYLOR (1960) Chapter 18,
DEL TORO and PARKER (1960) Chapter 10,
WILTS (1960) Chapter 5,
RAVEN (1961) Chapter 7,
BARBE (1963) Chapter 11,

LANGILL (1965) Chapter 10,
D'AZZO and HOUPIS (1966) Chapter 7,
DORF (1967) Chapter 6,
DRANSFIELD (1968) Chapter 7,
SENSICLE (1968) Chapter 7,
HARRISON and BOLLINGER (1969) Chapter 12,
MELSA and SCHULTZ (1969), Chapter 7,
TAYLOR (1969) Chapter 18,
WATKINS (1969) Chapter 7,
OGATA (1970) Chapter 8,
TOWILL (1970) Chapter 3,
EVELEIGH (1972) Chapter 6.
Discrete-time root-locus diagrams are discussed in
RAGAZZINI and FRANKLIN (1958) Chapter 5,
JURY (1958) Chapter 3,
TOU (1959) Section 9.8,
KUO (1963) Chapter 6,
LINDORFF (1965) Section 2.13,
GUPTA and HASDORFF (1970) Section 6.5.
Continuous-time and discrete-time root-locus diagrams are discussed together in
KUO (1967) Chapter 8,
SAUCEDO and SCHIRING (1968) Chapter 8.

4.7. Problems

(N.B. Problems 4.1–4.9 are about continuous-time systems.)

4.1. Sketch root-locus diagrams for closed-loop feedback systems having the following open-loop transfer functions:

(i) K/s^2

(ii) $\dfrac{K}{s^2(1 + sT)}$

(iii) $\dfrac{K(1 + aTs)}{s^2(1 + sT)}$; make two sketches, one for $a > 1$ and one for $a < 1$

(iv) $\dfrac{K(s^2 + s + 1)}{(s + 1)^4}$

Which of the five diagrams represents a stable system?

4.2. A simple position-control system positions a rotating load of inertia J and viscous friction coefficient F. The actuating signal is a torque proportional to position error. Sketch the system root-locus diagram, and hence determine the value of gain that will give the system a damping ratio $1/\sqrt{2}$.

   Check this result by deriving an expression for damping ratio directly from the closed-loop characteristic equation.

4.3. A closed-loop system has open-loop transfer function,

$$\frac{K}{s(1 + s)(1 + 0.5s)}.$$

Sketch a root-locus diagram for the system and hence determine the maximum value of K for which the system can be stable.

Check this result by a Nyquist diagram calculation.

4.4. The Bode diagram of Fig. 4.17(a) gives the open-loop frequency response of the system in Fig. 4.17(b). Sketch the relevant root-locus diagram, and locate the point at which the locus breaks away from the real axis. Hence determine whether or not there is an oscillating component in the transient response of the system.

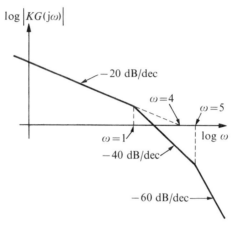

FIG. 4.17(a).

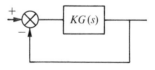

FIG. 4.17(b).

4.5. A certain servomechanism controls the position of a rotating load having inertia J and viscous friction of coefficient F. The control signal is the difference between an input signal and a signal $K_1 y + K_2 \, dy/dt$, where y is the output position: this signal is amplified by a power amplifier that has a time constant T_1. As a result of inductance in the motor field windings, the torque applied to the load lags behind the power amplifier with another time constant T_2.

Sketch root-locus diagrams for the system under each of the four conditions:

(i) $\dfrac{1}{T_1} > \dfrac{1}{T_2} > \dfrac{F}{J} > \dfrac{K_1}{K_2}$

(ii) $\dfrac{1}{T_1} > \dfrac{1}{T_2} > \dfrac{K_1}{K_2} > \dfrac{F}{J}$

(iii) $\dfrac{1}{T_1} > \dfrac{K_1}{K_2} > \dfrac{1}{T_2} > \dfrac{F}{J}$

(iv) $\dfrac{K_1}{K_2} > \dfrac{1}{T_1} > \dfrac{1}{T_2} > \dfrac{F}{J}$

4.6. A closed-loop system has open-loop transfer function

$$\frac{K(1 + 0\cdot3s)(1 + 25s)}{s^2(1 + s)}.$$

Sketch a root-locus diagram for the system and derive an equation for the point where the locus rejoins the real axis. Hence estimate the maximum value of K for which there is any oscillatory component in the transient response of the system.

4.7. Figure 4.18 shows a control system in which the gain K is to be chosen so that the oscillatory component of transient response has damping ratio $1/\sqrt{2}$. Make a sketch of the root-locus diagram of the system and use it to estimate the value for K. Also estimate the time constant of the slowest component of the transient response.

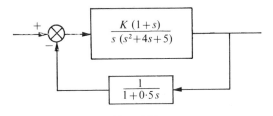

FIG. 4.18.

4.8. The controlled process in a certain feedback control system has transfer function

$$G(s) = \frac{10}{(1 + 0\cdot2s)^2(1 + 0\cdot05s)}.$$

A controller with transfer function $H(s)$ measures the system error and generates a control signal which is the input to $G(s)$. Draw closed-loop root-locus diagrams as K is varied for three forms of $H(s)$:

(i) $H(s) = K$
(ii) $H(s) = K(1 + 0\cdot2s)$
(iii) $H(s) = K(1 + 0\cdot2s)/(1 + 0\cdot02s)$

and use them to estimate what is the maximum value of K in each case that will give (a) stability, and (b) acceptable damping.

Discuss the merits of the three controllers: suggest why (ii) may be impracticable and compare the performance of the system with (i) and with (iii).

4.9. Figure 4.19 shows a sketch of the root-locus diagram for a closed-loop system having open-loop transfer function

$$\frac{K(1 + 5s)}{s^2(1 + s)}$$

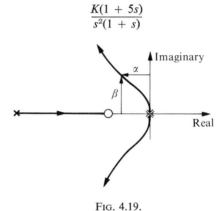

FIG. 4.19.

(i) Use the angle condition to show that the coordinates (α, β as shown) of the curved part of the locus are related by

$$\beta^2 = \frac{2\alpha^3 - 1 \cdot 6\alpha^2 + 0 \cdot 4\alpha}{0 \cdot 8 - 2\alpha}.$$

(ii) Compute values of β for $\alpha = 0 \cdot 1$, $0 \cdot 2$, $0 \cdot 3$, $0 \cdot 35$ and hence draw the root-locus diagram (use a scale of the order of $0 \cdot 02$ m to $0 \cdot 1$ unit).

(iii) Use the magnitude relationship to find the value of gain corresponding to a few points on the curve. Also derive an analytic expression to give values of gain along the locus on the real axis. Mark a few values of gain on the loci.

(iv) It is required to set the gain of the system so that all three closed-loop roots have the same real part. Estimate from the root-locus diagram (using a method of successive approximations if necessary) the appropriate value of K.

(v) With the value of K determined above, what is the time constant characterizing system transient response? What is the damping ratio of the oscillating part of the response?

(vi) What is the maximum damping ratio possible for the system? What is the corresponding value of K? What then is the largest time constant in the response of the system?

4.10. Sketch the root-locus diagram for the discrete-time system of Problem 1.11, and comment on the criterion used to choose a value for the gain K.

4.11. Sketch the root-locus diagram for the discrete-time system of Fig. 1.30 when the transfer functions are as specified in Problem 1.27. Hence find the maximum value of K for stability.

5. Controller transfer functions

5.1. Introduction

THE conflicting requirements in a control system for good steady-state performance and for good transient response cannot always be resolved simply by the choice of a value for gain, or of the number of integrations to be used.

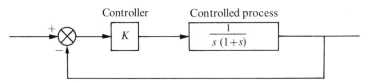

FIG. 5.1. Example of type 1 control system.

In the type 1 continuous-time system of Fig. 5.1, for example, the analysis of Section 3.5 shows that acceptable transient response requires the controller gain K to have the value

$$K = \tfrac{1}{2}.$$

This value of gain determines the steady-state performance in accordance with the table of Fig. 2.5, and any attempt to improve steady-state performance simply by increasing the value of K or by introducing an extra integration in the controller would lead to oscillatory response or to instability, as shown in Figs 5.2(i) and 5.2(ii).

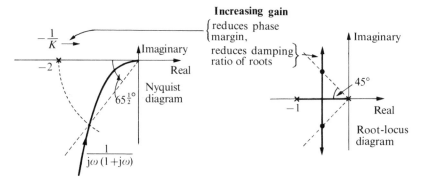

FIG. 5.2(i). Increasing the gain in the system of Fig. 5.1.

Unstable

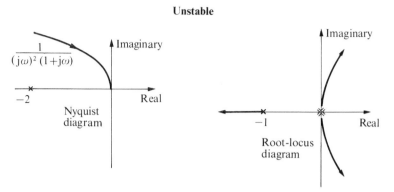

FIG. 5.2(ii). Extra integration in the system of Fig. 5.1.

Overall performance can often be improved by using a dynamic controller, with transfer function $KH(s)$ as shown in Fig. 5.3, to modify the dynamics of the system so that better performance can be achieved without deterioration of transient response. Controller transfer functions for continuous-time systems are often designed on the basis of approximate calculations relating to frequency response or root-locus diagrams; purely analytic design procedures have also been developed but are not so widely used for simple applications.

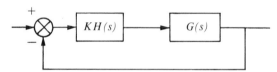

FIG. 5.3. Introducing transfer function $H(s)$ in the controller.

The approximate graphical methods are introduced in this chapter, and the analytic procedures in Chapter 6. Corresponding techniques for designing discrete-time controllers scarcely exist and are not discussed in these chapters, however the generalization of the analytic procedure in subsequent chapters applies to discrete-time as well as to continuous-time systems.

5.2. Phase advance and phase lag

A controller transfer function $H(s)$ to improve the overall performance of the system in Fig. 5.3 could be proposed on the basis of either of two lines of reasoning. Both possibilities are introduced here with reference to the Nyquist diagram for a system where the controlled process transfer function is

$$G(s) = \frac{1}{s(1 + s)},$$

as in Fig. 5.1. One way is to use the transfer function

$$H_1(s) = 1 + sT_1$$

which, as shown in Fig. 5.4, has little effect on the frequency response at low frequencies but modifies the response at high frequencies so that more gain can be used, with consequent improvement in steady-state performance.

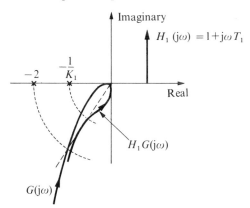

Fɪɢ. 5.4. Controller transfer function $H_1(s) = 1 + sT_1$ (phase advance).

Figure 5.4 shows that $H_1(s)$ makes it possible to increase the controller gain from the value $\frac{1}{2}$ to a larger value K_1 without reduction of phase margin; the steady-state errors will then be reduced by a factor $K_1/2$.

The other way is to use the transfer function

$$H_2(s) = 1 + \frac{1}{sT_2}$$

which, as shown in Fig. 5.5, has little effect on the frequency response at high frequencies but effectively introduces another integration at low frequencies so that the type-number of the system is increased, with consequent improve-

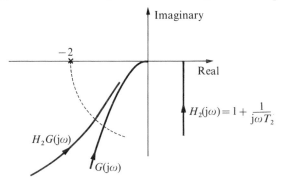

Fɪɢ. 5.5. Controller transfer function $H_2(s) = 1 + 1/sT_2$ (phase lag).

ment in steady-state performance. Figure 5.5 shows that $H_2(s)$ changes the type 1 system into a type 2 system, but with little reduction of phase margin.

The transfer functions $H_1(s)$ and $H_2(s)$ can be described as phase-advance and phase-lag controllers because the effect they have on the combined open-loop frequency-response function $HG(j\omega)$ is, respectively, to advance and to retard the phase angle, as in Figs 5.4 and 5.5. The effects of these transfer functions can also be illustrated on the root-locus diagram. The phase advance $H_1(s)$ introduces an extra zero into the open-loop transfer function; Fig. 5.6

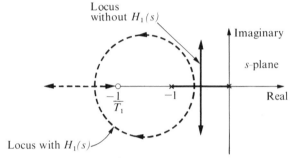

FIG. 5.6. Effect of phase advance $H_1(s)$ on root locus.

shows how the locus is drawn towards the zero and away from forbidden regions of the s-plane. Phase advance is thus equivalent to velocity feedback in position-control systems, which also has the effect of introducing an extra open-loop zero as can be seen from the block diagram of Figs 2.7, 2.8, or 3.7 or the example in Section 4.3.

The phase lag $H_2(s)$ can be written

$$H_2(s) = \frac{s + 1/T_2}{s}$$

to show that it introduces an open-loop zero to offset the destabilizing effect of the open-loop pole at the origin which increases the type number of the system. For the particular example discussed here the combined open-loop transfer function

$$KH_2G(s) = \frac{K(s + 1/T_2)}{s^2(s + 1)}$$

has the root-locus diagram shown in Fig. 4.10; the stabilizing time constant T_2 must be greater than the time constant of the controlled process, unity, for the closed-loop system to be stable. Phase lag is not normally usable in systems where the transfer function $G(s)$ of the controlled process has more than one integration, because its single open-loop zero cannot stabilize a system where the resulting combined $H_2G(s)$ would have more than two integrations.

Calculation of values for controller gains and time constants can become awkward when the controller includes either phase advance or phase lag, and it is usual to design to approximate, rather than to find exact, values. Examples of calculations for the two types of controller transfer function are as follows.

(i) Phase advance is to be used to give open-loop transfer function

$$K_1 H_1(s) G(s) = 10(1 + sT_1) \frac{1}{s(1 + s)};$$

what value of the phase-advance time constant T_1 will give the system a phase margin of about 60°? (The gain K_1 here is twenty times as large as what could be used in the system of Fig. 5.1, with consequent improvement in steady-state performance.)

An exact calculation of the phase margin here leads to a high-order polynomial in T_1 which cannot easily be solved; it is preferable to follow the approximate Nyquist diagram calculation illustrated in Fig. 5.7. First the

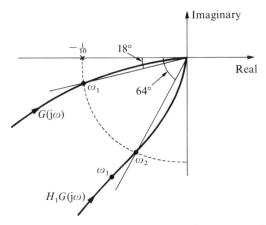

Fig. 5.7. Calculation of phase-advance time constant T_1.

phase margin is calculated assuming that no phase advance is used; the frequency ω_1 at which

$$|G(j\omega_1)| = \frac{1}{K_1}$$

is given by

$$\frac{1}{\omega_1 \sqrt{(1 + \omega_1^2)}} = \frac{1}{10}$$

and is found to be

$$\omega_1 = 3 \cdot 08,$$

which gives the phase margin

$$180° + \underline{/G(j\omega_1)} = 180° - 90° - \tan^{-1}\omega_1$$
$$\simeq 18°.$$

Next the time constant T_1 is chosen so that $H_1(s)$ will introduce $42°$ of phase advance (the difference between $60°$ and $18°$) at frequency ω_1;

$$\underline{/H(j\omega_1)} = \tan^{-1}\omega_1 T_1 = \tan^{-1} 3 \cdot 08 T_1 = 42°$$

or

$$T_1 = 0 \cdot 293.$$

The actual phase margin when T_1 has this value is not $60°$, because the appropriate frequency for calculating phase margin when the open-loop transfer function is $K_1 H_1 G(s)$ is not ω_1, but ω_2, as shown in Fig. 5.7. An approximate value for T_1 is then

$$T_1 = 0 \cdot 3.$$

The exact phase margin with this value of T_1 can be calculated by finding ω_2 from

$$|H_1 G(j\omega_2)| = \frac{\sqrt{(1 + 0 \cdot 09\omega_2^2)}}{\omega_2\sqrt{(1 + \omega_2^2)}} = \frac{1}{10}$$

which gives

$$\omega_2 = 3 \cdot 85,$$

and the phase margin is then

$$180° + \underline{/H_1 G(j\omega_2)} = 180° - 90° - \tan^{-1}\omega_2 + \tan^{-1}\omega_2 T_1 \simeq 64°$$

which is close enough to the specified value of $60°$.

(ii) Phase lag is to be used to give open-loop transfer function

$$K_2 H_2(s) G(s) = \frac{1}{2} \frac{1 + sT_2}{sT_2} \frac{1}{s(1 + s)};$$

what is the smallest value of the phase lag time constant T_2 that can be used without reducing the phase margin by more than $5°$? (the type-number here is two rather than one as in Fig. 5.1, with consequent improvement in steady-state performance. The gain in the loop increases as T_2 decreases, so T_2 should be as small as possible for further improvement in steady-state performance).

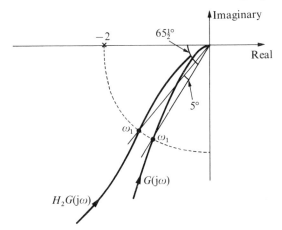

Fig. 5.8. Calculation of phase-lag time constant T_2.

Figure 5.8 shows the Nyquist diagram for this example. The phase margin, assuming no phase lag, is calculated first; this has been found in Section 3.5 to be $65\frac{1}{2}°$, and the relevant frequency is

$$\omega_1 = 0\cdot455.$$

Next the time constant T_2 is chosen so that $H_2(s)$ will introduce no more than $5°$ of phase lag at frequency ω_1;

$$\underline{/H_2(\mathrm{j}\omega_1)} = \underline{/1 - \dfrac{\mathrm{j}}{\omega_1 T_2}} = -\tan^{-1}\dfrac{1}{0\cdot455T_2} = -5°$$

or

$$T_2 \simeq 25.$$

The actual phase margin when T_2 has this value is very close to the specified value of $60\frac{1}{2}°$, because ω_1 is very close to the correct frequency for calculating the phase margin when the open-loop transfer function is $\frac{1}{2}H_2G(s)$; this can be shown by evaluating

$$|H_2G(\mathrm{j}\omega_1)| = \sqrt{(1 + (0\cdot0875)^2)}\times 2 \simeq 2 = \dfrac{1}{K_2}.$$

In most practical control systems it is not possible to use the ideal phase-advance transfer function $H_1(s)$ because the differentiation implied by the term sT_1 is unrealizable. Real controllers use transfer functions which approximate to the ideal over some working range of frequencies; similar approximating transfer functions are also often used in place of the ideal phase lag $H_2(s)$. Simple electrical networks with voltage transfer functions approximating to the ideals are shown in Fig. 5.9 together with their actual

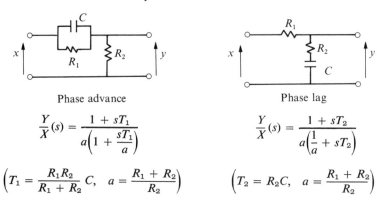

Phase advance

$$\frac{Y}{X}(s) = \frac{1 + sT_1}{a\left(1 + \dfrac{sT_1}{a}\right)}$$

$$\left(T_1 = \frac{R_1 R_2}{R_1 + R_2}\, C, \quad a = \frac{R_1 + R_2}{R_2}\right)$$

Phase lag

$$\frac{Y}{X}(s) = \frac{1 + sT_2}{a\left(\dfrac{1}{a} + sT_2\right)}$$

$$\left(T_2 = R_2 C, \quad a = \frac{R_1 + R_2}{R_2}\right)$$

Fig. 5.9. Electric circuits approximating ideal controller transfer functions.

transfer functions. In both cases the approximation improves as a increases towards infinity, but there is attenuation $1/a$ which must be removed by introducing extra amplification a; a common compromise value for a is about ten.

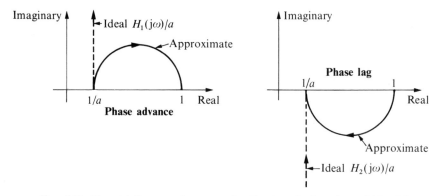

Fig. 5.10. Nyquist diagrams for approximating transfer functions of Fig. 5.9.

Figure 5.10 compares the shape of the frequency response of the approximating transfer functions with their ideals; it can be seen that for phase advance, approximation is good at low frequencies, and that for phase lag it is good at high frequencies.

The calculation of parameter values for controllers using the approximating transfer functions cannot normally be exact because of the high-order equations that tend to arise. Controller parameters are usually designed by approximate methods, such as were used in examples (i) and (ii) above, or by graphical methods using either open-loop frequency response or root-locus diagrams. The design may have to be iterated on a trial-and-error basis to allow for the differences between ideal and realizable controller transfer functions.

Problems 5.1–5.4

5.3. Comparison and combination of phase advance and phase lag

Phase-advance and phase-lag controller transfer functions improve overall performance in different ways. One is designed to improve the high frequency, or transient, response of the controlled process transfer function $G(s)$; the other is designed to improve the low frequency, or steady-state, response of $G(s)$. There is little interaction between the effects of the two and so both can be used together, in the same controller, with advantage. The effects of the ideal controller transfer functions are illustrated by considering the performance that can be achieved when the process

$$G(s) = \frac{1}{s(1 + s)}$$

of Section 5.2 is controlled by four different controller transfer functions.

(i) Controller without dynamics, as in Fig. 5.1 or Fig. 5.3,

$$KH(s) = \tfrac{1}{2}.$$

(ii) Ideal phase advance, as calculated in Section 5.2,

$$K_1 H_1(s) = 10(1 + 0{\cdot}3s).$$

(iii) Ideal phase lag, as calculated in Section 5.2,

$$K_2 H_2(s) = \tfrac{1}{50}(1 + 25s)/s.$$

(iv) Combination of ideal phase advance and ideal phase lag

$$K_1 K_2 H_1(s) H_2(s) = \tfrac{1}{5}(1 + 0{\cdot}3s)(1 + 25s)/s.$$

Figure 5.11 shows the open-loop frequency response of the system calculated from the transfer function for each of the four controllers; they all have phase margin of about 60°, which means satisfactory transient response.

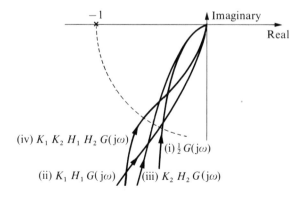

FIG. 5.11. Nyquist diagram for four different controllers when $G(s) = 1/s(1 + s)$.

Steady-state performances are compared in the table of Fig. 5.12 which shows the steady-state errors, calculated as in Fig. 2.5, produced by each controller in response to standard desired-value functions. The table shows that all the dynamic controllers improve steady-state performance and that the best performance is achieved by using both phase advance and phase lag together.

	Desired value $x(t)$		
	X	At	$\frac{1}{2}Bt^2$
(i) No controller dynamics	0	$2A$	∞
(ii) Ideal phase advance	0	$0\cdot1A$	∞
(iii) Ideal phase lag	0	0	$50B$
(iv) Phase advance and lag	0	0	$5B$

FIG. 5.12. Steady-state errors for four different controllers when $G(s) = 1/s(1 + s)$.

One further basis of comparison is the closed-loop frequency response of the four different systems. Figure 5.13 shows the closed-loop amplitude ratios M', calculated from the open-loop frequency response $M e^{j\phi}$ by the formula, from Section 1.6(ii),

$$M' e^{j\phi'} = \frac{M e^{j\phi}}{1 + M e^{j\phi}}. \qquad †$$

The range of frequencies for which M' is close to unity, sometimes called the bandwidth, gives an indication of the range of speeds that the controlled process can be made to follow without significant error. The figure shows that the phase-advance controller produces the largest bandwidth, and it therefore appears to be best from the point of view of speed of response. There are however practical disadvantages to the high speed response produced by the phase-advance controller: one disadvantage is that high-frequency noise, which exists in most practical systems, is not so well suppressed as by the other controllers with their steeper frequency response cut off at lower frequencies. Another disadvantage is that small dynamic lags in the system, which may have been negligible when the controlled process was originally

† This formula is sometimes presented graphically on a special diagram called a Nichols diagram which shows contours of constant M' and of constant ϕ' in a plane having rectangular axes $20 \log_{10} M$ and ϕ.

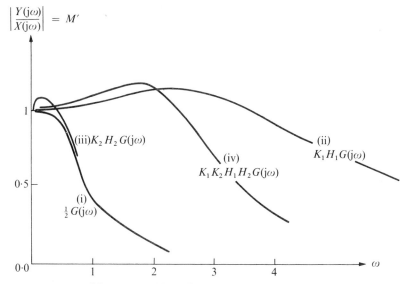

FIG. 5.13. Four different closed-loop frequency responses when $G(s) = 1/s(1 + s)$.

described by the transfer function $1/s(1 + s)$ can seriously affect stability at the higher speeds. For these reasons the controller with both phase advance and phase lag together, which increases bandwidth by a factor of about three, is generally regarded as giving the best closed-loop frequency response.

The above conclusions about how performance is improved by using dynamic controllers were derived from considering one specific controlled process $G(s)$, but they are generally applicable to most ordinary controlled processes. The principal conclusion is that it is desirable to use a controller that introduces both phase advance and phase lag. A controller transfer function of this nature can be written

$$KH(s) = K_1 K_2 (1 + sT_1)(1 + 1/sT_2)$$
$$= K(1 + sT_d + 1/sT_i),$$

where

$$K = K_1 K_2 (T_1 + T_2)/T_2, \qquad T_d = T_1 T_2/(T_1 + T_2), \qquad T_i = T_1 + T_2.$$

It is often used in the configuration of Fig. 5.3 and is known as a three-term controller because it produces as its output a controlling signal that can be regarded as the sum of three terms. One term, proportional to the controller input, is the basic feedback signal proportional to error; the second, derivative term has a stabilizing effect; and the third, integral term tends to eliminate steady-state errors.

Three-term controllers with adjustable parameters K, T_d, T_i are commercially available and are commonly used to control a great variety of industrial processes.† They are not necessarily suitable for all controlled processes, but the heuristic justification of the three terms, outlined above, is applicable to discrete-time as well as to continuous-time systems, and the discrete-time version of the three-term controller is recognized as a useful, practical control law; it has z-transfer function

$$KH(z) = K\big(1 + T_d(1 - z^{-1}) + 1/T_i(1 - z^{-1})\big).$$

Problems 5.5, 5.6

5.4. Bibliography

The design of continuous-time controller transfer functions is a central topic in almost every book on classical control engineering. It is discussed, often more extensively than here, in

JAMES, NICHOLS and PHILLIPS (1947) Chapter 4,
BROWN and CAMPBELL (1948) Chapters 7 and 8,
TRUXAL (1955) Chapters 4, 5, and 6,
SAVANT (1958) Chapter 6,
CHESTNUT and MAYER (1959) Chapters 12 and 13,
GILLE, PELEGRIN, and DECAULNE (1959) Chapter 18,
DEL TORO and PARKER (1960) Chapters 9 and 11,
THALER and BROWN (1960) Chapters 7–10,
WILTS (1960) Chapter 7,
RAVEN (1961) Chapter 10,
CLARK (1962) Chapters 7 and 9,
BARBE (1963) Chapter 13,
DOUCE (1963) Chapter 5,
HOROWITZ (1963),
SHINNERS (1964) Chapter 6; (1972) Chapter 7,
WEBB (1964) Chapter 8,
LANGILL (1965) Chapter 14,
D'AZZO and HOUPIS (1966) Chapters, 12, 13, and 14,
DISTEFANO, STUBBERUD, and WILLIAMS (1967) Chapters 12, 13, and 14,
DORF (1967) Chapter 10,
ELGERD (1967) Chapter 7,
DRANSFIELD (1968) Chapter 9,
SAUCEDO and SCHIRING (1968) Chapter 11,
SENSICLE (1968) Chapter 10,
MELSA and SCHULTZ (1969) Chapters 9 and 10,
WATKINS (1969) Chapter 9,
GUPTA and HASDORFF (1970) Chapter 8,
OGATA (1970) Chapter 10,
TAKAHASHI, RABINS, and AUSLANDER (1970) Chapters 8 and 9,

† See Ogata (1970) Chapter 5.

Towill (1970) Chapter 6,

Kwakernaak and Sivan (1972) Chapter 2.

The design of discrete-time controller transfer functions is discussed in

Ragazzini and Franklin (1958) Chapter 7,

Jury (1958) Chapters 5, 6, and 7,

Tou (1959) Chapter 9,

Monroe (1962) Chapter 14,

Kuo (1963) Chapters 8 and 9,

Shinners (1964) Chapter 9,

Lindorff (1965) Chapters 4 and 5,

Cadzow and Martens (1970) Sections 7.1–7.3,

Gupta and Hasdorff (1970) Chapter 9.

Discrete-time three-term controllers are discussed in

Harrison and Bollinger (1969) Section 16.5,

Cadzow and Martens (1970) Section 3.7,

Takahashi, Rabins and Auslander (1970) Sections 11.1, 11.2.

5.5. Problems

5.1. In the control system shown in Fig. 5.14 the gain K is to be so chosen that, when $H(s) = 1$, the system phase margin shall be 50°. An improved controller

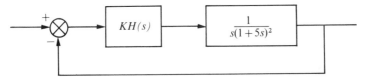

$$KH(s) \qquad \frac{1}{s(1+5s)^2}$$

Fig. 5.14.

transfer function $H(s) = 1 + 1/sT$ has time constant T chosen so that the phase margin of the system shall be approximately 45°.

(i) What should be the values of K and T?

(ii) For a constantly accelerating input $x = 0.001t^2$, what will be the steady-state error in the improved system?

5.2. It is proposed to improve the performance of the system of Problem 3.13 by introducing a phase-advance controller transfer function

$$KH(s) = K(1 + s).$$

(i) What value of K will give the resulting system a gain margin of 12 dB?

(ii) In what way will system performance have been improved?

5.3. Figure 5.15 shows a control system where the controlled object has transfer function

$$G(s) = \frac{2}{s(1 + 0.1s)(1 + 0.01s)}$$

and the phase margin is not to be less than 50°. *Sketch* the Bode diagram of the frequency response $G(j\omega)$ and hence make a *rough* estimate of the maximum value of K that can be achieved when the controller transfer function is

$$H(s) = K.$$

It is proposed to improve performance by redesigning the controller $H(s)$ to cancel one of the two non-zero poles of $G(s)$; show that it is preferable to cancel the factor $(1 + 0\cdot1s)$.

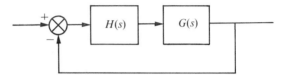

FIG. 5.15.

5.4. Sketch root-locus diagrams for the system of Problem 5.1 when

(i) $H(s) = 1$

(ii) $H(s) = 1 + \dfrac{1}{sT}$,

and T has the value calculated in Problem 5.1.

5.5. In a control system having the block diagram shown in Fig. 5.15 the controlled object has transfer function

$$G(s) = \frac{1}{(1 + sT)^3}.$$

Show by considering the root-locus diagram that a three-term controller for the system could have transfer function

$$H(s) = \frac{1}{4}\left(1 + \frac{T}{2}s + \frac{1}{2Ts}\right).$$

5.6. Figure 5.16 represents a control system in which the controller transfer function $H(s)$ is designed to make the output y follow a signal x in spite of disturbances z. The system is to have zero steady-state error when both x and z have constant values. Discuss what form of transfer function $H(s)$ will give satisfactory performance, and calculate values of its parameters.

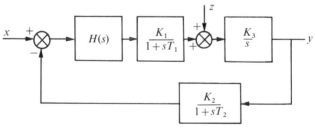

FIG. 5.16.

6. Analytic design of controllers

6.1. Performance criteria

ANALYTIC procedures for designing controllers are based on the idea that all aspects of the performance of a control system might be measured by a single, suitably defined, performance criterion. This contrasts with the approximate methods introduced in Chapter 5 where two aspects of performance, steady-state and transient, were considered separately and the controller transfer function was then chosen to compromise between their conflicting requirements. It was convenient to consider steady-state and transient performance separately because they correspond to the particular integral and complementary function of simple solutions to linear dynamic equations; the performance criteria used for analytic design are not so easily calculated but their introduction is justified, from the point of view of control theory, because they lead to a thorough-going mathematical theory of control systems. The procedures are introduced here for continuous-time systems; corresponding procedures, using results which are included in Section 6.2, exist for discrete-time systems.

The performance criterion for a continuous-time system is usually some functional of the time-dependent variables, $x(t)$ and $y(t)$ in Fig. 6.1(i), defined by an integral I of the form

$$I = \int_0^T h\big(x(t), y(t)\big)\, \mathrm{d}t,$$

where the function h specifies the nature of the criterion, and the range of integration

$$0 \leqslant t \leqslant T$$

includes all the time during which the system operates. An example of such a performance criterion is the *integral* of *absolute value* of *error* (IAE), illustrated in Fig. 6.1, specified by setting the function h equal to the absolute value of error,

$$h = |x - y|,$$

and the upper limit of integration

$$T = \infty.$$

(i) Typical control
    system

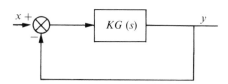

(ii) Typical step
     responses

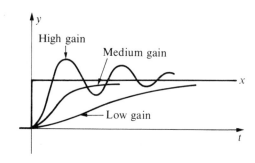

(iii) Absolute value
      of error

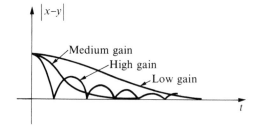

(iv) $\text{IAE} = \int_0^{\infty} \left| x - y \right| \, dt$

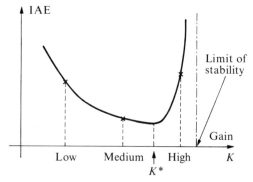

FIG. 6.1. The IAE-criterion.

The figure shows how a control system (i) produces different step responses (ii) for different values of its gain K; each step response gives one value of the IAE-criterion which is the area under the corresponding graph (iii) of absolute value of error against time. These values are marked on a typical graph (iv) showing how the IAE-criterion varies with gain K: the criterion is large for low values of gain because of large steady-state errors, and it is large for high values of gain because of oscillatory transient response; in between these two extremes small values of the criterion correspond to values of gain which make a good compromise between steady-state and transient performance. There is one particular value of gain, K^* in Fig. 6.1(iv), which minimizes the IAE-criterion for step responses; when the criterion is used as the sole measure of performance, K^* is the best, or 'optimum', value for the gain in the system.

The performance criteria used for analytic design always have an optimum point, like the minimum in Fig. 6.1(iv), and the procedure is to design the controller so that the system will operate at that optimum. The behaviour of the resulting system depends on what performance criterion is used and on what features of the controller transfer function are at the designer's disposal; a system in which the gain has been chosen so as to minimize the IAE-criterion for step responses may behave very differently from one in which some other parameter, say a velocity-feedback coefficient, has been chosen to optimize some other criterion, specified by some other function h or with some other input function $x(t)$ than a step.

Various forms for the function h measure various aspects of system performance. Some examples are given below

(i) If

$$h = (x - y)^2,$$

then

$$I = \int_0^\infty (x - y)^2 \, dt$$

is the *integral* of the *square* of the *error* (ISE). This ISE-criterion measures errors in the same way as does the IAE-criterion, but has the advantage that it can sometimes be calculated analytically, as discussed in Section 6.2.

(ii) If the controlling variable is u, as in Fig. 2.1, and

$$h = u^2,$$

then

$$I = \int_0^\infty u^2 \, dt$$

measures the amount of effort expended by the controller. There are many practical applications where u represents a flow of energy, for example a heat-flow controlling a temperature, so that it is economically desirable to minimize the amount of effort expended.

(iii) If the controlling variable u is a force controlling the position of a load and if

$$h = u\dot{y},$$

then

$$I = \int_0^\infty u\dot{y} \, \mathrm{d}t$$

measures the work done by the controller. This again is a measure of effort expended but it may not be so easy to calculate as the measure in (ii) above.

(iv) The function h can include time t as an explicit variable to weight performance in time. For example if

$$h = t(x - y)^2,$$

then

$$I = \int_0^\infty t(x - y)^2 \, \mathrm{d}t$$

attaches little weight to transient errors while t is small but penalizes residual errors with increasing severity as t becomes large. This criterion is sometimes called the *integral* of the product of *time* and the *square* of *error* (ITSE) criterion.

(v) Sometimes the design objective is to take the controlled process from one steady state to another steady state as rapidly as possible. If

$$h = 1,$$

then

$$I = \int_0^T 1 \, \mathrm{d}t = T$$

measures the time taken. Problems where the controller is designed to optimize this criterion are called minimum-time control problems; they are discussed in Section 9.4.

Whatever performance criterion is used, it must have certain properties:

(i) It must be monotonic with respect to quality of performance; it must increase (or decrease) as performance deteriorates (or improves), so that it can provide a basis for optimization by minimization (or maximization). Plain integral of error lacks this property; Fig. 6.2 shows the error of an

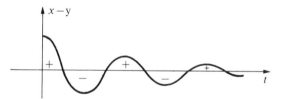

FIG. 6.2. Oscillatory sign of integral of error.

oscillatory step response which, although very poor, would give a low value
to the integral

$$I = \int_0^\infty (x - y) \, dt$$

because of its oscillating sign. Criteria like the IAE and ISE use functions h
that are non-linear in the variables of the system in order to avoid this sort
of anomaly.

(ii) It must be finite. When the range of the integral defining I is infinite, as
in most of the examples given above, the integral may not converge. If the
input function $x(t)$ and the system dynamics are such as to cause a finite steady-
state error, as discussed in Chapter 2, then the integrals of the IAE- and ISE-
criteria will not converge. Figure 6.3 shows typical errors in step response for
type zero and type one systems; the finite steady-state error of the type zero

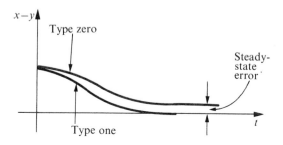

FIG. 6.3. Error in step responses.

response would cause the values of both IAE- and ISE-criteria to be infinite.
Finite performance criteria for type zero systems can be generated by considering the response to an impulse, which gives zero steady-state error.

(iii) For analytic design it must be possible to evaluate the integral I so as
to give an analytic expression relating the performance of the system to
whatever features of the controller are to be optimized. The only class of
problem for which this is generally possible is the class of linear systems having
criterion function h quadratic in the system variables, like the ISE-criterion.

Analytic design procedures are therefore concerned with designing controllers to optimize quadratic performance criteria. These quadratic criteria
normally correlate closely to the subjective ideas of designers and users of
control systems about what is 'good' performance; however the main reason
for concentrating on them is because the resulting integrals can be analytically evaluated. The method for calculating the integrals is introduced in
Section 6.2.

The design procedure itself depends on what feature of the controller is
to be optimized. There are two classes of problem.

(i) Fixed configuration, where the form of the controller transfer function
is given and it is required to optimize the value of some parameter, like the
gain in Fig. 6.1.

(ii) Semi-free configuration, where the form of the controller is not specified
and it is required to find whatever controller will produce optimum performance of a given controlled process. The design methods discussed in
Chapter 5 were examples of attempts to match the form of the controller to
a given controlled process, but no attempt was made to find an optimum
form.

The procedures for fixed configuration design are based on transfer function
analysis and are discussed here in Chapter 6 and again in Section 13.2. The
semi-free problem is more complicated; design procedures can be, and
originally were,† based on transfer function analysis but they are more conveniently presented in the state-space language of optimal control theory in
Chapters 8, 9, and 13.

6.2. Evaluating quadratic performance criteria

Analytic design is concerned with optimizing quadratic performance criteria
which in continuous-time have the form

$$I = \int_0^\infty x^2(t)\, \mathrm{dt}, \qquad (6.1)$$

and in discrete-time

$$I = \sum_{i=0}^\infty x^2(i). \qquad (6.2)$$

These can be evaluated using the transformations of Section 1.4.

Evaluation of the discrete-time summation (6.2) uses the inverse z-trans-
form, which is derived by considering the polynomial function

$$z^{i-1}X(z) \equiv x(0)z^{i-1} + x(1)z^{i-2} + \cdots$$
$$+ x(i-1) + x(i)z^{-1} + x(i+1)z^{-2} + \cdots$$

† See Newton, Gould and Kaiser (1957) or Chang (1961).

This expansion can be regarded as a Laurent series† in the complex variable z. Its contour integral around any closed path encircling the origin,

$$\oint z^{i-1} X(z)\, dz = 2\pi j x(i),$$

provides an expression for $x(i)$:

$$x(i) = \frac{1}{2\pi j} \oint z^{i-1} X(z)\, dz,$$

which is the inverse of the z-transformation defined in Section 1.4. The sum I can be written

$$I = \sum_{i=0}^{\infty} x(i) \frac{1}{2\pi j} \oint z^{i-1} X(z)\, dz,$$

and changing the order of summation and integration gives

$$I = \frac{1}{2\pi j} \oint X(z) \sum_{i=0}^{\infty} x(i)\, z^{i-1}\, dz$$

$$= \frac{1}{2\pi j} \oint X(z) \sum_{i=0}^{\infty} x(i) z^{i}\, \frac{dz}{z},$$

which can be written, using the definition of the z-transformation,

$$I = \frac{1}{2\pi j} \oint X(z) X(z^{-1})\, \frac{dz}{z}. \qquad (6.3)$$

Equation (6.3) is useful because the contour integral can be evaluated analytically when the z-transform $X(z)$ is the ratio of two polynomials in z:

$$X(z) = \frac{C(z)}{D(z)}$$

as is the case for variables in ordinary linear systems. The table of Fig. 6.4 gives values of the integrals

$$I_n = \frac{1}{2\pi j} \oint \frac{C(z)C(z^{-1})}{D(z)D(z^{-1})}\, \frac{dz}{z}$$

for small values of n, the order of the polynomials‡

$$D(z) = d_0 + d_1 z + \cdots + d_n z^n,$$

$$C(z) = c_0 + c_1 z + \cdots + c_n z^n.$$

The integrals converge only if the order of the numerator polynomial $C(z)$ is no higher than that of the denominator polynomial $D(z)$ and if $D(z)$ has all

---

† See, for example, Kreysig (1972) Chapters 15 and 16.
‡ Values of the integrals for higher values of n can be computed using a recursive algorithm given in Åström (1970) Section 5.2.

$$I_0 = \frac{c_0{}^2}{d_0{}^2}$$

$$I_1 = \frac{c_1{}^2 - 2c_0c_1d_0/d_1 + c_0{}^2}{d_1{}^2 - d_0{}^2}$$

$$I_2 = \frac{\left\{ \begin{array}{l} ((d_0 + d_2)(c_2d_2 - c_1d_0) - d_1(c_2d_2 - c_0d_1))^2 + \\[4pt] + ((d_0 + d_2)^2 - d_1{}^2)(c_1d_2 - c_0d_1)^2 + \\[4pt] + c_0{}^2(d_2{}^2 - d_0{}^2)((d_0 + d_2)^2 - d_1{}^2) \end{array} \right\}}{d_2{}^2(d_2{}^2 - d_0{}^2)((d_0 + d_2)^2 - d_1{}^2)}$$

Fig. 6.4. Table of summations I_0, I_1, I_2.

its zeroes within the unit circle; these conditions are normally satisfied by the z-transforms of variables in stable systems.

Evaluation of the continuous-time integral (6.1) uses the inverse Laplace transform in the form of a line integral

$$x(t) = \frac{1}{2\pi j} \int_{-j\infty}^{j\infty} X(s)\, e^{st}\, ds.$$

The integral I can be written

$$I = \int_0^\infty x(t) \left(\frac{1}{2\pi j} \int_{-j\infty}^{j\infty} X(s)\, e^{st}\, ds \right) dt,$$

and changing the order of integration gives

$$I = \frac{1}{2\pi j} \int_{-j\infty}^{j\infty} X(s) \int_0^\infty x(t)\, e^{st}\, dt\, ds,$$

which can be written, using the definition of the Laplace transformation,

$$I = \frac{1}{2\pi j} \int_{-j\infty}^{j\infty} X(s)X(-s)\, ds. \qquad (6.4)$$

Equation (6.4), giving the transformed expression for the integral I, is known as Parseval's theorem.

Parseval's theorem is useful because the line integral can be evaluated analytically provided that the Laplace transform $X(s)$ can be expressed as the ratio of two finite polynomials in s.

$$X(s) = \frac{C(s)}{D(s)}.$$

In linear control systems the variables have Laplace transforms that can be so expressed, and the integrals arising in analytic design can therefore be evaluated.

The table of Fig. 6.5 gives values of the integrals

$$I_n = \frac{1}{2\pi j} \int_{-j\infty}^{j\infty} \frac{C(s)C(-s)}{D(s)D(-s)} \, ds$$

$$I_1 = \frac{c_0^2}{2d_0 d_1}$$

$$I_2 = \frac{c_1^2 d_0 + c_0^2 d_2}{2d_0 d_1 d_2}$$

$$I_3 = \frac{c_2^2 d_0 d_1 + (c_1^2 - 2c_0 c_2)d_0 d_3 + c_0^2 d_2 d_3}{2d_0 d_3(-d_0 d_3 + d_1 d_2)}$$

$$I_4 = \frac{\left\{ \begin{array}{l} c_3^2(-d_0^2 d_3 + d_0 d_1 d_2) + (c_2^2 - 2c_1 c_3)\,d_0 d_1 d_4 + \\ + (c_1^2 - 2c_0 c_2)d_0 d_3 d_4 + c_0^2(-d_1 d_4^2 + d_2 d_3 d_4) \end{array} \right\}}{2d_0 d_4(-d_0 d_3^2 - d_1^2 d_4^2 + d_1 d_2 d_3)}$$

FIG. 6.5. Table of integrals $I_1 - I_4$.

for small values of n, the order of the denominator polynomial.† The values are expressed in terms of the coefficients of the denominator polynomial

$$D(s) = d_0 + d_1 s + \cdots + d_n s^n$$

and of the numerator polynomial

$$C(s) = c_0 + c_1 s + \cdots + c_{n-1} s^{n-1}.$$

The integrals converge only if $C(s)$ is of lower order than $D(s)$ and if all the

† Tabulation of integrals for higher values of n (up to 10) and details of the method of integration are to be found in various textbooks, for example James, Nichols, and Phillips (1947), Section 7.9 and Appendix A, or Newton, Gould, and Kaiser (1957), Appendix E. The integrals can also be computed using a recursive algorithm given in Åström (1970) Section 5.3.

zeros of $D(s)$ have negative real parts; these conditions are normally satisfied by the Laplace transform of variables in stable systems.

Problems 6.1, 6.2

6.3. Fixed-configuration design procedure for continuous-time systems

The procedure for fixed-configuration design is introduced by considering the ISE-criterion for some continuous-time examples with step inputs.

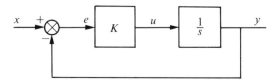

FIG. 6.6. A simple linear system.

In the system of Fig. 6.6 the Laplace transform of the error is

$$E(s) = \frac{1}{1 + K/s} X(s)$$

and when the input function $x(t)$ is a unit step with Laplace transform, from the table of Fig. 1.13,

$$X(s) = \frac{1}{s},$$

this becomes

$$E(s) = \frac{1}{s + K}.$$

The ISE-criterion can then be found using Parseval's theorem to write

$$I = \frac{1}{2\pi_j} \int_{-j\infty}^{j\infty} \frac{1}{s + K} \frac{1}{-s + K} \, ds,$$

which corresponds to I_1 in the table of Fig. 6.5 and has the value

$$I = \frac{1}{2K}.$$

The optimum value of the gain K, which minimizes the criterion I, is infinite here. This is what would be expected of a first-order system where there is no possibility of the transient response becoming oscillatory and the gain can be chosen with reference to steady-state performance only.

Figure 6.7 is a block diagram of the third-order position-control system of

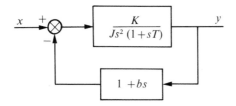

FIG. 6.7. Third-order position-control system.

Fig. 3.8, discussed in previous chapters. The Laplace transform of the error is as in Section 2.3,

$$E(s) = \frac{1 + \dfrac{K(1 + bs)}{Js^2(1 + sT)} - \dfrac{K}{Js^2(1 + sT)}}{1 + \dfrac{K(1 + bs)}{Js^2(1 + sT)}} X(s),$$

and when the input function $x(t)$ is a unit step this becomes

$$E(s) = \frac{TJs^2 + Js + Kb}{TJs^3 + Js^2 + Kbs + K}.$$

The ISE-criterion can then be found, as above, from the integral I_3 of the table; it is

$$I = \frac{1}{2}\left(b + \frac{J/K}{b - T}\right).$$

This criterion has a local minimum with respect to the velocity feedback coefficient b which can be found by calculus, but the minimizing value of gain K is infinite. The optimum value for b is found by setting

$$\frac{\partial I}{\partial b} = \frac{1}{2}\left(1 - \frac{J/K}{(b - T)^2}\right) = 0,$$

which gives the minimizing value

$$b^* = T + \sqrt{(J/K)},$$

and when K is infinite this becomes

$$b^{**} = T.$$

The significance of this is that the value of b should be chosen so that the open-loop zero due to the phase advance of velocity feedback exactly cancels the open-loop pole due to the phase lag of the amplifier dynamics; then the system becomes second order and the gain can be made infinite without danger of instability. In practice there are always additional small lags in the system, as discussed at Fig. 4.11, and so infinite values of gain would result in an unstable system. In this example the design procedure gives a practical value for the velocity feedback coefficient b, equal to the time constant T, but not for the gain K.

Practical controllers do not have infinite values of gain, so the two examples given above seem to show that the analytic procedure is not useful for designing controller gains. The infinite values of gain resulted because the mathematical models, that is the differential equations or transfer functions used to describe the system dynamics, were not entirely accurate in that they neglected small lags which have a limiting effect on stability. No mathematical design procedure can be expected to give useful practical results unless all relevant practical features are described in the mathematical models on which analysis is based; one way to make the analytic design procedure here more usable would be to ensure that all dynamic effects were accurately described in the mathematical model. However accurate information about small, but limiting, dynamic effects is not always available and so it is necessary to find an alternative way to make the design procedure more usable.

This alternative is to modify the performance criterion so that it penalizes high values of gain for reasons other than those of stability. The ISE-criterion is modified by adding a term which measures the amount of effort expended by the controller, so that

$$I = \int_0^\infty (x - y)^2 \, \mathrm{d}t + q \int_0^\infty u^2 \, \mathrm{d}t,$$

where u is the controlling variable, as in Section 6.1, and q is a positive constant specifying the importance of control effort relative to errors. This criterion can be evaluated by the method of Section 6.2. It penalizes high values of controller gain by penalizing high values of u, the variable at the controller output; if the constant q is properly chosen it leads to designs where the destabilizing effect of small lags is not critical because the gain is not made impracticably high. The modified criterion also reflects other features of practical control systems which are that control effort is expensive and that most real control variables u saturate, that is the range of values which they can take is limited. For example, if u were a heat-flow rate in a temperature-control system, upper and lower limits on the rate at which heat could be supplied or rejected would be imposed by physical limitations of the equipment available. If the constant q is properly chosen the modified criterion can lead to designs where the control variable u is not expected to exceed its saturation limits.†

The constant q which weights the control effort is similar to the damping

† In minimum-time control problems it is desirable that a saturating control variable always takes one or other of its limiting values (see Section 9.4) but the linear analysis in Chapter 6 assumes that no saturation occurs.

ratio and the phase margin in that a desirable value for it must be specified before any designing can be done. It differs from the damping ratio and the phase margin in that there are no general-purpose values, like $1/\sqrt{2}$ or $50°$, which can reasonably be expected to lead to acceptable performance of almost any control system. The difficulty of knowing what value to assign to q is a major practical disadvantage of the analytic design procedure.

In the system of Fig. 6.6, for example, the term measuring controller effort is

$$\int_0^\infty u^2 \, dt = K^2 \int_0^\infty e^2 \, dt,$$

and when the input $x(t)$ is a unit step this is

$$\int_0^\infty u^2 \, dt = K/2,$$

and the modified performance criterion is

$$I = \frac{1}{2}\left(\frac{1}{K} + qK\right).$$

This criterion has a local minimum with respect to the gain K which can be found by calculus,

$$\frac{\partial I}{\partial K} = \frac{1}{2}\left(-\frac{1}{K^2} + q\right) = 0,$$

to give the minimizing value

$$K^* = 1/\sqrt{q}.$$

The optimum value of gain is thus entirely dependent on the constant q. In other examples the optimum gain may not be entirely dependent on q, but it is scarcely ever totally independent of it.

Problems 6.3–6.7

6.4. Bibliography

Analytic design procedures in terms of transfer functions were originally introduced for semi-free stochastic problems and the bibliography for this is given in Section 13.6. Fixed-configuration deterministic problems are discussed in

NEWTON, GOULD and KAISER (1957) Chapter 2,
TOU (1959) Section 10.4,
CHANG (1961) Chapter 2,
MERRIAM (1964) Chapter 2,
D'AZZO and HOUPIS (1966) Chapter 17,

EVELEIGH (1967) Chapters 2 and 3,
MCCAUSLAND (1969) Chapter 2,
GUPTA and HASDORFF (1970) Chapter 11,
OGATA (1970) Chapter 7.

6.5. Problems

6.1. A discrete-time system has z-transfer function

$$\frac{Y(z)}{X(z)} = \frac{1}{z + a}.$$

Use the table of Fig. 6.4 to prove that when the input $x(i)$ is the unit impulse
($X(z) = 1$) the sum of the squared values of the outputs is

$$\sum_{i=0}^{\infty} y(i)^2 = \frac{1}{1 - a^2}.$$

Confirm this result using the impulse-response function discussed in Section 1.5.

6.2. A continuous-time system has transfer function

$$\frac{Y(s)}{X(s)} = \frac{1}{s^2 + 2\zeta s + 1}.$$

Use the table of Fig. 6.5 to evaluate the integral of the square of the output y when
the input x is a unit impulse.

6.3. If the input x to the system of Problem 6.2 is a step function, what value of
the damping ratio ζ will minimize the integral of the square of the difference
between x and y?

6.4. Use the result of Problem 6.3 to find the value of b to minimize the ISE-
performance criterion for the system shown in the Fig. 6.8 when the input x is a
step function.

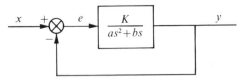

FIG. 6.8.

Explain why it is not practicable to use the same criterion to find values for a
and K.

6.5. What value of gain K minimizes the ISE-performance criterion for the system
shown in Fig. 6.9 when the input x is a step function?
What is the gain margin of the resulting system?

6.6. Derive an expression giving the ISE-performance criterion for the discrete-
time system of Problem 1.11 when the input x is an impulse function. It can be

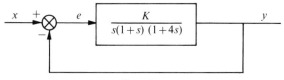

FIG. 6.9.

shown (numerically) that the criterion is minimized when $K = 0 \cdot 4/a$; find the roots of the resulting characteristic equation and show them on the root-locus diagram of Problem 4.10.

6.7. What values of K and b in the system of Fig. 2.7 minimize the performance criterion

$$I = \int_0^\infty \left((x - y)^2 + qu^2 \right) dt$$

when the input x is a step function? What is the damping ratio of the resulting system?

7. State space

7.1. Introduction

THE high-order dynamic equations (1.1) and (1.2) introduced in Chapter 1 are not the only way of describing the dynamics of controlled processes; more general mathematical methods for analysing and designing control systems are based on an alternative description of process dynamics using a set of simultaneous first-order dynamic equations. For example, the nth-order differential equation in a single variable y;

$$a_n \frac{d^n y}{dt^n} + \cdots + a_1 \frac{dy}{dt} + a_0 y = u$$

is equivalent to n first-order differential equations, in variables $x_1, \ldots, x_n$ defined by

$$x_1 \equiv y, \quad x_2 \equiv \frac{dy}{dt}, \quad \cdots, \quad x_n \equiv \frac{d^{n-1} y}{dt^{n-1}},$$

$$\dot{x}_1 = x_2$$

$$\dot{x}_2 = x_3$$

$$\vdots$$

$$\dot{x}_{n-1} = x_n$$

$$\dot{x}_n = -\frac{a_{n-1}}{a_n} x_n - \cdots - \frac{a_0}{a_n} x_1 + \frac{1}{a_n} u.$$

Similarly, the nth-order difference equation

$$a_n y(i) + a_{n-1} y(i-1) + \cdots + a_1 y(i-n+1) + a_0 y(i-n) = u(i)$$

is equivalent to n first-order difference equations in variables $x_1, \ldots, x_n$ defined by

$$x_1(i) \equiv y(i - n), \quad x_2(i) \equiv y(i - n + 1), \quad \ldots, \quad x_n(i) \equiv y(i - 1),$$

$$x_1(i + 1) = x_2(i)$$

$$x_2(i + 1) = x_3(i)$$

$$\vdots$$

$$x_{n-1}(i + 1) = x_n(i)$$

$$x_n(i + 1) = -\frac{a_{n-1}}{a_n} x_n(i) - \cdots - \frac{a_0}{a_n} x_1(i) + \frac{1}{a_n} u(i).$$

Such sets of simultaneous equations are conveniently written in vector notation, and when the dynamic equations are linear, matrices also can be used. Thus continuous-time, linear dynamics are written

$$\dot{\mathbf{x}} = \mathbf{A}\mathbf{x} + \mathbf{B}\mathbf{u}, \tag{7.1}$$

and discrete-time, linear dynamics

$$\mathbf{x}(i + 1) = \mathbf{A}\mathbf{x}(i) + \mathbf{B}\mathbf{u}(i). \tag{7.2}$$

Points about this notation are as follows.

(i) $\mathbf{x}$ is a vector with n components $x_1, \ldots, x_n$, where n is the order of the dynamic system.

(ii) $\mathbf{u}$ is a vector with m components, representing up to m separate inputs. Usually m is not greater than n.

(iii) $\mathbf{A}$ and $\mathbf{B}$ are matrices of order $n \times n$ and $n \times m$ respectively. In both the examples above it was the case that

$$\mathbf{A} = \begin{bmatrix} 0 & 1 & & \\ \vdots & & \ddots & \\ 0 & & & 1 \\ \hline -\dfrac{a_0}{a_n} & -\dfrac{a_1}{a_n} & \cdots & -\dfrac{a_{n-1}}{a_n} \end{bmatrix}, \qquad \mathbf{B} = \begin{bmatrix} 0 \\ \vdots \\ 0 \\ \vdots \\ \dfrac{1}{a_n} \end{bmatrix}.$$

Matrices having the form of $\mathbf{A}$ here are known as companion matrices.

(iv) Eigenvalues† of the matrix $\mathbf{A}$ are roots of an equation

$$|\mathbf{A} - s\mathbf{I}| = 0,$$

† Eigenvalues of a matrix $\mathbf{A}$ are values of the scalar quantity s for which the n simultaneous equations represented by

$$\mathbf{A}\mathbf{x} = s\mathbf{x}$$

have non-trivial solutions. In general there are n eigenvalues $s_1, \ldots, s_n$ and n corresponding solutions $\mathbf{x}_1, \ldots, \mathbf{x}_n$, known as eigenvectors. See Kreysig (1972) Section 6.4.

which is the matrix form of the characteristic equation (1.3), and the eigenvalues are thus identical with the roots of the characteristic equation. This can be further demonstrated by setting the inputs $\mathbf{u}$ to zero and seeking complementary-function solutions to the resulting homogeneous equations in the general form

$$\mathbf{x}(t) = \mathbf{x}(0)\,e^{st} \tag{7.3}$$

for the continuous-time equations, or

$$\mathbf{x}(i) = \mathbf{x}(0)s^i \tag{7.4}$$

for the discrete-time equations. Substituting equations (7.3) and (7.4) into the homogeneous forms of equations (7.1) and (7.2) leads to two equations,

$$s\mathbf{x}(0)\,e^{st} = \mathbf{A}\mathbf{x}(0)\,e^{st}$$

and

$$\mathbf{x}(0)s^{i+1} = \mathbf{A}\mathbf{x}(0)s^i,$$

which both reduce to the standard eigenvalue equation

$$s\mathbf{x}(0) = \mathbf{A}\mathbf{x}(0).$$

It can therefore be seen that the number s which determines the dynamic behaviour of the response in equations (7.3) and (7.4) is an eigenvalue of $\mathbf{A}$. It can also be seen that the response can only have the form of equations (7.3) and (7.4) if the initial state $\mathbf{x}(0)$ is an eigenvector of $\mathbf{A}$; transient responses from other, more general, initial states are discussed in terms of state space in Section 7.2.

The main reason why it is so convenient to describe dynamics by equations (7.1) or (7.2) is that first-order dynamic equations are easier to handle than higher-order equations. An additional reason is that the restriction to single-variable systems, which was implicit throughout the preceding chapters, disappears; equations (7.1) and (7.2) can describe multi-variable dynamic systems with many inputs and outputs, like that shown in Fig. 7.1, as well as

FIG. 7.1. Multi-variable dynamic system.

single-variable systems like that introduced in Fig. 1.2. In either case it is the matrices $\mathbf{A}$ and $\mathbf{B}$ which specify a particular linear system, and so the dynamic properties of systems can be investigated by investigating properties of the matrices.

Problems 7.1, 7.2

7.2. State space

In equations (7.1) and (7.2) the dynamic state of a system is represented by a vector $\mathbf{x}$. This vector $\mathbf{x}$ is known as the state vector and its n components are known as state variables. It is sometimes helpful to regard the n state variables as coordinate axes defining an n-dimensional state space. An example of a two-dimensional state space, which can be illustrated on a two-dimensional sheet of paper, is provided by the typical second-order differential equation

$$\ddot{y} + 2\zeta\omega_0\dot{y} + \omega_0^2 y = u;$$

using state variables

$$x_1 \equiv y, \qquad x_2 \equiv \dot{y},$$

this second-order equation is equivalent to two first-order equations

$$\dot{x}_1 = x_2,$$
$$\dot{x}_2 = -\omega_0^2 x_1 - 2\zeta\omega_0 x_2 + u.$$

Assuming that the input u is zero, the transient response can be described by trajectories in a state space having x_1 and x_2 as coordinate axes; the slope of such trajectories is given by

$$\frac{\mathrm{d}x_2}{\mathrm{d}x_1} = \frac{\dot{x}_2}{\dot{x}_1} = -\left(2\zeta\omega_0 + \frac{\omega_0^2 x_1}{x_2}\right).$$

Figure 7.2 shows how the shape of these trajectories depends on the damping ratio ζ†: for zero damping the closed trajectories (i) correspond to simple

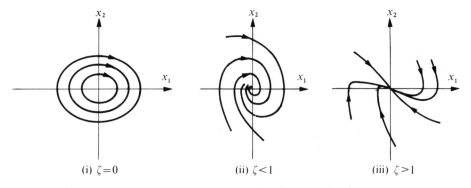

(i) $\zeta=0$ (ii) $\zeta<1$ (iii) $\zeta>1$

FIG. 7.2. Transient responses in state plane for second-order system.

harmonic motion; for underdamping the spiralling trajectories (ii) correspond to oscillatory response; and for overdamping the trajectories (iii) show no overshoot. The arrows on the trajectories indicate the direction of motion in

† Derivation of Fig. 7.2 is given in Section 10.1.

time; for this example, where x_2 is equal to $\dot{x}_1$, the arrows show how x_1 always increases for positive x_2 and decreases for negative x_2.

In general every point in state space corresponds to a value of the state vector $\mathbf{x}$, and the initial state for any solution of the dynamic equations (7.1) or (7.2) can be regarded as a point in state space. The eigenvectors $\mathbf{x}_1, \ldots, \mathbf{x}_n$ of the matrix $\mathbf{A}$ can be regarded as defining a set of n directions in state space, and provided they are linearly independent any point in state space can be written in the form of a weighted sum of the eigenvectors:

$$\mathbf{x} = c_1\mathbf{x}_1 + \cdots + c_n\mathbf{x}_n = \mathbf{Xc},$$

where the weighting coefficients $\mathbf{c}$ are sometimes known as normal coordinates and the matrix $\mathbf{X}$ having eigenvector direction cosines for its columns is known as the modal matrix.† Any initial state can therefore be written in the form

$$\mathbf{x}(0) = \mathbf{Xc}(0),$$

and the general form of the complementary function (7.3) for continuous-time equations is

$$\mathbf{x}(t) = c_1(0)\mathbf{x}_1\, e^{s_1 t} + \cdots + c_n(0)\mathbf{x}_n\, e^{s_n t}, \qquad (7.5)$$

and of the complementary function (7.4) for discrete-time equations

$$\mathbf{x}(i) = c_1(0)\mathbf{x}_1(s_1)^i + \cdots + c_n(0)\mathbf{x}_n(s_n)^i. \qquad (7.6)$$

Every solution of a dynamic equation can be represented by a trajectory in state space. The shape of trajectories representing the complementary functions, or transient responses, of equation (7.5) and (7.6) depends on the matrix $\mathbf{A}$ only. When the inputs $\mathbf{u}$ are not zero the particular integral solution, which depends partly on the matrix $\mathbf{B}$, can affect trajectories in state space. For example, if a discrete-time process has only a single control input u it is possible to determine the number of discrete-time intervals needed to transfer its dynamic state from an arbitrary initial state $\mathbf{x}$ to a specified final state, say the origin of state space, which is often a point of equilibrium. The dynamics of such a single-input process are represented by equation (7.2),

$$\mathbf{x}(i + 1) = \mathbf{Ax}(i) + u(i)\mathbf{b},$$

where $\mathbf{b}$ is an n-vector. Then the set $\mathbf{p}_1$ of points from which it is possible to get to the origin in a single step can be found by setting $\mathbf{x}(i)$ equal to $\mathbf{p}_1$ and $\mathbf{x}(i + 1)$ equal to zero to give

$$\mathbf{Ap}_1 + u_1\mathbf{b} = 0;$$

---

† A set of n n-vectors $\mathbf{x}_1, \ldots, \mathbf{x}_n$ is linearly independent provided that the determinant of the square matrix

$$\mathbf{X} = [\mathbf{x}_1 \,\vdots\, \mathbf{x}_2 \,\vdots\, \cdots \,\vdots\, \mathbf{x}_n]$$

is non-zero. Any square matrix satisfying this condition is said to be non-singular. See Kreysig (1972) Section 6.10.

Non-singular matrices have linearly independent eigenvectors. See Zadeh and Desoer (1963) Sections 5.4, 5.5.

this is the vector form of the equation of a straight line in state space, so the set $\mathbf{p}_1$ is the set of all points in that straight line. Similarly the set $\mathbf{p}_2$ of points from which it is possible to get to the origin in two steps, or to $\mathbf{p}_1$ in one step, is found by setting $\mathbf{x}(i)$ equal to $\mathbf{p}_2$ and $\mathbf{x}(i + 1)$ equal to $\mathbf{p}_1$ to give

$$\mathbf{p}_1 = \mathbf{A}\mathbf{p}_2 + u_2\mathbf{b}$$

or

$$\mathbf{A}^2\mathbf{p}_2 + u_2\mathbf{A}\mathbf{b} + u_1\mathbf{b} = \mathbf{0}.$$

Provided that the vectors $\mathbf{b}$, $\mathbf{A}\mathbf{b}$ are linearly independent, this is the vector equation of a plane in state space, and so the set $\mathbf{p}_2$ is the set of all points in that plane. This argument can be repeated n times to show that the set $\mathbf{p}_n$ of points from which it is possible to get to the origin in n steps is given by

$$\mathbf{A}^n\mathbf{p}_n + u_n\mathbf{A}^{n-1}\mathbf{b} + \cdots + u_1\mathbf{b} = \mathbf{0}.$$

Provided that the set of vectors $\mathbf{b}$, $\mathbf{A}\mathbf{b}$, ..., $\mathbf{A}^{n-1}\mathbf{b}$ is linearly independent, the set of points $\mathbf{p}_n$ satisfying this equation includes the whole of n-dimensional state space. It follows that n is the maximum number of steps needed to get from any arbitrary initial state to the origin or, with a shift of coordinate zeros, to any other specified final state.

The state-space description of dynamics is essential for most of the control theory presented in subsequent chapters; it provides the framework for discussing continuous-time and discrete-time systems, with linear or non-linear equations, whether deterministic or with uncertainty. A particular problem that comes to light when the state-space description is used is the question, introduced in the above discrete-time example, of whether or not a system is controllable in the sense that the available control signals can produce any desired change of its dynamic state.

7.3. Controllability

A set of dynamic equations relating a state vector $\mathbf{x}$ to controls $\mathbf{u}$ is said to be controllable if there exist controls $\mathbf{u}$ which, in a finite time, will transfer the dynamic state $\mathbf{x}$ between any two arbitrarily chosen points in state space. When the equations are linear, (7.1) or (7.2), this question of controllability is a question about the interconnections between the controls and the state variables. For example, the interconnections in a set of continuous-time equations (7.1) having distinct eigenvalues† can be clarified by using the modal matrix $\mathbf{X}$ to transform them into equations in the normal coordinates $\mathbf{c}$ defined by

$$\mathbf{x} = \mathbf{X}\mathbf{c}.$$

† When $\mathbf{A}$ has repeated eigenvalues there is no transformation to a diagonal matrix $\mathbf{S}$ as here but $\mathbf{A}$ can instead be represented in the Jordan canonical form; see Saucedo and Schiring (1968) Section 5.9 or Bronson (1969) Chapter 8.

Premultiplying equation (7.1) by X^{-1} gives

$$\dot{c} = X^{-1}AXc + X^{-1}Bu,$$

and it follows from the definitions of eigenvalues and eigenvectors that the product $X^{-1}AX$, where X is the modal matrix of A, simplifies to

$$X^{-1}AX = \begin{bmatrix} s_1 & & 0 \\ & \cdot & \\ 0 & & \cdot \\ & & s_n \end{bmatrix} = S,$$

a diagonal matrix S of the eigenvalues of A. So the differential equations in the normal coordinates are

$$\dot{c} = Sc + B^*u,$$

where

$$B^* = X^{-1}B.$$

This transformation to the normal coordinates gives a form of the differential equations in which the dynamic variables c are decoupled, and the interconnections between the controls u and variables c depends only on the matrix B^*. The equations in c can be written out,

$$\dot{c}_1 = s_1c_1 + b_{11}^*u_1 + \cdots + b_{1m}^*u_m$$

$$\cdot \quad \cdot \quad \cdot \quad \cdot \quad \cdot \quad \cdot \quad \cdot \quad \cdot \quad \cdot$$

$$\dot{c}_n = s_nc_n + b_{n1}^*u_1 + \cdots + b_{nm}^*u_m$$

to show that if there is any zero row in the matrix B^* the corresponding normal coordinate cannot be affected by any of the controls u and, since it is decoupled from all other normal coordinates, the equations will be uncontrollable.

The matrix B^* which appears in the transformation to normal coordinates can be used to investigate controllability of both continuous-time and discrete-time equations having distinct eigenvalues. However, when the eigenvalues of A are not all distinct the transformation breaks down and the condition for controllability cannot be expressed in terms of B^*. It has been shown† that the general condition for controllability, applicable whether the eigenvalues of A are distinct or not, is a generalization of the result derived for the discrete-time example in Section 7.2. The general condition states that equations (7.1) or (7.2) are controllable provided that the set of n $n \times m$-matrices

† Kalman (1960*b*).

B, AB, ..., A^{n-1}B is linearly independent.† This condition applies both to continuous-time and to discrete-time equations. It can be used to confirm that any of the controlled processes in the examples of previous chapters is controllable; for example, the rotating inertia introduced in Section 1.2 had position y related to applied torque u by a second-order differential equation

$$J\ddot{y} = u$$

which, with state variables

$$x_1 \equiv y, \qquad x_2 \equiv \dot{y},$$

is equivalent to equations (7.1) with

$$\mathbf{A} = \begin{bmatrix} 0 & 1 \\ 0 & 0 \end{bmatrix}, \qquad \mathbf{B} = \begin{bmatrix} 0 \\ 1/J \end{bmatrix}.$$

The matrix **M** to determine whether or not **B** and **AB** are linearly independent here is

$$\mathbf{M} = \begin{bmatrix} 0 & 1/J \\ 1/J & 0 \end{bmatrix},$$

which is square and non-singular, so that the system is controllable.

An example of an uncontrollable system is illustrated in Fig. 7.3, where the same torque u is applied to two rotating inertias which are not otherwise

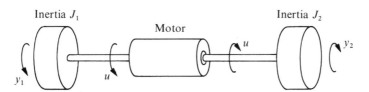

FIG. 7.3. Example of an uncontrollable system.

mechanically connected; this might happen if one intertia J_1 were connected to the stator and the other inertia J_2 to the rotor of a motor developing torque u. The differential equations of this system would be

$$J_1\ddot{y}_1 = u$$
$$J_2\ddot{y}_2 = u$$

† A set of $n \times m$-matrices **B, AB, ..., A^{n-1}B** is linearly independent provided that the $n \times nm$-matrix

$$\mathbf{M} \equiv [\mathbf{B} \, \vdots \, \mathbf{AB} \, \vdots \, \cdots \, \vdots \, \mathbf{A}^{n-1}\mathbf{B}]$$

has rank n.

The rank of a matrix is the order of the largest non-zero determinant that can be obtained from the elements of the matrix. See Kreysig (1972) Section 6.10.

and, with state variables

$$x_1 \equiv y_1, \qquad x_2 \equiv \dot{y}_1, \qquad x_3 \equiv y_2, \qquad x_4 \equiv \dot{y}_2,$$

these are equivalent to equations (7.1) with

$$\mathbf{A} = \begin{bmatrix} 0 & 1 & 0 & 0 \\ 0 & 0 & 0 & 0 \\ 0 & 0 & 0 & 1 \\ 0 & 0 & 0 & 0 \end{bmatrix}, \qquad \mathbf{B} = \begin{bmatrix} 0 \\ 1/J_1 \\ 0 \\ 1/J_2 \end{bmatrix}.$$

Self-products of the matrix $\mathbf{A}$ here are zero and the matrix $\mathbf{M}$ to determine whether or not the vectors $\mathbf{B}, \mathbf{AB}, \mathbf{A}^2\mathbf{B}, \mathbf{A}^3\mathbf{B}$ are linearly independent is

$$\mathbf{M} = \begin{bmatrix} 0 & 1/J_1 & 0 & 0 \\ 1/J_1 & 0 & 0 & 0 \\ 0 & 1/J_2 & 0 & 0 \\ 1/J_2 & 0 & 0 & 0 \end{bmatrix},$$

which is square and singular, so that the vectors are not linearly independent and the system is uncontrollable. The interpretation of this uncontrollability is that it would not be possible to position the two inertias independently using the one torque.

The foregoing test for controllability gives no indication of what control law should be used to achieve any given objective; it only answers the fundamental question of whether control is possible at all. Controllability of non-linear dynamic equations may depend not only on interconnections but also on the nature of the non-linearities in the equations and on the allowable magnitudes of the variables; there are no general results about controllability of non-linear equations.

Problems 7.3–7.6

7.4. Observability

Most control laws are feedback laws, for the reasons introduced in Section 2.1, and the state-space description of dynamics makes it clear that the information required for feedback control is the dynamic state $\mathbf{x}$. In most real controlled processes the dynamic state $\mathbf{x}$ is not identical with the observable outputs; in the examples given so far in Chapter 7, the observable output was assumed to be a single variable y which was just one component of the state vector $\mathbf{x}$. A linear equation to account for differences between the state $\mathbf{x}$ and the output $\mathbf{y}$ is

$$\mathbf{y} = \mathbf{Cx}, \tag{7.7}$$

where $\mathbf{y}$ is a vector with p components representing the observable outputs shown in Fig. 7.1; p is usually not greater than the order n of the equations, and $\mathbf{C}$ is a $p \times n$-matrix.

When the output of a process is not its state the question arises of whether or not it is possible to evaluate the state from observations of the output. A set of dynamic equations together with an output equation is said to be observable if the output $\mathbf{y}$ contains sufficient information for the state $\mathbf{x}$ to be evaluated in a finite time. When the equations are linear, (7.1) or (7.2) and (7.7), this question of observability is, like the question of controllability, about interconnections between the state variables and the outputs. For example, using the modal matrix $\mathbf{X}$ to transform the output equations (7.7) into equations in the normal coordinates $\mathbf{c}$ gives

$$\mathbf{y} = \mathbf{C}^*\mathbf{c},$$

where

$$\mathbf{C}^* = \mathbf{CX},$$

and so the output equations can be written†

$$y_1 = c_{11}^* c_1 + \cdots + c_{1n}^* c_n$$

$$\cdot \quad \cdot \quad \cdot \quad \cdot \quad \cdot \quad \cdot$$

$$y_p = c_{p1}^* c_1 + \cdots + c_{pn}^* c_n$$

to show that if there is a zero column in the matrix $\mathbf{C}^*$ the corresponding normal coordinate affects none of the outputs $\mathbf{y}$ and the equations are therefore unobservable.

The matrix $\mathbf{C}^*$ can be used to investigate observability of both continuous-time and discrete-time equations but, as with the matrix $\mathbf{B}^*$ for controllability, this method cannot be used if the eigenvalues of $\mathbf{A}$ are not all distinct. The general condition for observability of equations (7.1) or (7.2) and (7.7), corresponding to the general condition for controllability, and applicable to both continuous-time and discrete-time equations, is that the set of n $n \times p$-matrices $\mathbf{C}'$, $\mathbf{A}'\mathbf{C}'$, ..., $(\mathbf{A}')^{n-1}\mathbf{C}'$ (where $\mathbf{A}'$, $\mathbf{C}'$, etc. are the transposes of $\mathbf{A}$, $\mathbf{C}$, etc.) be linearly independent. It can be used to confirm that any of the controlled processes, with high-order dynamic equations and a single output, in the examples of previous chapters is observable. For example, the

† The symbol c is used here with a single subscript to represent normal coordinates and with a double subscript to represent elements of the matrix $\mathbf{C}^*$.

rotating inertia discussed in Section 7.3 has

$$\mathbf{A} = \begin{bmatrix} 0 & 1 \\ 0 & 0 \end{bmatrix}, \qquad \mathbf{C} = [1 \quad 0],$$

and the matrix $\mathbf{M}$ to determine whether or not $\mathbf{C}'$ and $\mathbf{A}'\mathbf{C}'$ are linearly independent here is

$$\mathbf{M} = \begin{bmatrix} 1 & 0 \\ 0 & 1 \end{bmatrix},$$

which is square and non-singular, so that the system is observable.

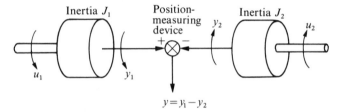

$$y = y_1 - y_2$$

FIG. 7.4. Example of an unobservable system.

An example of an unobservable system is illustrated in Fig. 7.4 where two rotating inertias have their positions y_1 and y_2 controlled by separate torques u_1 and u_2, but the only observable output y is produced by a device which measures the difference $y_1 - y_2$; this could happen with any differential position-measuring device. Using the same state variables

$$x_1 \equiv y_1, \qquad x_2 \equiv \dot{y}_1, \qquad x_3 \equiv y_2, \qquad x_4 \equiv \dot{y}_2,$$

as in the example of Section 7.3 the matrix $\mathbf{A}$ is

$$\mathbf{A} = \begin{bmatrix} 0 & 1 & 0 & 0 \\ 0 & 0 & 0 & 0 \\ 0 & 0 & 0 & 1 \\ 0 & 0 & 0 & 0 \end{bmatrix},$$

and the matrix $\mathbf{C}$ of equation (7.7) for the output y is

$$\mathbf{C} = [1 \quad 0 \quad -1 \quad 0].$$

Self-products of the matrix $\mathbf{A}'$ here are zero, and the matrix $\mathbf{M}$ to determine whether or not the vectors $\mathbf{C}', \mathbf{A}'\mathbf{C}', (\mathbf{A}')^2\mathbf{C}', (\mathbf{A}')^3\mathbf{C}'$ are linearly independent is

$$\mathbf{M} = \begin{bmatrix} 1 & 0 & 0 & 0 \\ 0 & 1 & 0 & 0 \\ -1 & 0 & 0 & 0 \\ 0 & -1 & 0 & 0 \end{bmatrix},$$

which is square and singular, so that the vectors are not linearly independent and the system is unobservable. The interpretation of this unobservability is that it would not be possible to evaluate the independent positions y_1 and y_2 from the output y, and it follows that y would not, by itself, be an adequate feedback signal for controlling the positions of the inertias.

The foregoing test for observability gives no indication of how to evaluate the state x from observations of the output y; it only answers the fundamental question of whether such evaluation is possible at all. The design of special dynamic systems, known as *observers*, to perform the evaluation is beyond the scope of this book;† methods for estimating states in the presence of uncertainty are discussed in Chapter 14. As with controllability, there are no general results about observability of non-linear equations.

Problem 7.7

7.5. General descriptions of linear systems

The state-space description of dynamics shows that the nature of a linear system depends on properties of the matrices $\mathbf{A}$, $\mathbf{B}$ and $\mathbf{C}$ of equations (7.1) or (7.2) and (7.7). The important properties are:

(i) the order n of the matrix $\mathbf{A}$, and of the system. This number, which is the number of state variables needed to describe the dynamic state of the system, is a measure of the complexity of the system.

(ii) the eigenvalues and eigenvectors of the matrix $\mathbf{A}$. These determine the dynamic behaviour of the system.

(iii) the rank of a controllability matrix

$$\mathbf{M_C} \equiv [\mathbf{B} \ \vdots \ \mathbf{AB} \ \vdots \cdots \vdots \ \mathbf{A}^{n-1}\mathbf{B}] \qquad (7.8)$$

which must be n if the system is to be controllable.

(iv) the rank of an observability matrix

$$\mathbf{M_O} \equiv [\mathbf{C}' \ \vdots \ \mathbf{A}'\mathbf{C}' \ \vdots \cdots \vdots \ (\mathbf{A}')^{n-1}\mathbf{C}'] \qquad (7.9)$$

which must be n if the system is to be observable.

All these properties are invariant under linear transformations of the state variables x: one such transformation was that used in previous sections to find the normal coordinates and to diagonalize the dynamic equations,

$$x = \mathbf{X}c \quad \text{or} \quad c = \mathbf{X}^{-1}x,$$

but any other linear transformation, for example

$$x_T = \mathbf{T}x,$$

provided that the transformation matrix $\mathbf{T}$ is non-singular, would also preserve the properties of the equations. This sort of linear transformation can

† See Luenberger (1964) or Dorf (1965) Section 5.3 or Sage (1968) Section 11.3 or Åström (1970) Sections 5.4 and 5.5 or Anderson and Moore (1971) Sections 8.1–8.3 or Kwakernaak and Sivan (1972) Section 4.2.

be regarded as a rotation of the coordinate axes in state space; it alters the values of the coordinate, or state, variables but not the shape of the trajectories described. It follows that there is no unique set of state variables **x** for any linear system; the number of state variables must be the same as the order n of the system but any linearly independent combination of the normal coordinates, or of any other set of state variables, can be used.

For the general class of single-variable continuous-time or discrete-time systems with dynamic equations,†

$$\frac{d^n y}{dt^n} + \cdots + a_1 \frac{dy}{dt} + a_0 y = b_n \frac{d^n u}{dt^n} + \cdots + b_0 u, \qquad (7.10a)$$

$$y(i) + \cdots + a_1 y(i - n + 1) + a_0 y(i - n) = b_n u(i) + \cdots + b_0 u(i - n) \qquad (7.10b)$$

convenient sets of state variables are:

$$\text{continuous-time } \mathbf{x}(t) \equiv \mathbf{T}_1 \begin{bmatrix} y \\ \dfrac{dy}{dt} \\ \vdots \\ \dfrac{d^{n-1}y}{dt^{n-1}} \end{bmatrix} + \mathbf{T}_2 \begin{bmatrix} u \\ \dfrac{du}{dt} \\ \vdots \\ \dfrac{d^{n-1}u}{dt^{n-1}} \end{bmatrix}, \qquad (7.11a)$$

$$\text{discrete-time } \mathbf{x}(i) \equiv \mathbf{T}_3 \begin{bmatrix} y(i) \\ y(i-1) \\ \vdots \\ y(i-n+1) \end{bmatrix} + \mathbf{T}_4 \begin{bmatrix} u(i) \\ u(i-1) \\ \vdots \\ u(i-n+1) \end{bmatrix}, \qquad (7.11b)$$

where $\mathbf{T}_1, \mathbf{T}_2, \mathbf{T}_3, \mathbf{T}_4$ are $n \times n$-matrices defined by

$$\mathbf{T}_1 \equiv \begin{bmatrix} a_1 & a_2 & . & a_{n-1} & 1 \\ a_2 & & . & 1 & \\ . & & . & . & \\ a_{n-1} & 1 & & & \\ 1 & & & & \end{bmatrix}, \qquad \mathbf{T}_2 \equiv \begin{bmatrix} -b_1 & -b_2 & . & & -b_n \\ -b_2 & & . & -b_n & \\ . & & . & . & \\ . & & -b_n & & \\ -b_n & & & & \end{bmatrix},$$

† There is no loss of generality in setting the leading coefficient a_n on the left-hand side of the equations to unity, nor in assuming the same number of terms on both sides of the equations. Systems with different numbers of terms are represented by setting appropriate coefficients to zero.

$$
\mathbf{T}_3 \equiv \left[\begin{array}{cccc}
-a_0 & & & \\
-a_1 & -a_0 & & \\
\cdot & \cdot & \cdot & \\
-a_{n-2} & \cdot & \cdot & -a_0 \\
\hline
1 & & &
\end{array}\right], \quad
\mathbf{T}_4 \equiv \left[\begin{array}{cccc}
b_0 & & & \\
b_1 & b_0 & & \\
\cdot & \cdot & \cdot & \\
b_{n-2} & \cdot & \cdot & b_0 \\
\hline
-b_n & & &
\end{array}\right].
$$

These state variables are convenient because they lead to canonical forms of the dynamic equations in n-dimensional state space:

$$
\dot{\mathbf{x}} = \mathbf{A}\mathbf{x}(t) + \mathbf{B}u(t), \tag{7.12a}
$$

$$
\mathbf{x}(i+1) = \mathbf{A}\mathbf{x}(i) + \mathbf{B}u(i), \tag{7.12b}
$$

where in both of equations (7.12)

$$
\mathbf{A} = \left[\begin{array}{ccccc}
0 & \cdots & 0 & -a_0 \\
\hline
1 & & & -a_1 \\
& \cdot & & \cdot \\
& & \cdot & \cdot \\
& & 1 & -a_{n-1}
\end{array}\right], \quad
\mathbf{B} = \left[\begin{array}{c}
b_0 - a_0 b_n \\
b_1 - a_1 b_n \\
\cdot \\
\cdot \\
b_{n-1} - a_{n-1} b_n
\end{array}\right].
$$

The principal significance of this canonical form is that it provides a state-space representation for any single-variable system. In this representation the matrix $\mathbf{A}$ is a companion matrix similar to that in equations (7.1) and (7.2).

The current value of the output variable y in both continuous-time and discrete-time systems is related to current values of the control u and the state $\mathbf{x}$ of equations (7.11) by an equation like (7.7):

$$
y = \mathbf{C}\mathbf{x} + Du, \tag{7.13}
$$

where

$$
\mathbf{C} = [0 \quad \cdots \quad 0 \quad 1], \qquad D = b_n.
$$

The matrix $\mathbf{C}$ here can be used to form the observability matrix $\mathbf{M}_0$ of equation (7.9) for single-variable systems:

$$
\mathbf{M}_0 = \left[\begin{array}{ccccc}
& & & & 1 \\
& & & \cdot & -a_{n-1} \\
& & 1 & \cdot & a_{n-1}^2 - a_{n-2} \\
& 1 & -a_{n-1} & & \cdot \\
1 & -a_{n-1} & -a_{n-1}^2 - a_{n-2} & \cdot &
\end{array}\right].
$$

This matrix $\mathbf{M}_0$ has determinant ± 1 and rank n, and therefore shows that single-variable systems specified by dynamic equations (7.10) never exhibit unobservability.

It can similarly be shown† that equations (7.10) never exhibit uncontrollability, and that uncontrollability and unobservability are features of state-space equations (7.1) or (7.2) and (7.7) rather than of input–output equations (7.10). Any set of state variables used to represent a dynamic system can, in general, be subdivided into four subsets depending on whether they are controllable (C) or observable (O). Figure 7.5 shows how this subdivision can be used to describe interconnections in the general linear system of Fig. 7.1; it

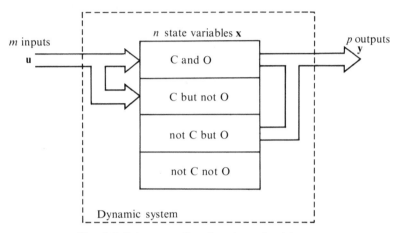

FIG. 7.5. Interconnections in a dynamic system.

shows how only the subset of state variables which are both controllable and observable affects the response of outputs **y** to inputs **u**. Therefore, although for any given set of state-space equations there is a unique relationship between inputs **u** and outputs **y**, for any given input–output equations there can be an infinite number of possible sets of state-space equations. Uncontrollability and unobservability arise when the number of state variables exceeds the order of the dynamic equations specifying the input–output relationship.

The question of what is meant by the order of a multi-variable system is beyond the scope of this book, as also is the specification of canonical forms for the general class of multi-variable systems.‡

Problems 7.8–7.11

7.6. Bibliography

The development of state-space methods for description and analysis of linear dynamic systems is, at the time of writing, an active area of research in mathematical system theory. Introductions to state space, similar to that given here are in

Tou (1964) Chapter 3,

De Russo, Roy, and Close (1965) Chapters 5 and 6,

† See Kalman (1963*a*).   ‡ See Kalman (1966).

DORF (1965) Chapters 2 and 3,
FREEMAN (1965) Chapter 2,
LINDORFF (1965) Chapter 7,
GUPTA (1966) Chapters 7 and 8,
ELGERD (1967) Chapters 3 and 4,
KUO (1967) Chapter 3; (1970) Chapter 4,
OGATA (1967) Sections 1.1–1.3; (1970) Chapter 14,
SCHULTZ and MELSA (1967) Chapter 2,
CHEN and HAAS (1968) Chapter 6,
KOPPEL (1968) Chapters 3 and 4,
SAUCEDO and SCHIRING (1968) Chapters 12 and 13,
TIMOTHY and BONA (1968) Chapter 4,
BRYSON and HO (1969) Appendix B,
MCCAUSLAND (1969) Chapter 4,
PRIME (1969) Chapter 2,
WATKINS (1969) Chapter 5,
CADZOW and MARTENS (1970) Chapter 5,
GUPTA and HASDORFF (1970) Chapter 7,
TAKAHASHI, RABINS and AUSLANDER (1970) Chapter 2,
EVELEIGH (1972) Chapter 13,
SMYTH (1972) Chapter 4.

Discussion of controllability and observability is in
LEE (1964) Section 4.1.2,
TOU (1964) Section 4.4,
DORF (1965) Chapter 5,
ATHANS and FALB (1966) Section 4.15 *et seq.*,
ELGERD (1967) Chapter 5,
KUO (1967) Chapter 11,
LEE and MARKUS (1967) Section 2.4,
OGATA (1967) Chapter 7; (1970) Sections 16.2 and 16.3,
CHEN and HAAS (1968) Section 8.V,
SAGE (1968) Sections 11.1 and 11.2,
TIMOTHY and BONA (1968) Chapter 9,
BRYSON and HO (1969) Appendix B,
MCCAUSLAND (1969) Section 4.10,
PRIME (1969) Sections 3.5 and 3.6,
CADZOW and MARTENS (1970) Section 7.4,
ROSENBROCK (1970) Sections 5.2 and 5.3,
TAKAHASHI, RABINS, and AUSLANDER (1970) Section 3.4,
KWAKERNAAK and SIVAN (1972) Section 1.6–1.8.

Canonical forms for single-variable systems are described in
LEE (1964) Section 4.13,
MERRIAM (1964) Section 1.4,
ATHANS and FALB (1966) Section 4.10,
OGATA (1967) Section 4.2; (1970) pp. 675–677, 697–698,

Hsu and Meyer (1968) Chapter 2,
McCausland (1969) Section 4.2,
Desoer (1970) Section IV.6,
Kwakernaak and Sivan (1972) Section 1.9.
Further aspects of this whole branch of mathematical system theory are discussed in
Zadeh and Desoer (1963),
Johnson and Wonham (1966),
Kalman, Falb and Arbib (1969),
Brockett (1970),
Chen (1970),
D'Angelo (1970),
Desoer (1970),
Rosenbrock (1970),
Rubio (1971),
Barnett (1971).

7.7. Problems

7.1. Write state equations in the canonical form

$$\dot{\mathbf{x}} = \mathbf{A}\mathbf{x} + \mathbf{B}u$$

to describe the dynamics of single-variable continuous-time systems having transfer functions $Y(s)/U(s)$ as follows:

(i) $\dfrac{1}{1 + sT}$

(ii) $\dfrac{K}{(1 + sT_1)(1 + sT_2)}$

(iii) $\dfrac{10}{s^4}$

(iv) $\dfrac{4}{s^3 + 7s^2 + 14s + 8}$

(v) $\dfrac{1}{s^9 - 1}$

7.2. What are the eigenvalues of the matrix $\mathbf{A}$ for each of the systems in Problem 7.1?

7.3. Write state equations in the diagonal form

$$\dot{\mathbf{c}} = \mathbf{S}\mathbf{c} + \mathbf{B}^*u,$$

where $\mathbf{S}$ is diagonal, for as many as possible of the systems (i), (ii), (iii) of Problem 7.1.

7.4. In the discrete-time position-control system of Problem 1.12 the position y

and the velocity v can be taken as state variables x_1 and x_2. Find the sequence of controls $u(1)$, $u(2)$ which will take an arbitrary initial state $x(1)$ to the origin in two steps.

7.5. Use the result of Problem 7.3(ii) to investigate the controllability of a discrete-time system having z-transfer function

$$\frac{Y(z)}{U(z)} = \frac{k}{(a_1 + z^{-1})(a_2 + z^{-1})}.$$

7.6. Use the result of Problem 7.1(iii) to investigate the controllability of a discrete-time system having difference equation

$$y(i) = 10u(i - 4).$$

7.7. Investigate the controllability and observability of the continuous-time system having input u, states x_1, x_2 and output y as shown in Fig. 7.6.

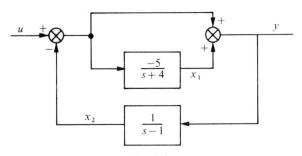

FIG. 7.6.

7.8. Substitute equations (7.11) into equations (7.12) and hence verify that they represent equations (7.10).

7.9. For a discrete-time system having z-transfer function

$$\frac{Y(z)}{U(z)} = \frac{5}{1 + 2z^{-1} + 3z^{-2} + 4z^{-3}}$$

specify variables $x(i)$ and corresponding equations

$$x(i + 1) = Ax(i) + Bu(i)$$

(i) in the canonical form of equation (7.2),
(ii) in the canonical form of equation (7.10b).

7.10. Write state equations in the canonical form

$$x(i + 1) = Ax(i) + Bu(i)$$

to describe the dynamics of a discrete-time system having z-transfer function

$$\frac{Y(z)}{U(z)} = \frac{3z^3 + z + 3}{z^3 - z^2 - 4z + 1}.$$

7.11. Find the controllability matrix for a continuous-time system having transfer function

$$\frac{Y(s)}{U(s)} = \frac{b_2 s^2}{s^2 + a_1 s + a_0}.$$

Comment on the circumstances under which the system may be uncontrollable.

8. Optimal LQP-control

8.1. Introduction

THE semi-free design problem, where a controlled process and a performance criterion are specified and it is required to find the control law to optimize performance, was introduced in Section 6.1 and can now be reformulated in terms of state space. The state-space formulation, with its set of simultaneous first-order dynamic equations, makes a very general class of semi-free problems amenable to mathematical techniques of optimization. This class includes all problems where the controlled process can be described by ordinary, as opposed to partial, dynamic equations written

$$\dot{\mathbf{x}} = \mathbf{g}(\mathbf{x}, \mathbf{u}, t) \qquad \text{(continuous time)} \qquad (8.1a)$$
$$\mathbf{x}(i + 1) = \mathbf{G}\big(\mathbf{x}(i), \mathbf{u}(i), i\big) \qquad \text{(discrete time)}, \qquad (8.1b)$$

and where the performance criterion I is of the form

$$I = \int_{t_1}^{t_2} h(\mathbf{x}, \mathbf{u}, t)\, \mathrm{d}t \qquad \text{(continuous time)} \qquad (8.2a)$$

$$I = \sum_{i=i_1}^{i_2} H\big(\mathbf{x}(i), \mathbf{u}(i), i\big) \qquad \text{(discrete time)}. \qquad (8.2b)$$

In equations (8.1) $\mathbf{g}$ and $\mathbf{G}$ are not necessarily linear, vector functions, and in equations (8.2) h and H are scalar functions, of the state $\mathbf{x}$, the control $\mathbf{u}$, and of time t or i. The question of what is the optimal control law for a given situation, as specified by equations (8.1) and (8.2), can be regarded as a fundamental question in control theory; the value of the state-space formulation is that it translates this fundamental question into a mathematical problem which can be attacked by mathematical techniques.

When the dynamic equations (8.1) are linear and the performance criterion (8.2) is quadratic the mathematical problem has an explicit solution giving the optimal control law. Such problems with *linear* dynamics and *quadratic* performance criteria are known as LQP-problems; the analytic design problems solved in Chapter 6 were examples of LQP-problems. Chapter 8 introduces the mathematical techniques of optimal control theory and discusses LQP-

problems; other deterministic problems are discussed in Chapter 9, and
stochastic problems, where equations (8.1) and (8.2) are modified to account
for uncertainty, are discussed in Chapters 13 and 15. The mathematical
technique used to attack all these optimal control problems is dynamic
programming.

8.2. Dynamic programming

Dynamic programming is a slightly misleading name for what is really a
mathematical theory of multistage decision processes. The discrete-time op-
timal-control problem, where it is required to make a series of decisions about
the controls $u(1), u(2), \ldots$, is an obvious example of a multistage decision
process; other examples are to be found in most subjects that make use of
applied mathematics, such as economics, design of industrial plant, operations research, learning theory, games theory. The name dynamic programming, with its suggestion of computer techniques, was introduced because
the theory leads in general to equations which are well conditioned for numerical solution using a digital computer; the general usefulness of dynamic
programming is limited, as with other techniques where solutions must be
computed numerically, to problems of low dimensionality. In spite of this
general practical difficulty, the theory of multistage decision processes
provides a very convenient theoretical framework for discussions of the
fundamental question about optimal control of both discrete-time and continuous-time processes. For LQP-problems the dynamic programming equations have analytic solutions.

   The basic theory of dynamic programming is introduced by considering a
general, stationary, single-variable, discrete-time problem with dynamic
equation

$$x(i + 1) = G\big(x(i), u(i)\big)$$

and performance criterion

$$I = \sum_{i=1}^{N} H\big(x(i), u(i)\big),$$

which could be written out

$$I = H\big(x(1), u(1)\big) + H\big\{G\big(x(1), u(1)\big), u(2)\big\} +$$
$$+ H\big[G\{G\big(x(1), u(1)\big), u(2)\}, u(3)\big] + \cdots + H(N)$$
$$= I\big(x(1), u(1), \ldots, u(N)\big)$$

to show that it depends on the initial state $x(1)$ and the series of controls
$u(1), \ldots, u(N)$. The multistage decision problem of finding values of the controls $u(1), \ldots, u(N)$ to minimize the performance criterion I thus appears to

be the sort of problem solved by the classical calculus method of setting partial derivatives to zero to find a stationary value. There are, however, several reasons why this method is unsuitable as a general procedure for multistage decision processes: the partial derivatives $\partial I/\partial u(1), \ldots, \partial I/\partial u(N)$ may not exist; the simultaneous equations derived by setting all the partial derivatives to zero may not be easy to solve; there may be more than one stationary value of I; the minimum of I may not be a stationary value at all. These difficulties are all avoided by dynamic programming.

The point of departure for dynamic programming is the above observation that, for a given problem, the value of I depends only on the initial state $x(1)$ and the controls $u(1), \ldots, u(N)$. It follows that the optimal value of I, minimized with respect to $u(1), \ldots, u(N)$, depends only on the initial state $x(1)$. This dependence is emphasized by defining a function

$f_N(x) \equiv$ Value of the performance criterion I over N stages using the optimal control and starting from the initial state $x(1) = x$

$$= \min_{u(1),\ldots,\,u(N)} \{I(x, u(1), \ldots, u(N))\}.$$

Now, regarding I purely as a mathematical function of its arguments x, $u(1), \ldots, u(N)$ there is no reason why the N minimizations of the definition should not be performed in any arbitrary order. In particular, the minimization with respect to $u(1)$ could be done last to give

$$f_N(x) = \min_{u(1)} \{ \min_{u(2),\ldots,\,u(N)} \{I\} \},$$

and since the performance criterion I has its first term $H(x, u(1))$ independent of $u(2), \ldots, u(N)$ the inner minimization can be written

$$\min_{u(2),\ldots,\,u(N)} \{I\} = H(x, u(1)) + \min_{u(2),\ldots\,u(N)} \{H(G(x, u(1)), u(2)) + \cdots + H(N)\}.$$

It follows from the notation used to define the function $f_N(x)$ that the second term in this expression is

$$\min_{u(2),\ldots\,u(N)} \{H(G(x, u(1)), u(2)) + \cdots + H(N)\} \equiv f_{N-1}(G(x, u(1))),$$

so that the expression for $f_N(x)$ becomes

$$f_N(x) = \min_{u(1)} \{H(x, u(1)) + f_{N-1}(G(x, u(1)))\}, \tag{8.3}$$

a functional recurrence equation relating the function f_N for an N-stage process to the function f_{N-1} for an $(N - 1)$-stage process. This recurrence equation can be solved for successive values of N provided that the solution function $f_N(x)$ is known for some one value of N. The required initial solution can be found for a single-stage process, $N = 1$, where the definitions give

$$f_1(x) = \min_{u(1)} \{H(x(1), u(1))\}. \tag{8.4}$$

The functional equations (8.3) and (8.4) are the fundamental equations of dynamic programming. Their solution consists of a set of functions $f_N(x)$ describing optimal values of the performance criterion I and also a set of functions of x giving the minimizing, or optimal, values of the control u. It is convenient to describe this optimal control with a notation similar to that used for the optimal values of I,

$u_N(x) \equiv$ The optimal value of $u(1)$ in a process with N stages-to-go and with state $x(1) = x$,

so that the minimizing value of $u(1)$ in equation (8.3) is written $u_N(x)$. The equations cannot generally be solved analytically but the recursive nature of equation (8.3) makes it suitable for numerical solution using a digital computer.

Numerical aspects of dynamic programming are beyond the scope of this book.† For the purposes of control theory the importance of dynamic programming is that it can be used to formulate a wide range of problems and that it leads to analytic solutions to LQP-problems.

An example of a single-variable discrete-time LQP-problem has dynamic equation

$$x(i + 1) = ax(i) + bu(i)$$

and quadratic criterion function

$$H(i) = x^2(i) + qu^2(i).$$

Then substituting into equation (8.4) gives

$$f_1(x) = \min_{u(1)} \{x^2 + qu^2(1)\}$$

and the minimizing value of $u(1)$ is zero so that

$$u_1(x) = 0$$

and

$$f_1(x) = x^2.$$

Now this solution for $f_1(x)$ can be substituted into equation (8.3) to give

$$f_2(x) = \min_{u(1)} \{x^2 + qu^2(1) + f_1(ax + bu(1))\}$$

$$= \min_{u(1)} \{x^2 + a^2x^2 + 2abxu(1) + (q + b^2)u^2(1)\},$$

and the minimizing value of $u(1)$ can be found by calculus

$$\frac{d\{\ \}}{du(1)} = 2abx + 2(q + b^2)u(1) = 0$$

† See Bellman and Dreyfus (1962), Larson (1968), and Jacobson and Mayne (1970).

which gives

$$u_2(x) = -\frac{ab}{q + b^2}\, x,$$

$$f_2(x) = \left(1 + \frac{qa^2}{q + b^2}\right) x^2.$$

This solution for $f_2(x)$ can be substituted into equation (8.3) and the minimizing value of $u(1)$ found by calculus, which gives

$$u_3(x) = -\frac{ab(q + b^2 + qa^2)}{(q + b^2)^2 + \dot{q}a^2 b^4}\, x,$$

and so on. It is the case that the optimal control here is always linear in the state x and that the solution functions $f_N(x)$ are all quadratic in x, so the analysis could proceed indefinitely.

If the LQP-problem in the example were to run for three discrete-time stages, the optimal control $u^*(1)$ at the first stage would be found from the solution for $u_3(x)$, the optimal first control when there are three stages-to-go,

$$u^*(1) = u_3\big(x(1)\big) = -\frac{ab(q + b^2 + qa^2)}{(q + b^2)^2 + qa^2 b^4}\, x(1).$$

The optimal control $u^*(2)$ at the second stage would be found from $u_2(x)$, the optimal first control when there are two stages-to-go,

$$u^*(2) = u_2\big(x(2)\big) = -\frac{ab}{q + b^2}\, x(2),$$

and similarly the optimal control $u^*(3)$ at the third and final stage would be

$$u^*(3) = u_1\big(x(3)\big) = 0.$$

The optimal control law can thus be derived from the functional recurrence equations as a series of optimal first controls for a series of processes with decreasing number of stages-to-go.

This property of the optimal control law, that it can be regarded as a series of optimal first controls, holds for all problems which can be written in the general form of equations (8.1) and (8.2). It is known as the Principle of Optimality, expressed by the general statement that 'an optimal control sequence has the property that, whatever the initial state and the optimal first control may be, the remaining controls constitute an optimal control sequence with regard to the state resulting from the first control'. The fundamental functional equations of dynamic programming are usually derived directly from the Principle of Optimality.

For example, in the general single-variable problem the state $x(2)$ after the first decision $u(1)$ will be

$$x(2) = G(x(1), u(1))$$

and there will remain $N - 1$ stages-to-go. The Principle of Optimality implies that when the optimal control law is used the contribution to the performance criterion I from the last $N - 1$ stages will be $f_{N-1}(x(2))$, where

$$f_{N-1}(x(2)) = f_{N-1}(G(x(1), u(1))) \cdot$$

The contribution to I from the first stage is $H(x(1), u(1))$ and so the performance criterion when the optimal control is used for N stages will be

$$I = H(x(1), u(1)) + f_{N-1}(G(x(1), u(1))) \cdot$$

But the optimal control minimizes I with respect to N controls, including the first control $u(1)$, so the optimal value of I over N stages is

$$f_N(x(1)) = \min_{u(1)}\{H(x(1), u(1)) + f_{N-1}(G(x(1), u(1)))\} ,$$

which is equation (8.3) again.

The functional recurrence equations of dynamic programming, derived for discrete-time processes, can be carried over for continuous-time processes. For a general single-variable continuous-time problem with dynamic equation

$$\dot{x} = g(x, u)$$

and performance criterion

$$I = \int_0^T h(x, u)\, dt,$$

the optimal values of I are expressed by a function

$f(x, T) \equiv$ Value of I over time T using the optimal control and starting from state x,

and the optimal control is described by a similar notation:

$u(x, T) \equiv$ The optimal value of u when the time-to-go is T and the state is x.

The time interval $(0, T)$ of the performance criterion I can be subdivided into two component intervals $(0, \Delta)$ and (Δ, T) and I can then be expressed as the sum of two components. If Δ is so small that the increment δx in x over the first interval $(0, \Delta)$ is infinitesimal, the first component is

$$\int_0^\Delta h(x, u)\, dt \simeq h(x, u)\Delta,$$

and the second component is given by the Principle of Optimality as

$$\int_{\Delta}^{T} h(x, u) \, dt = f(x + \delta x, T - \Delta),$$

where

$$x + \delta x \simeq x(t = 0) + \dot{x}(t = 0)\Delta$$
$$= x + g(x, u)\Delta.$$

Provided that the solution function $f(x, T)$ is differentiable, the second component can be approximated by a first-order Taylor series expansion,

$$\int_{\Delta}^{T} h(x, u) \, dt \simeq f(x, T) + \frac{\partial f}{\partial x} g(x, u)\Delta - \frac{\partial f}{\partial T} \Delta.$$

Summing the two components of I gives

$$I \simeq h(x, u)\Delta + f(x, T) + \frac{\partial f}{\partial x} g(x, u)\Delta - \frac{\partial f}{\partial T} \Delta,$$

and minimizing with respect to the initial control u over $(0, \Delta)$ gives

$$f(x, T) \simeq \min_{u} \left\{ h(x, u)\Delta + f(x, T) + \frac{\partial f}{\partial x} g(x, u)\Delta - \frac{\partial f}{\partial T} \Delta \right\}.$$

In the limit as Δ goes to zero the approximations above become equalities giving a functional differential equation†

$$\frac{\partial f}{\partial T} = \min_{u} \left\{ h(x, u) + \frac{\partial f}{\partial x} g(x, u) \right\}. \tag{8.5}$$

The minimizing value of u in equation (8.5) is the optimal control $u(x, T)$.

Problems 8.1, 8.2

8.3. Multi-dimensional processes

The arguments used to derive the single-variable dynamic programming equations (8.3) and (8.5) are in no way restricted to single-variable processes. For multi-variable control problems, described by equations (8.1) and (8.2), the same arguments lead to a discrete-time functional recurrence equation

$$f_N(\mathbf{x}) = \min_{\mathbf{u}} \left\{ H(\mathbf{x}, \mathbf{u}) + f_{N-1}(\mathbf{G}(\mathbf{x}, \mathbf{u})) \right\} \tag{8.6a}$$

† A fuller discussion of limiting processes leading to functional differential equations for continuous-time systems is in Jacobs (1967) Chapter 3.

and to a continuous-time functional differential equation

$$\frac{\partial f}{\partial T} = \min_{\mathbf{u}} \left\{ h(\mathbf{x}, \mathbf{u}) + \sum_{j=1}^{n} \frac{\partial f}{\partial x_j} g_j(\mathbf{x}, \mathbf{u}) \right\}. \tag{8.6b}$$

These equations (8.6) are similar to the single-variable equations (8.3) and (8.5) except that the argument of the solution functions is the state vector $\mathbf{x}$ instead of the single state variable x.

One class of multi-dimensional LQP-problems has the dynamic equations (7.1), (7.2) introduced in Section 7.1 and repeated here as (8.7):

$$\mathbf{x}(i + 1) = \mathbf{A}\mathbf{x}(i) + \mathbf{B}\mathbf{u}(i) \quad \text{(discrete time)} \tag{8.7a}$$

$$\dot{\mathbf{x}} = \mathbf{A}\mathbf{x} + \mathbf{B}\mathbf{u} \quad \text{(continuous time)} \tag{8.7b}$$

together with quadratic performance criteria specified by putting

$$H = h = \mathbf{x}'\mathbf{P}\mathbf{x} + \mathbf{u}'\mathbf{Q}\mathbf{u} \tag{8.8}$$

in equations (8.2). This class of problems could be described as stationary regulators. They are stationary in that the matrices $\mathbf{A}$, $\mathbf{B}$, $\mathbf{P}$, and $\mathbf{Q}$ are not time-varying, and regulators in the sense that the performance criterion specified by equation (8.8) describes problems where the objective is to regulate the state $\mathbf{x}$ to a constant value chosen to coincide with the origin of state space. Non-stationary problems are discussed in Section 8.4 and tracking, as opposed to regulator, problems are discussed in Section 8.5.

Provided that the matrices $\mathbf{P}$ and $\mathbf{Q}$ in equation (8.8) are symmetric, which is not usually a serious restriction, solution functions f to the dynamic programming equations (8.6) specified by equations (8.7), (8.2), and (8.8) can be written in terms of symmetric matrices $\mathbf{V}$:

$$f_N(\mathbf{x}) = \mathbf{x}'\mathbf{V}_N\mathbf{x} \quad \text{(discrete time)}$$
$$f(\mathbf{x}, T) = \mathbf{x}'\mathbf{V}(T)\mathbf{x} \quad \text{(continuous time)}.$$

The matrices $\mathbf{V}$ here are functions of time-to-go and they can be found by substituting into the dynamic programming equations; the discrete-time functional recurrence equation (8.6a) gives

$$\mathbf{x}'\mathbf{V}_N\mathbf{x} = \min_{\mathbf{u}} \left\{ \mathbf{x}'\mathbf{P}\mathbf{x} + \mathbf{u}'\mathbf{Q}\mathbf{u} + (\mathbf{A}\mathbf{x} + \mathbf{B}\mathbf{u})'\mathbf{V}_{N-1}(\mathbf{A}\mathbf{x} + \mathbf{B}\mathbf{u}) \right\},$$

and the continuous-time functional differential equation (8.6b)† gives

$$\mathbf{x}'\frac{d\mathbf{V}}{dT}\mathbf{x} = \min_{\mathbf{u}} \left\{ \mathbf{x}'\mathbf{P}\mathbf{x} + \mathbf{u}'\mathbf{Q}\mathbf{u} + 2\mathbf{x}'\mathbf{V}(\mathbf{A}\mathbf{x} + \mathbf{B}\mathbf{u}) \right\}.$$

† In Equation (8.6b) the sum

$$\sum_{j=1}^{n} \frac{\partial f}{\partial x_j} g_j$$

The minimizing values of **u** can be found from calculus conditions:

$$\frac{d\{\ \}}{d\mathbf{u}} = 2\mathbf{u}'\mathbf{Q} + 2(\mathbf{A}\mathbf{x} + \mathbf{B}\mathbf{u})'\mathbf{V}_{N-1}\mathbf{B} = \mathbf{0} \quad \text{(discrete time)}$$

$$\frac{d\{\ \}}{d\mathbf{u}} = 2\mathbf{u}'\mathbf{Q} + 2\mathbf{x}'\mathbf{V}\mathbf{B} = \mathbf{0} \qquad\qquad \text{(continuous time)}$$

and give optimal controls which are linear in the state **x**,

where
$$\left.\begin{aligned}\mathbf{u}_N(\mathbf{x}) &= -\mathbf{K}_N\mathbf{x}, \\ \mathbf{K}_N &= (\mathbf{Q} + \mathbf{B}'\mathbf{V}_{N-1}\mathbf{B})^{-1}\mathbf{B}'\mathbf{V}_{N-1}\mathbf{A};\end{aligned}\right\} \text{(discrete time)} \qquad (8.9a)$$

and

where
$$\left.\begin{aligned}\mathbf{u}(\mathbf{x}, T) &= -\mathbf{K}(T)\mathbf{x}, \\ \mathbf{K}(T) &= \mathbf{Q}^{-1}\mathbf{B}'\mathbf{V}.\end{aligned}\right\} \text{(continuous time)} \qquad (8.9b)$$

Substituting the discrete-time optimal control of equation (8.9a) into the discrete-time functional recurrence equation gives a quadratic expression

$$\mathbf{x}'\mathbf{V}_N\mathbf{x} = \mathbf{x}'\left(\mathbf{P} + \mathbf{K}_N'\mathbf{Q}\mathbf{K}_N + (\mathbf{A} - \mathbf{B}\mathbf{K}_N)'\mathbf{V}_{N-1}(\mathbf{A} - \mathbf{B}\mathbf{K}_N)\right)\mathbf{x}$$

in which the bracketed term ( ) is symmetric provided that $\mathbf{V}_{N-1}$ is symmetric. When this term ( ) is symmetric, the expression reduces to a difference equation for the symmetric matrix $\mathbf{V}_N$:

---

can be written

$$\sum_{j=1}^{n}\frac{\partial f}{\partial x_j}g_j = \left[\frac{\partial f}{\partial x_1}, \ldots, \frac{\partial f}{\partial x_n}\right]\begin{bmatrix}g_1 \\ \vdots \\ g_n\end{bmatrix} = \begin{bmatrix}\dfrac{\partial f}{\partial x_1} \\ \vdots \\ \dfrac{\partial f}{\partial x_n}\end{bmatrix}'\mathbf{g}.$$

When $f = \mathbf{x}'\mathbf{V}\mathbf{x}$ the partial derivatives can be written out as

$$\begin{bmatrix}\dfrac{\partial f}{\partial x_1} \\ \vdots \\ \dfrac{\partial f}{\partial x_n}\end{bmatrix} = \begin{bmatrix}2(x_1 v_{11} + \cdots + x_n v_{1n}) \\ \cdot \quad \cdot \quad \cdot \quad \cdot \quad \cdot \\ 2(x_1 v_{n1} + \cdots + x_n v_{nn})\end{bmatrix} = 2\mathbf{V}\mathbf{x}.$$

Then the sum becomes
$$\sum_{j=1}^{n}\frac{\partial f}{\partial x_j}g_j = 2(\mathbf{V}\mathbf{x})'\mathbf{g} = 2\mathbf{x}'\mathbf{V}'\mathbf{g},$$

and if the matrix **V** is symmetric the sum can be written
$$\sum_{j=1}^{n}\frac{\partial f}{\partial x_j}g_j = 2\mathbf{x}'\mathbf{V}\mathbf{g} = 2\mathbf{x}'\mathbf{V}(\mathbf{A}\mathbf{x} + \mathbf{B}\mathbf{u}).$$

$$\mathbf{V}_N = \mathbf{P} + \mathbf{K}'_N \mathbf{Q} \mathbf{K}_N + (\mathbf{A} - \mathbf{B}\mathbf{K}_N)' \mathbf{V}_{N-1}(\mathbf{A} - \mathbf{B}\mathbf{K}_N)$$

$$= \mathbf{P} + \mathbf{K}'_N(\mathbf{Q} + \mathbf{B}'\mathbf{V}_{N-1}\mathbf{B})\mathbf{K}_N + \mathbf{A}'\mathbf{V}_{N-1}\mathbf{A} - \mathbf{A}'\mathbf{V}_{N-1}\mathbf{B}\mathbf{K}_N - \mathbf{K}'_N\mathbf{B}'\mathbf{V}_{N-1}\mathbf{A},$$

and using equation (8.9a), which defines $\mathbf{K}_N$, this becomes

$$\mathbf{V}_N = \mathbf{P} + \mathbf{A}'\mathbf{V}_{N-1}\mathbf{A} - \mathbf{A}'\mathbf{V}_{N-1}\mathbf{B}(\mathbf{Q} + \mathbf{B}'\mathbf{V}_{N-1}\mathbf{B})^{-1}\mathbf{B}'\mathbf{V}_{N-1}\mathbf{A}. \quad (8.10a)$$

Similarly, substituting the continuous-time optimal control of equation (8.9b) into the continuous-time functional differential equation gives a quadratic expression

$$\mathbf{x}' \frac{d\mathbf{V}}{dT} \mathbf{x} = \mathbf{x}'(\mathbf{P} + 2\mathbf{V}\mathbf{A} - \mathbf{V}\mathbf{B}\mathbf{Q}^{-1}\mathbf{B}'\mathbf{V})\mathbf{x}$$

in which the bracketed term ( ) can be made symmetric by using the identity

$$2\mathbf{x}'\mathbf{V}\mathbf{A}\mathbf{x} \equiv \mathbf{x}'\mathbf{A}'\mathbf{V}'\mathbf{x} + \mathbf{x}'\mathbf{V}\mathbf{A}\mathbf{x} = \mathbf{x}'(\mathbf{A}'\mathbf{V} + \mathbf{V}\mathbf{A})\mathbf{x},$$

where $\mathbf{V}$ and $(\mathbf{A}'\mathbf{V} + \mathbf{V}\mathbf{A})$ are symmetric. The resulting differential equation for the symmetric matrix $\mathbf{V}(T)$ is

$$\frac{d\mathbf{V}}{dT} = \mathbf{P} + \mathbf{A}'\mathbf{V} + \mathbf{V}\mathbf{A} - \mathbf{V}\mathbf{B}\mathbf{Q}^{-1}\mathbf{B}'\mathbf{V} \quad (8.10b)$$

and belongs to a class of non-linear differential equations known as Riccati equations. Boundary conditions, which are needed to solve equations (8.10), are provided in discrete time by considering a single-stage process, $N = 1$, where it follows from the definitions that

$$\mathbf{u}_1(\mathbf{x}) = \mathbf{0}, \qquad \mathbf{K}_1 = \mathbf{0},$$

giving

$$\mathbf{V}_1 = \mathbf{P}; \quad (8.11a)$$

and in continuous time by the corresponding result for a process with zero time-to-go, $T = 0$,

$$\mathbf{V}(0) = \mathbf{0}.$$

$$(8.11b)$$

It can be shown† that unique stable solutions to the dynamic equations (8.10) exist when

(i) the controlled process specified by matrices $\mathbf{A}$ and $\mathbf{B}$ of equations (8.7) is controllable in the sense of Section 7.3,

(ii) the symmetric matrices $\mathbf{P}$, and for continuous-time systems the diagonal matrix $\mathbf{Q}$, in the performance criteria of equations (8.8) are positive definite.‡

The fact that, subject to conditions (i) and (ii), solutions exist justifies the assumption used to derive equations (8.10), that the functions f are quadratic

---

† See Kalman (1960c).

‡ A square matrix $\mathbf{P}$ is positive definite if the scalar quantity $\mathbf{x}'\mathbf{P}\mathbf{x}$ is positive for all non-zero vectors $\mathbf{x}$.

in **x**. The conditions are often satisfied in control problems where the objective is to regulate process variables to constant desired values. They are, however, not always satisfied where the objective is to make process variables follow time-varying desired values, as discussed in Chapter 2: optimal control laws for such problems are discussed in Section 8.5.

The condition about positive definiteness of the matrix **Q** is necessary for continuous-time systems but not for discrete-time systems. This reflects a fundamental difference between continuous-time and discrete-time systems which can be demonstrated by considering what happens in equations (8.9) and (8.10) when the matrix **Q**, instead of being positive definite, is null, that is when cost-of-control is excluded from the performance criterion I. When **Q** is null the inverse $\mathbf{Q}^{-1}$, which appears in the continuous-time equations (8.9b) and (8.10b), becomes infinite in a way which implies that the optimal feedback gain **K** and the optimal control **u*** also become infinite. The meaninglessness of this result has already been discussed in Chapter 6 where it was seen to be one of the reasons for introducing cost-of-control terms into performance criteria for fixed-configuration analytic design problems. Cost-of-control terms specified by positive definite matrices **Q** must similarly be included in performance criteria for semi-free, continuous-time LQP-problems. The discrete-time equations (8.9a) and (8.10a) differ from the continuous-time equations in that they do not contain $\mathbf{Q}^{-1}$ and that they therefore specify finite optimal control laws when **Q** is null: for example when **Q** is null and **B** is square and non-singular the discrete-time equations (8.9a) specify an optimal control law

$$\mathbf{u}^*(i) = -\mathbf{B}^{-1}\mathbf{A}\mathbf{x}(i),$$

which takes the state **x** to the origin of state space in a single step. This difference between continuous-time and discrete-time LQP-problems arises because the discretization in time constitutes a constraint on discrete-time systems which has no equivalent in the continuous-time systems of equations (8.7b) and (8.8). Some continuous-time systems are subject to a related constraint in the form of a constraint on the range of values of the control **u**; this makes the control problem non-linear and is discussed in Section 9.4.

Equations (8.9), (8.10), and (8.11) can be used to find the optimal control for any LQP-problem specified by matrices **A**, **B**, **P**, **Q**, such that solutions to equations (8.10) exist. An example is the general second-order, continuous-time system having differential equation

$$a_2\ddot{y} + a_1\dot{y} + a_0 y = u$$

and performance criterion

$$I = \int_0^T (y^2 + qu^2)\, dt.$$

Using a state vector

$$\mathbf{x} \equiv \begin{bmatrix} y \\ \dot{y} \end{bmatrix},$$

the matrices would be

$$\mathbf{A} = \begin{bmatrix} 0 & 1 \\ -\dfrac{a_0}{a_2} & -\dfrac{a_1}{a_2} \end{bmatrix}, \quad \mathbf{B} = \begin{bmatrix} 0 \\ \dfrac{1}{a_2} \end{bmatrix},$$

$$\mathbf{P} = \begin{bmatrix} 1 & 0 \\ 0 & 0 \end{bmatrix}. \quad \mathbf{Q} = q,$$

and substituting these into the matrix Riccati equation (8.10b) gives

$$
\frac{\mathrm{d}\mathbf{V}}{\mathrm{d}T} = \begin{bmatrix} 1 & 0 \\ 0 & 0 \end{bmatrix} + \begin{bmatrix} 0 & -\dfrac{a_0}{a_2} \\ 1 & -\dfrac{a_1}{a_2} \end{bmatrix} \begin{bmatrix} v_{11} & v_{12} \\ v_{12} & v_{22} \end{bmatrix} + \begin{bmatrix} v_{11} & v_{12} \\ v_{12} & v_{22} \end{bmatrix} \begin{bmatrix} 0 & 1 \\ -\dfrac{a_0}{a_2} & -\dfrac{a_1}{a_2} \end{bmatrix}
$$

$$
- \begin{bmatrix} v_{11} & v_{12} \\ v_{12} & v_{22} \end{bmatrix} \begin{bmatrix} 0 \\ \dfrac{1}{a_2} \end{bmatrix} q^{-1} \begin{bmatrix} 0 & \dfrac{1}{a_2} \end{bmatrix} \begin{bmatrix} v_{11} & v_{12} \\ v_{12} & v_{22} \end{bmatrix},
$$

which is the matrix form of three simultaneous Riccati differential equations
for the three elements v_{11}, v_{12}, v_{22} of the symmetric matrix $\mathbf{V}$:

$$\frac{\mathrm{d}v_{11}}{\mathrm{d}T} = 1 - 2\frac{a_0}{a_2}v_{12} - \frac{(v_{12})^2}{qa_2^2},$$

$$\frac{\mathrm{d}v_{12}}{\mathrm{d}T} = -\frac{a_0}{a_2}v_{22} + v_{11} - \frac{a_1}{a_2}v_{12} - \frac{v_{12}v_{22}}{qa_2^2},$$

$$\frac{\mathrm{d}v_{22}}{\mathrm{d}T} = 2v_{12} - 2\frac{a_1}{a_2}v_{22} - \frac{(v_{22})^2}{qa_2^2}.$$

The optimal control of equation (8.9b),

$$u(x, T) = -q^{-1} \begin{bmatrix} 0 & \dfrac{1}{a_2} \end{bmatrix} \begin{bmatrix} v_{11} & v_{12} \\ v_{12} & v_{22} \end{bmatrix} \begin{bmatrix} y \\ \dot{y} \end{bmatrix}$$

$$= -\frac{v_{12}}{qa_2}y - \frac{v_{22}}{qa_2}\dot{y},$$

depends on the solution of the differential equations for v_{11}, v_{12}, and v_{22}. These non-linear differential equations do not have analytic solutions but they are in a convenient form to be solved by analogue computation; Fig. 8.1 shows a solution generated on an analogue computer for one particular case.

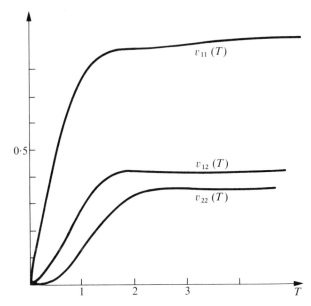

FIG. 8.1. Solution of Riccati equations when $a_0 = 1$, $a_1 = 1$, $a_2 = 1$, $q = 1$.

The figure shows that as the time-to-go T becomes large the elements of V converge to steady values so that both the optimal value f of the performance criterion and the optimal control law given by substituting V into equation (8.9b) become independent of time-to-go. This convergence of the solution to equations (8.10) as the time-to-go becomes infinite corresponds to the statement that stable solutions exist and is typical of all controllable systems. In general the steady-state solutions for problems with infinite time-to-go are the unique solutions to steady-state versions of the dynamic equations (8.10),

$$V = P + A'VA - A'VB(Q + B'VB)^{-1}B'VA \quad \text{(discrete time)} \quad (8.12a)$$

and

$$0 = P + A'V + VA - VBQ^{-1}B'V \quad \text{(continuous time)} \quad (8.12b)$$

but these equations (8.12) are non-linear and not well conditioned for direct solution, and it can be more convenient to find steady-state solutions V by simulating the dynamic equations (8.10) and letting them converge to the steady state as illustrated in Fig. 8.1. The existence of unique steady-state solutions to the dynamic equations (8.10) for V has two general implications.

(i) The value of $\mathbf{K}$ in equations (8.9) specifying the optimal controls depends on $\mathbf{V}$. A steady-state value of $\mathbf{V}$ therefore implies a steady-state value of $\mathbf{K}$. It follows that in stationary problems with infinite time-to-go the optimal control law is not time-varying.

(ii) The matrix $\mathbf{V}$ specifies the optimal value f of the performance criterion I. When $\mathbf{V}$ is independent of time-to-go I is independent of the period of time over which the quadratic function H or h of equation (8.8)

$$H = h = \mathbf{x'Px} + \mathbf{u'Qu} = \mathbf{x'(P + K'QK)x}$$

is summed or integrated to generate I. This can only happen if H goes to zero, and therefore implies that the state $\mathbf{x}$ is driven to the origin of state space. Thus a control system using the steady-state value of $\mathbf{K}$ for a controllable process will always drive the state to the origin.

For the second-order continuous-time example discussed above, equations (8.12b) give

$$0 = 1 - 2\frac{a_0}{a_2}v_{12} - \frac{(v_{12})^2}{qa_2^2},$$

$$0 = -\frac{a_0}{a_2}v_{22} + v_{11} - \frac{a_1}{a_2}v_{12} - \frac{v_{12}v_{22}}{qa_2^2},$$

$$0 = 2v_{12} - 2\frac{a_1}{a_2}v_{22} - \frac{(v_{22})^2}{qa_2^2},$$

equations which are sufficiently simple that they can be solved directly. For the particular case of Fig. 8.1 the solutions are

$$v_{11} = 0{\cdot}91, \qquad v_{12} = 0{\cdot}41, \qquad v_{22} = 0{\cdot}35.$$

Also the general expression for the optimal control, which depends only on v_{12} and v_{22}, is found to be

$$u(\mathbf{x}, \infty) = -\left\{\sqrt{\left(a_0^2 + \frac{1}{q}\right)} - a_0\right\} y$$

$$- \left\{\sqrt{\left(a_1^2 + 2a_2\left\{\sqrt{\left(a_0^2 + \frac{1}{q}\right)} - a_0\right\}\right)} - a_1\right\} \dot{y}.$$

Combining this equation for the optimal control with the second-order differential equation of the controlled process shows that the resulting optimal control system is a second-order system with differential equation

$$\ddot{y} + 2\zeta\omega_0\dot{y} + \omega_0^2 y = 0,$$

having natural frequency

$$\omega_0 = (a_2)^{-1/2}\left(a_0 + \frac{1}{q}\right)^{1/4}$$

and damping ratio

$$\zeta = \frac{1}{\sqrt{2}}\left(1 + \frac{a_1^2 - 2a_0a_2}{2a_2\sqrt{(a_0^2 + 1/q)}}\right)^{1/2}.$$

This optimal value of damping ratio can be compared with the value $1/\sqrt{2}$ suggested in Section 3.1 as usually giving acceptable transient response; it can be seen that the value $1/\sqrt{2}$ is optimal whenever the coefficients of the differential equation of the controlled process satisfy the condition

$$a_1^2 - 2a_0a_2 = 0.$$

One process satisfying this condition is the rotating load, first introduced in Section 1.2, having differential equation

$$J\ddot{y} = u$$

with coefficients

$$a_2 = J, \qquad a_1 = 0, \qquad a_0 = 0.$$

Thus LQP-theory justifies the common engineering practice of adjusting damping ratio in position-control systems to a value of about $1/\sqrt{2}$.

A more general conclusion, which follows from the general equations (8.9) for optimal LQP-control, is that the optimal control $\mathbf{u}^*$ is always a linear function of the dynamic state $\mathbf{x}$ of the controlled process. This justifies the widespread use of linear, feedback-control laws in practice. In real processes the state variables are not always available for feedback, usually for two reasons.

(i) Observable outputs are not identical with states or are fewer in number, as discussed in Section 7.4, and dynamic observers may be needed to evaluate states.

(ii) Observation of process variables is subject to measurement noise which introduces uncertainty about the exact values of the variables. This can make evaluation of time derivatives, which are often defined as state variables, particularly difficult. Techniques for estimating values of state variables from noisy observations are discussed in Chapter 14; the use of such estimates in place of the true values $\mathbf{x}$ in the linear, feedback-control laws of equations (8.9) is discussed in Chapter 13.

Problems 8.3–8.9

8.4. Non-stationary problems

Generalization of the results in Section 8.3 to non-stationary problems, where the matrices $\mathbf{A}$, $\mathbf{B}$, $\mathbf{P}$, and $\mathbf{Q}$ may be time-varying, follows from generalization of the dynamic programming equations (8.6) which are the point of departure for the analysis of LQP-problems. This can be achieved by using the Principle of Optimality to derive dynamic programming equations for the general class of non-stationary processes where the dynamics and the performance-criterion function are time-varying as indicated in the notation of equations (8.1) and (8.2). The resulting equations are similar to equations (8.6) for stationary problems, but the notation becomes more complicated because it becomes necessary to distinguish between the time-to-go to the end of the process and the absolute value of time. It is convenient to make the distinction by relating absolute values of time to the time when the process finishes, i_2 for discrete-time and t_2 for continuous-time processes; then the time-to-go, N or T, can be regarded as measuring time backwards from when the process finishes. Figure 8.2 shows the distinction for discrete-time processes where

$$N = i_2 - i + 1.$$

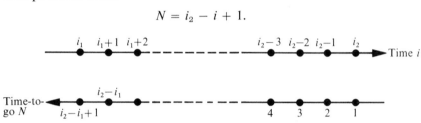

FIG. 8.2. The distinction between time and time-to-go.

For continuous-time processes the corresponding relationship is

$$T = t_2 - t.$$

Appropriate definitions of solution functions for optimal values of I and of the control $\mathbf{u}$ are for discrete-time:

$f_N(\mathbf{x}, i_2) \equiv$ Value of I using the optimal control over N stages, finishing at time i_2 and starting from state $\mathbf{x}(i_2 - N + 1) = \mathbf{x}$,

$\mathbf{u}_N(\mathbf{x}, i_2) \equiv$ Optimal control $\mathbf{u}^*(i_2 - N + 1)$ for a process finishing at time i_2 and starting from state $\mathbf{x}(i_2 - N + 1) = \mathbf{x}$;

and for continuous-time:

$f(\mathbf{x}, T, t_2) \equiv$ Value of I using the optimal control over time T, finishing at time t_2 and starting from state $\mathbf{x}(t_2 - T) = \mathbf{x}$,

$\mathbf{u}(\mathbf{x}, T, t_2) \equiv$ Optimal control $\mathbf{u}^*(t_2 - T)$ for a process finishing at time t_2 and starting from state $\mathbf{x}(t_2 - T) = \mathbf{x}$.

With this notation the general dynamic programming equations can be shown to be the discrete-time functional recurrence equation

$$f_N(\mathbf{x}, i_2) = \min_{\mathbf{u}} \left\{ H(\mathbf{x}, \mathbf{u}, i_2 - N + 1) + f_{N-1}\left(\mathbf{G}(\mathbf{x}, \mathbf{u}, i_2 - N + 1), i_2\right) \right\},$$

$$(8.13a)$$

and the continuous-time, functional differential equation

$$\frac{\partial f}{\partial T} = \min_{\mathbf{u}} \left\{ h(\mathbf{x}, \mathbf{u}, t_2 - T) + \sum_{j=1}^{n} \frac{\partial f}{\partial x_j} g_j(\mathbf{x}, \mathbf{u}, t_2 - T) \right\}. \quad (8.13b)$$

For non-stationary LQP-problems the solution functions f to these equations (8.13) are quadratic in $\mathbf{x}$, as for stationary problems, and can be written in terms of matrices $\mathbf{V}$ which depend on the absolute value of time:

$$f_N(\mathbf{x}, i_2) = \mathbf{x}'\mathbf{V}_{i_2 - N + 1}\mathbf{x} = \mathbf{x}'\mathbf{V}_i\mathbf{x} \qquad \text{(discrete time)}$$

$$f(\mathbf{x}, T, t_2) = \mathbf{x}'\mathbf{V}(t_2 - T)\mathbf{x} = \mathbf{x}'\mathbf{V}(t)\mathbf{x} \quad \text{(continuous time)}.$$

Substituting these expressions for f into the LQP-versions of equations (8.13) leads to non-stationary solutions similar to those of equations (8.9) and (8.10) for stationary problems:

$$\mathbf{u}(i) = -\mathbf{K}_i\mathbf{x}(i),$$
where
$$\mathbf{K}_i = \left(\mathbf{Q}(i) + \mathbf{B}'(i)\mathbf{V}_{i+1}\mathbf{B}(i)\right)^{-1}\mathbf{B}'(i)\mathbf{V}_{i+1}\mathbf{A}(i)$$ (discrete time)

(8.14a)

and

$$\mathbf{u}(t) = -\mathbf{K}(t)\mathbf{x}(t),$$
where
$$\mathbf{K}(t) = \mathbf{Q}^{-1}(t)\mathbf{B}'(t)\mathbf{V}(t)$$ (continuous time).   (8.14b)

The discrete-time matrices $\mathbf{V}_i$ are given by difference equations,

$$\mathbf{V}_i = \mathbf{P}(i) + \mathbf{A}'(i)\mathbf{V}_{i+1}\mathbf{A}(i)$$
$$- \mathbf{A}'(i)\mathbf{V}_{i+1}\mathbf{B}(i)\left(\mathbf{Q}(i) + \mathbf{B}'(i)\mathbf{V}_{i+1}\mathbf{B}(i)\right)^{-1}\mathbf{B}(i)\mathbf{V}_{i+1}\mathbf{A}(i), \quad (8.15a)$$

the continuous-time matrices $\mathbf{V}(t)$ are given by Riccati equations

$$-\frac{d\mathbf{V}}{dt} = \mathbf{P}(t) + \mathbf{A}'(t)\mathbf{V}(t) + \mathbf{V}(t)\mathbf{A}(t) - \mathbf{V}(t)\mathbf{B}(t)\mathbf{Q}^{-1}(t)\mathbf{B}'(t)\mathbf{V}(t),$$

(8.15b)

and the boundary conditions, corresponding to equations (8.11), are

$$\mathbf{V}_{i_2} = \mathbf{P}(i_2) \quad \text{(discrete time)}, \quad\quad\quad (8.16a)$$

$$\mathbf{V}(t_2) = \mathbf{0} \quad\quad \text{(continuous time)}. \quad\quad\quad (8.16b)$$

Equations (8.14), (8.15), and (8.16) specify the solution for any LQP-problem with known, time-varying matrices $\mathbf{A}$, $\mathbf{B}$, $\mathbf{P}$, $\mathbf{Q}$.

A common class of non-stationary problems is where the objective is to drive the state $\mathbf{x}$ to a specified value, often the origin of state space, by the time the control process terminates. Such problems are known as terminal-control problems and can be specified by performance criteria with time-varying matrices $\mathbf{P}$,

$$I = \mathbf{x}'(i_2)\mathbf{P}_1\mathbf{x}(i_2) + \sum_{i=i_1}^{i_2-1} (\mathbf{x}'\mathbf{Px} + \mathbf{u}'\mathbf{Qu}), \quad \text{(discrete time)}$$

$$I = \mathbf{x}'(t_2)\mathbf{P}_1\mathbf{x}(t_2) + \int_{t_1}^{t_2} (\mathbf{x}'\mathbf{Px} + \mathbf{u}'\mathbf{Qu})dt, \text{ (continuous time)}$$

where the matrix $\mathbf{P}_1$ measures the cost of terminal errors and the matrices $\mathbf{P}$ and $\mathbf{Q}$ may be time-varying, constant, or zero. When the matrices $\mathbf{A}$, $\mathbf{B}$, $\mathbf{P}$, $\mathbf{Q}$ are all constant and the only non-stationary feature is the cost of terminal errors, the non-stationary equations (8.14) and (8.15) reduce to the equations (8.9) and (8.10) for stationary problems but with the boundary conditions of equations (8.11) replaced by

$$\mathbf{V}_1 = \mathbf{P}_1 + \mathbf{P}, \quad \text{(discrete time)} \tag{8.17a}$$

$$\mathbf{V}(0) = \mathbf{P}_1 \qquad \text{(continuous time).} \tag{8.17b}$$

Problems 8.10, 8.11

8.5. Tracking problems

The LQP-theory developed in Section 8.3 for regulator problems can be generalized to problems where outputs $\mathbf{y}$ of the controlled process, specified by equations (7.7) repeated here as (8.18),

$$\mathbf{y} = \mathbf{Cx} \tag{8.18}$$

are required to follow time-varying desired values $\mathbf{z}(i)$ or $\mathbf{z}(t)$. This class of problems includes those introduced in Chapter 2; it is known as the class of tracking, or servo, problems. Tracking performance can be measured by the magnitude of errors between the outputs $\mathbf{y}$ of equation (8.18) and their desired values $\mathbf{z}$, so the appropriate quadratic performance criterion I for tracking problems is given by equations (8.2) with

$$H = h = (\mathbf{z} - \mathbf{Cx})'\mathbf{P}(\mathbf{z} - \mathbf{Cx}) + \mathbf{u}'\mathbf{Qu}. \tag{8.19}$$

The generalization of LQP-theory to problems having the performance criteria of equation (8.19) takes either of two forms depending on whether or not the desired values $\mathbf{z}$ are known in advance.

When the values of $\mathbf{z}$ are known in advance the tracking problem is a non-stationary LQP-problem. The quadratic functions H and h of equation (8.19) can be rewritten

$$H = h = \mathbf{x}'\mathbf{C}'\mathbf{PCx} + \mathbf{u}'\mathbf{Qu} - 2\mathbf{z}'\mathbf{PCx} + \mathbf{z}'\mathbf{Pz}$$

to show that they contain two additional terms, $-2\mathbf{z}'\mathbf{PCx}$ and $\mathbf{z}'\mathbf{Pz}$, which do not appear in the corresponding functions of equations (8.8) for regulator problems and which are time-varying, because they depend on $\mathbf{z}$, so that the problem must be regarded as non-stationary. In spite of the additional terms in H and h the tracking problem can be solved by analysis similar to that used for regulator problems in Sections 8.3 and 8.4. The analysis is based on the assumption that solution functions f to the functional equations (8.13) for LQP-tracking problems have additional terms corresponding to those in H and h and take the form

$$f = \mathbf{x}'\mathbf{V}\mathbf{x} - 2\mathbf{v}'\mathbf{x} + w,$$

where $\mathbf{V}$ is a symmetric matrix which depends on the absolute value of time as in Section 8.4, and $\mathbf{v}$ and w are a vector and a scalar which also depend on the absolute value of time. Substituting this form of f into versions of the functional equations (8.13) specified by equations (8.7), (8.2), and (8.19) leads to optimal control laws which are similar to those of equations (8.14) for non-stationary regulators except that they have an additional term proportional to $\mathbf{v}$.

$$\begin{array}{l} \mathbf{u}(i) = -\mathbf{K}_i\mathbf{x}(i) + (\mathbf{Q} + \mathbf{B}'\mathbf{V}_{i+1}\mathbf{B})^{-1}\mathbf{B}'\mathbf{v}(i), \\[2mm] \mathbf{K}_i = (\mathbf{Q} + \mathbf{B}'\mathbf{V}_{i+1}\mathbf{B})^{-1}\mathbf{B}'\mathbf{V}_{i+1}\mathbf{A}, \end{array} \left. \vphantom{\begin{array}{l} a \\ b \end{array}} \right\} \quad \text{(discrete time)}$$

where   (8.20a)

and

$$\begin{array}{l} \mathbf{u}(t) = -\mathbf{K}(t)\mathbf{x}(t) + \mathbf{Q}^{-1}\mathbf{B}'\mathbf{v}(t), \\[2mm] \mathbf{K}(t) = \mathbf{Q}^{-1}\mathbf{B}'\mathbf{V}(t). \end{array} \left. \vphantom{\begin{array}{l} a \\ b \end{array}} \right\} \quad \text{(continuous time).} \quad \text{(8.20b)}$$

where

The matrices $\mathbf{V}$ here are given by dynamic equations similar to (8.15). For discrete-time systems,

$$\mathbf{V}_i = \mathbf{C}'\mathbf{P}\mathbf{C} + \mathbf{A}'\mathbf{V}_{i+1}\mathbf{A} - \mathbf{A}'\mathbf{V}_{i+1}\mathbf{B}(\mathbf{Q} + \mathbf{B}'\mathbf{V}_{i+1}\mathbf{B})^{-1}\mathbf{B}\mathbf{V}_{i+1}\mathbf{A}$$

$$\text{(8.21a)}$$

with boundary condition

$$\mathbf{V}_{t_2} = \mathbf{C}'\mathbf{P}\mathbf{C};$$

and for continuous-time systems,

$$-\frac{d\mathbf{V}}{dt} = \mathbf{C}'\mathbf{P}\mathbf{C} + \mathbf{A}'\mathbf{V}(t) + \mathbf{V}(t)\mathbf{A} - \mathbf{V}(t)\mathbf{B}\mathbf{Q}^{-1}\mathbf{B}'\mathbf{V}(t) \quad \text{(8.21b)}$$

with boundary condition

$$\mathbf{V}(t_2) = 0.$$

The additional vectors $\mathbf{v}$ in equations (8.20) are given by linear dynamic equations. For discrete-time systems,

$$\mathbf{v}(i) = (\mathbf{A} - \mathbf{B}\mathbf{K}_i)'\mathbf{v}(i + 1) + \mathbf{C}'\mathbf{P}\mathbf{z}(i) \quad \text{(8.22a)}$$

with boundary condition

$$\mathbf{v}(i_2) = \mathbf{C}'\mathbf{P}\mathbf{z}(i_2);$$

and for continuous-time systems,

$$-\frac{d\mathbf{v}}{dt} = \left(\mathbf{A} - \mathbf{B\bar{K}}(t)\right)'\mathbf{v}(t) + \mathbf{C}'\mathbf{Pz}(t) \tag{8.22b}$$

with boundary condition

$$\mathbf{v}(t_2) = 0.$$

These equations (8.22) can be regarded as specifying dynamic systems characterized by matrices $(\mathbf{A} - \mathbf{BK})'$ and driven in reverse time by the known desired values $\mathbf{z}$. They are closely related to the optimal, feedback-control systems specified by combining equations (8.7) and (8.20) to give

$$\mathbf{x}(i + 1) = (\mathbf{A} - \mathbf{BK}_i)\mathbf{x}(i) + \mathbf{B}(\mathbf{Q} + \mathbf{B}'\mathbf{V}_{i+1}\mathbf{B})^{-1}\mathbf{B}'\mathbf{v}(i), \quad \text{(discrete time)}$$

$$\dot{\mathbf{x}} = \left(\mathbf{A} - \mathbf{BK}(t)\right)\mathbf{x} + \mathbf{BQ}^{-1}\mathbf{B}'\mathbf{v}, \qquad\qquad \text{(continuous time)}$$

equations of dynamic systems characterized by matrices $(\mathbf{A} - \mathbf{BK})$ and driven by the variables $\mathbf{v}$. The systems, specified by equations (8.22), having state $\mathbf{v}$ are said to be adjoint† to the optimal feedback-control systems and the variables $\mathbf{v}$ are described as adjoint variables.

The scalar w which appears in the performance criterion does not affect the control laws of equations (8.20). It satisfies the dynamic equations of an integrator driven in reverse time by the adjoint variables $\mathbf{v}$ and the known desired values $\mathbf{z}$. For discrete-time systems,

$$w(i) = w(i + 1) - \mathbf{v}'(i + 1)\mathbf{B}(\mathbf{Q} + \mathbf{B}'\mathbf{V}_{i+1}\mathbf{B})^{-1}\mathbf{B}'\mathbf{v}(i + 1) + \mathbf{z}'(i)\mathbf{Pz}(i) \tag{8.23a}$$

with boundary condition

$$w(i_2) = \mathbf{z}'(i_2)\mathbf{Pz}(i_2);$$

and for continuous-time systems,

$$-\frac{dw}{dt} = -\mathbf{v}'(t)\mathbf{BQ}^{-1}\mathbf{B}'\mathbf{v}(t) + \mathbf{z}'(t)\mathbf{Pz}(t) \tag{8.23b}$$

with boundary condition

$$w(t_2) = 0.$$

An example of a tracking problem with desired values known in advance is that of making a prespecified step change in the position y of a rotating load having differential equation

$$J\ddot{y} = u \tag{8.24}$$

---

† Pairs of dynamic systems related according to

$$\mathbf{x}(i + 1) = \mathbf{Mx}(i); \qquad \mathbf{v}(i - 1) = \mathbf{M}'\mathbf{v}(i) \qquad \text{(discrete time)}$$

$$\dot{\mathbf{x}} = \mathbf{Mx}; \qquad\qquad \dot{\mathbf{v}} = -\mathbf{M}'\mathbf{v} \qquad \text{(continuous time)}$$

are said to be adjoint.

specified by matrices

$$A = \begin{bmatrix} 0 & 1 \\ 0 & 0 \end{bmatrix}, \quad B = \begin{bmatrix} 0 \\ 1/J \end{bmatrix}, \quad C = [1 \ 0].$$

The equation (8.19) specifying tracking performance becomes

$$h = p(z - y)^2 + qu^2,$$

where p and q are constants and z is a variable representing the desired position. Assuming that the system will continue to operate for a long time into the future the optimal control is given by the steady-state solution of the Riccati equation (8.21b) and is

$$u(t) = -\sqrt{(p/q)}x_1(t) - \sqrt{\left(2J\sqrt{(p/q)}\right)}x_2(t) + \frac{1}{Jq}v_2(t), \qquad (8.25)$$

where the state variables x_1 and x_2 are position and velocity of the load, as before, and v_2 is an adjoint variable. Eliminating u from equations (8.24) and (8.25) gives

$$\ddot{y} + 2(1/\sqrt{2})\sqrt{\left(\frac{1}{J}\sqrt{(p/q)}\right)}\dot{y} + \frac{1}{J}\sqrt{(p/q)}y = \frac{1}{J^2q}v_2,$$

and shows that the optimal control system has damping ratio $1/\sqrt{2}$, as in Section 8.3, and is driven by v_2. This variable v_2 satisfies the adjoint equations (8.22b) which give

$$\ddot{v}_2 - 2(1/\sqrt{2})\sqrt{\left(\frac{1}{J}\sqrt{(p/q)}\right)}\dot{v}_2 + \frac{1}{J}\sqrt{(p/q)}v_2 = pz,$$

a similar differential equation in reverse time driven by the desired value z. Figure 8.3 illustrates what the adjoint variable v_2 and the position y would be if it were known in advance that the desired position z was to change by a prespecified amount Z at a prespecified time τ. The value f of the quadratic performance criterion using the optimal control of equation (8.25) for the prespecified step response is†

$$f = 0{\cdot}75Z^2 \sqrt{\left(2Jp\sqrt{(pq)}\right)}.$$

This shows an improvement of 25% over the value

$$Z^2\sqrt{\left(2Jp\sqrt{(pq)}\right)}$$

which would be achieved by the feedback-control law of the optimal regulator in Section 8.3, in the absence of prior information about the step.

---

† f is calculated here assuming the initial state a long time before the step to be zero and using Parseval's theorem (Fig. 6.5) to evaluate the integral of equation (8.23b) for w.

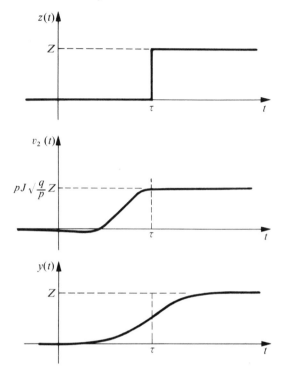

FIG. 8.3. Optimal response to a pre-specified step.

The improvement in performance due to using adjoint variables **v** which account for prespecified variables **z** cannot generally be realized in practice. The principal reason for this is that prespecified variables **z** are scarcely ever encountered in practical problems; uncertainty about future desired values is a fundamental feature of most control problems, as mentioned in Section 2.1 and discussed further in Part III. However even where values of **z** can be prespecified, as in the step-response example above, it is not easy to generate, in real time, adjoint variables **v** specified by dynamic equations in reverse time. The adjoint variable $v_2(t)$ in Fig. 8.3 is an example of a variable which would not be easy to realize in a practical control system. The generalization of LQP-theory to tracking problems where desired values **z** are known in advance is thus a topic of theoretical rather than of practical interest.

Another topic of theoretical interest is the possibility of generalizing LQP-theory to problems where desired values **z** are not completely known in advance but where a class of desired-value functions is prespecified. Examples of such classes are the classes of constant, ramp, and quadratic functions introduced in Chapter 2. For purposes of LQP-theory classes of functions are specified by regarding them as solutions to dynamic equations. For example, the class of constant, continuous-time functions,

$$z = Z,$$

where Z may take any constant value, are particular integrals to the first-order differential equation

$$\dot{z} = 0,$$

and can be represented by an additional state variable

$$x_{n+1} \equiv z$$

satisfying an additional dynamic equation

$$\dot{x}_{n+1} = 0.$$

Similarly the class of discrete-time, constant-plus-ramp functions

$$z = Z + Ai,$$

where Z and A may take any constant values, are particular integrals to a second-order difference equation

$$z(i) - 2z(i-1) + z(i-2) = 0,$$

which can be represented by additional state variables

$$x_{n+1}(i) \equiv z(i), \qquad x_{n+2}(i) \equiv z(i-1)$$

satisfying additional dynamic equations

$$x_{n+1}(i+1) = 2x_{n+1}(i) - x_{n+2}(i)$$
$$x_{n+2}(i+1) = x_{n+1}(i).$$

Thus any polynomial desired-value function can be regarded as the particular integral solution to a dynamic equation of order one greater than the order of the polynomial, and can be represented by additional state variables and equations. Exponential desired-value functions can be regarded as complementary function solutions to dynamic equations, and can also be represented by additional state variables and equations; in particular the sinusoidal function

$$z = A \cos \omega t$$

is a complementary function to the second-order differential equation

$$\ddot{z} + \omega^2 z = 0,$$

and can be represented by additional state variables

$$x_{n+1} \equiv z, \qquad x_{n+2} \equiv \dot{z}$$

satisfying additional dynamic equations

$$\dot{x}_{n+1} = x_{n+2},$$
$$\dot{x}_{n+2} = -\omega^2 x_{n+1}.$$

In general a set of m desired values z could be identified with m additional state variables each of which might be regarded as the solution to a dynamic equation represented by an appropriate further number of additional state variables. The resulting set of first-order dynamic equations, using states x_2 to represent m desired-value functions, is written

$$x_2(i + 1) = A_2 x_2(i) \quad \text{(discrete time)} \tag{8.26a}$$

$$\dot{x}_2 = A_2 x_2 \quad \text{(continuous time).} \tag{8.26b}$$

Tracking problems where desired-value functions are known to belong to a class which can be specified by equations (8.26) can be put into the form of regulator problems like those in Section 8.3. For this purpose the previous notation is modified so that the equations (8.7) of the controlled process are written

$$x_1(i + 1) = A_1 x_1(i) + B_1 u(i) \quad \text{(discrete time),}$$

$$\dot{x}_1 = A_1 x_1 + B_1 u \quad \text{(continuous time);}$$

the output equation (8.18) is written

$$y = C_1 x_1;$$

the desired values z are written

$$z = C_2 x_2;$$

and the performance measure of equation (8.19) is written

$$H = h = (C_2 x_2 - C_1 x_1)' P_1 (C_2 x_2 - C_1 x_1) + u' Q u.$$

Now if an augmented state vector is defined

$$x \equiv \begin{bmatrix} x_1 \\ x_2 \end{bmatrix}$$

with augmented matrices

$$A = \left[\begin{array}{c|c} A_1 & 0 \\ \hline 0 & A_2 \end{array} \right], \quad B = \begin{bmatrix} B_1 \\ \hline 0 \end{bmatrix}, \tag{8.27}$$

$$P = \begin{bmatrix} -C_1' \\ \hline C_2' \end{bmatrix} P_1 [-C_1 \mid C_2]$$

the tracking problem is reduced to a regulator problem having state vector x and specified by matrices A, B, P, Q as in Section 8.3.

A feature of this regulator problem is that there is no interconnection between the controls u and the state variables x_2 describing desired values, so the system specified by matrices A and B of equations (8.27) is not controllable in the sense of Section 7.3. The existence of unique, steady-state solutions to the dynamic equations (8.10) specifying optimal controls is therefore not necessarily guaranteed. Solutions are however known to exist for the class of uncontrollable systems in which all uncontrollable modes are stable: such systems are said to be stabilizable† and are characterized by equations (8.27) when the matrices A_1 and B_1 form a controllable pair and all eigenvalues of the matrix A_2 satisfy the stability conditions of Fig. 1.4.

Tracking problems where the desired-value function belongs to the class of constant, ramp, or quadratic functions introduced in Chapter 2 are generally not controllable and not stabilizable because such desired-value functions cannot be represented as solutions to stable dynamic equations and eigenvalues of the corresponding matrices A_2 do not satisfy the stability conditions.

A special class of tracking problem which reduces to a controllable regulator problem is where the controlled process and the desired-value functions are matched in the sense that the matrix A_1 specifying dynamics of the controlled process is the same as the matrix A_2 specifying desired values:

$$A_1 = A_2.$$

Then a corresponding regulator problem can be specified using a state vector

$$X \equiv x_1 - x_2$$

which satisfies dynamic equations having matrices

$$A = A_1 = A_2, \qquad B = B_1$$

representing a system which is controllable provided that the original controlled process, specified by matrices A_1 and B_1, is controllable. An example of such a controllable tracking problem is that of a single-input, single-output continuous-time controlled process consisting of n integrators having differential equation

$$\frac{d^n y}{dt^n} = u$$

with a polynomial desired-value function of order $n - 1$

$$z = Z + At + Bt^2 + \cdots + Mt^{n-1}$$

satisfying a similar differential equation:

$$\frac{d^n z}{dt^n} = 0.$$

† See Wonham (1967) and (1968) where a similar property, called detectability, related to observability is also discussed. An unobservable system is said to be detectable when all unobservable modes are stable.

The fact that unique, steady-state solutions to the dynamic equations (8.10) for **V** exist in this controllable example where the performance measure h of equation (8.19) can be of the form

$$h = (z - y)^2 + qu^2$$

implies, as discussed in Section 8.3, that the difference $(z - y)$ between desired and actual output is driven to zero by the optimal control. The example thus confirms the results derived in Section 2.2 about the relationship between type-number and zero steady-state errors.

Tracking problems with finite, steady-state errors were discussed in Chapter 2. These are problems where the performance measure h remains finite causing the performance criterion I of equations (8.2) to increase indefinitely so that there can be no steady-state solution to the dynamic equations (8.10) specifying the optimal value

$$f = \mathbf{x}'\mathbf{V}\mathbf{x}$$

of I. Generalization of LQP-theory for such problems is beyond the scope of this book.†

The methods described above for generalizing LQP-theory to tracking problems can also be used to develop LQP-theory for problems where external disturbances affect the controlled process as discussed in Section 2.4. These are problems having dynamic equations

$$\mathbf{x}_1(i + 1) = \mathbf{A}_1\mathbf{x}_1(i) + \mathbf{B}_1\mathbf{u}(i) + \mathbf{B}_2\mathbf{z}_2(i), \quad \text{(discrete time)} \quad \text{(8.28a)}$$

$$\dot{\mathbf{x}}_1 = \mathbf{A}_1\mathbf{x}_1 + \mathbf{B}_1\mathbf{u} + \mathbf{B}_2\mathbf{z}_2, \quad \text{(continuous time) (8.28b)}$$

where $\mathbf{z}_2$ are the external disturbances which may be known in advance or may belong to a prespecified class of functions represented by additional state variables $\mathbf{x}_3$ according to

$$\mathbf{z}_2 = \mathbf{C}_3\mathbf{x}_3$$

and satisfying dynamic equations

$$\mathbf{x}_3(i + 1) = \mathbf{A}_3\mathbf{x}_3(i) \quad \text{(discrete time)}$$

$$\dot{\mathbf{x}}_3 = \mathbf{A}_3\mathbf{x}_3 \quad \text{(continuous time)}.$$

An augmented state vector can be defined

$$\mathbf{x} \equiv \begin{bmatrix} \mathbf{x}_1 \\ \mathbf{x}_3 \end{bmatrix}$$

with augmented matrices

$$\mathbf{A} = \begin{bmatrix} \mathbf{A}_1 & \mathbf{B}_2\mathbf{C}_3 \\ \hline \mathbf{0} & \mathbf{A}_3 \end{bmatrix}, \quad \mathbf{B} = \begin{bmatrix} \mathbf{B}_1 \\ \hline \mathbf{0} \end{bmatrix},$$

† At the time of writing this appears to be a matter for further research.

which represent an uncontrollable and possibly unstabilizable system. Conditions of matching between the disturbance and the controlled process, which would reduce the problem to a controllable regulator problem and confirm the results of Section 2.4, are not as easy to derive here as for the previous tracking problem.†

Problems 8.12–8.15

8.6. Bibliography

Dynamic programming is the subject of several books,

BELLMAN (1957),
BELLMAN and DREYFUS (1962),
ARIS (1964),
ROBERTS (1964),
BELLMAN and KALABA (1965),
NEMHAUSER (1966),
JACOBS (1967),
BOUDAREL, DELMOS, and GUICHET (1971).

Its application to LQP-problems is discussed in

TOU (1963),
DORF (1965) Sections 10.4 and 10.5,
NOTON (1965) Chapters 2 and 3; (1972) Section 4.6,
MCCAUSLAND (1967) Chapter 7,
OGATA (1967) Chapter 9,
HSU and MEYER (1968) Chapter 15,
MEDITCH (1969) Section 9.3,
PRIME (1969) Chapter 5,
KIRK (1970) Chapter 3,
KUO (1970) Chapter 7,
KWAKERNAAK and SIVAN (1972) Section 6.4.

LQP regulator problems are discussed in

ATHANS and FALB (1966) Chapter 9,
LAPIDUS and LUUS (1967) Chapter 3,
LEE and MARKUS (1967) Chapter 3,
SAGE (1968) Section 5.1,
BRYSON and HO (1969) Chapter 5,
CITRON (1969) Chapter 8,
KIRK (1970) Section 5.2,
SPEEDY, BROWN, and GOODWIN (1970) Sections 7.8 and 8.3
ANDERSON and MOORE (1971),
KWAKERNAAK and SIVAN (1972) Chapter 3.

The tracking problem is discussed in

ATHANS and FALB (1966) Sections 9.9–9.12,
ANDERSON and MOORE (1971) Chapter 11,
BHATTACHARYA and PEARSON (1972).

† At the time of writing this appears to be a matter for further research.

8.7. Problems

8.1. The problem of investing money over a period of N years can be represented as a multistage decision process where each year £x is subdivided into £u which are invested and £$(x - u)$ which are spent. The process is assumed to have dynamic equation

$$x(i + 1) = au(i)$$

where the integer i counts the years and the growth rate a is constant and greater than unity. The satisfaction from money spent in any year is specified by a function H,

$$H(i) = H\big(x(i) - u(i)\big)$$

and the objective is to maximize the total satisfaction I over N years:

$$I = \sum_{i=1}^{N} H(i).$$

Define an optimal return function and write the dynamic programming functional recurrence equation for the process. Hence find the optimal investment policy for the two cases:

(i) diminishing returns $H(x - u) = \sqrt{(x - u)}$,

(ii) linear return $H(x - u) = x - u$.

8.2. A discrete-time controlled process has dynamic equation

$$x(i + 1) = 0\!\cdot\!74x(i) + 0\!\cdot\!26u(i).$$

Use a dynamic programming functional recurrence equation to compute values of $u(i)$ that minimize

$$I = \sum_{i=1}^{3} \big(x^2(i) + 0\!\cdot\!2u^2(i)\big)$$

for two cases:

(i) When $x(4)$ is unspecified,
(ii) when $x(4)$ must be zero.

8.3. A continuous-time, controlled process has differential equation

$$\dot{x} = u$$

and performance criterion

$$I = \int_{0}^{T} (x^2 + u^2)\, \mathrm{d}t.$$

Write down the Riccati equation for optimal control and hence find the optimal control law.

8.4. Work through the derivation of equations (8.9) and (8.10).

8.5. A single-variable continuous-time controlled process has transfer function

$$\frac{Y(s)}{U(s)} = \frac{s + 1}{s^2 + s + 1},$$

and performance criterion

$$I = \int_0^\infty (y^2 + u^2)\ \mathrm{dt}.$$

Express the transfer function in a canonical state-space form, write the steady-state Riccati equations for optimal control, and hence find the transfer function of the optimal feedback-control law. What is the damping ratio of the resulting control system?

8.6. In the discrete-time, position-control system of Problem 1.12 take the position y and the velocity v to be states x_1 and x_2, take the values of the coefficients a and b both to be unity, and iterate equations (8.9a) and (8.10a) to find the optimal control law to minimize the performance criterion

$$I = \sum_{i=1}^{N} \left(y^2(i) + v^2(i)\right).$$

What are the roots of the characteristic equation of the resulting control system as N goes to infinity?

8.7. Using the results of Problem 8.6,
  (i) make a sketch showing how the optimal feedback gain varies with time over the final stages of the process (as i goes to N);
  (ii) compare the optimal values of I with the value achieved using the control of Problem 7.4 which takes any initial state to the origin in two steps.

8.8. Find the optimal one-step-ahead control law for a discrete-time system where the controlled process is specified by equation (8.7a) and the performance criterion

$$I = \mathbf{x}'(i + 1)\mathbf{P}\mathbf{x}(i + 1) + \mathbf{u}'(i)\mathbf{Q}\mathbf{u}(i)$$

is evaluated instantaneously rather than as a summation. Under what circumstances does the optimal control take any initial state $\mathbf{x}$ to the origin of state space in a single step?

8.9. Show by mathematical induction that in discrete-time systems having the controlled process of equation (8.7a), and with a constant linear feedback-control law

$$\mathbf{u} = -\mathbf{K}\mathbf{x},$$

the value of the performance criterion I of equation (8.8a) is given by

$$\mathbf{x}'(1)\mathbf{V}_N\mathbf{x}(1),$$

where

$$V_N = P + K'QK + (A - BK)'V_{N-1}(A - BK),$$
$$V_1 = P + K'QK.$$

What is the corresponding expression for continuous-time systems?

8.10. Problem 8.3 can be reformulated as a terminal-control problem with performance criterion

$$I = x^2(T) + \int_0^T u^2 \, dt.$$

What is the resulting optimal control law?

8.11. Show that if the control system of Problem 8.10 starts from initial state $x(0) = X$ the resulting trajectory is

$$x(t) = X\left(1 - \frac{t}{1+T}\right).$$

Make a sketch comparing this with the corresponding trajectory

$$x(t) = X\left(\frac{e^t}{1+e^{2T}} + \frac{e^{-t}}{1-e^{-2T}}\right)$$

for Problem 8.3.

8.12. Work through the derivation of equations (8.20), (8.21), (8.22), and (8.23).

8.13. Derive equation (8.25) and check that it leads to a step response which is 25% better than could be achieved by the feedback-control law of an optimal regulator having no prior information about the step.

8.14. Prove that the LQP-optimal control for a single-variable tracking problem where the controlled process has transfer function

$$G(s) = \frac{K(1 + \beta_1 s + \cdots)}{s^N(1 + \alpha_1 s + \cdots)}$$

can produce zero steady-state error for all polynomial desired-value functions of order $N - 1$.

8.15. Derive equations specifying the optimal control for regulator problems subject to disturbances, specified by equations (8.8) and (8.28), when it can be assumed that values of the disturbances are known in advance.

Part II
Deterministic non-linear systems

9. Optimal control theory

9.1. General equations of optimal control

IN Chapter 8 analytic solutions to LQP-problems were obtained by deriving dynamic programming equations of optimal control for a general class of problems and solving the equations for the special class of LQP-problems. The general class of problems, specified by dynamic equations (8.1), repeated here as (9.1),

$$\mathbf{x}(i + 1) = \mathbf{G}(\mathbf{x}(i), \mathbf{u}(i), i), \qquad \text{(discrete time)} \qquad (9.1a)$$

$$\dot{\mathbf{x}} = \mathbf{g}(\mathbf{x}, \mathbf{u}, t), \qquad \text{(continuous time)} \quad (9.1b)$$

and by performance criteria (8.2), repeated here as (9.2),

$$I = \sum_{i = i_1}^{i_2} H\left(\mathbf{x}(i), \mathbf{u}(i), i\right), \qquad \text{(discrete time)} \qquad (9.2a)$$

$$I = \int_{t_1}^{t_2} h(\mathbf{x}, \mathbf{u}, t)\, dt, \qquad \text{(continuous time)} \quad (9.2b)$$

includes all non-linear, deterministic, optimal control problems having ordinary, as opposed to partial, dynamic equations. The dynamic programming analysis was in terms of solution functions

$f_N(\mathbf{x}, i_2) \equiv$ Discrete-time value of I using optimal control over N stages, finishing at time i_2 and starting from state $\mathbf{x}(i_2 - N + 1) = \mathbf{x}$,

$f(\mathbf{x}, T, t_2) \equiv$ Continuous-time value of I using optimal control over time T, finishing at time t_2 and starting from state $\mathbf{x}(t_2 - T) = \mathbf{x}$,

and the resulting equations are the discrete-time functional recurrence equation (8.13a), repeated here as (9.3),

$$f_N(\mathbf{x}, i_2) = \min_{\mathbf{u}} \left\{ H(\mathbf{x}, \mathbf{u}, i_2 - N + 1) + f_{N-1}\left(\mathbf{G}(\mathbf{x}, \mathbf{u}, i_2 - N + 1), i_2\right) \right\},$$

$$(9.3)$$

and the continuous-time, functional differential equation (8.13b), repeated here as (9.4),

$$\frac{\partial f}{\partial T} = \min_{\mathbf{u}} \left\{ h(\mathbf{x}, \mathbf{u}, t_2 - T) + \sum_{j=1}^{n} \frac{\partial f}{\partial x_j} g_j(\mathbf{x}, \mathbf{u}, t_2 - T) \right\}. \tag{9.4}$$

Initial solutions for the general, discrete-time, recurrence equation (9.3) usually depend on how the problem is constrained at its finishing time i_2. In terminal-control problems, where it is specified that the last control $\mathbf{u}(i_2)$ must† take the state to a fixed value $\mathbf{x}(i_2 + 1)$, the control $\mathbf{u}(i_2)$ must be found from the dynamic equation (9.1a).

$$\mathbf{x}(i_2 + 1) = \mathbf{G}(\mathbf{x}(i_2), \mathbf{u}(i_2), i_2)$$

rather than from any minimization, and the initial solution will then be given by

$$f_1(\mathbf{x}, i_2) = H(\mathbf{x}, \mathbf{u}(i_2), i_2). \tag{9.5a}$$

If the terminal state is not constrained to any particular value the last control $\mathbf{u}(i_2)$ is given by a minimization and the initial solution is

$$f_1(\mathbf{x}, i_2) = \min_{\mathbf{u}} \left\{ H(\mathbf{x}, \mathbf{u}, i_2) \right\}. \tag{9.5b}$$

Boundary values for the general, continuous-time differential equation (9.4) are found, in a similar way, by considering the behaviour of the solution function f at the terminal time t_2. The definition of f can be written

$$f(\mathbf{x}, T, t_2) \equiv \min_{\mathbf{u}(t)} \left\{ \int_{t_2 - T}^{t_2} h(\mathbf{x}(t), \mathbf{u}(t), t) \, dt \right\}$$

and it follows that when the range of integration T is zero

$$f|_{T=0} = f(\mathbf{x}, 0, t_2) = 0.\ddagger \tag{9.6}$$

Equation (9.6) specifies the value of an arbitrary additive constant, but is of little help in determining the shape of a solution function. A more useful

---

† The terminal-control problem here, where the terminal state is fixed, is related to, but is not identical with, the terminal-control problem discussed in Section 8.4, where terminal errors were heavily weighted rather than forbidden.

‡ Equation (9.6) is derived for performance criteria like (9.2b) with no special term weighting terminal errors. When such a term is included, as discussed in Section 8.4, and the performance criterion has the form

$$I = h_1(\mathbf{x}(t_2)) + \int_{t_1}^{t_2} h(\mathbf{x}, \mathbf{u}, t) \, dt,$$

equation (9.6) must be replaced by

$$f|_{T=0} = h_1(\mathbf{x}(t_2)). \tag{9.6a}$$

boundary condition, which also follows from the definition of f, is

$$\left.\frac{\partial f}{\partial T}\right|_{T=0} = \min_{\mathbf{u}}\{h(\mathbf{x}, \mathbf{u}, t_2)\}. \tag{9.7}$$

Equation (9.7) corresponds to the boundary conditions (9.5) for discrete-time equations, except that the distinction between problems where the terminal state is fixed and problems where it is free is not so clear. Equation (9.7) is the boundary condition to use when the terminal state is free; when the terminal state is fixed the fixed value $\mathbf{x}(t_2)$ provides a boundary value on the solution to the differential equation (9.4).

The general dynamic programming equations (9.3) and (9.4), together with their boundary conditions, specify optimal controls for the general class of deterministic non-linear control problems, and are therefore of considerable importance. They cannot usually be solved analytically for non-LQP-problems, so their practical usefulness is limited, and specific solutions can usually only be obtained numerically with a digital computer.

The dynamic programming equations can be regarded as generalizations of equations derived by other mathematical techniques of optimization. In particular the continuous-time equation (9.4) corresponds to the Hamilton–Jacobi equation of the calculus of variations. If it is assumed that the control $\mathbf{u}$ is taking its optimal value, so that the minimization need not be written, and if time is represented by its absolute value,

$$t = t_2 - T,$$

equation (9.4) becomes

$$\frac{\partial f}{\partial t} + h(\mathbf{x}, \mathbf{u}, t) + \sum_{j=1}^{n} \frac{\partial f}{\partial x_j} g_j(\mathbf{x}, \mathbf{u}, t) = 0 \tag{9.8}$$

and is sometimes called the Hamilton–Jacobi–Bellman equation. The Euler equation of the calculus of variations and the Pontriagin Maximum Principle can both be derived from the continuous-time dynamic programming equation.

9.2. The Euler equation of the calculus of variations

The continuous-time, dynamic programming equation (9.4) leads to the Euler equation of the calculus of variations whenever the problem is such that the minimization of the bracketed term { } can be achieved by calculus, that is by additional equations†

† The notation here is that $\partial h/\partial \mathbf{u}$ represents the gradient $(\partial h/\partial u_1, \partial h/\partial u_2, \ldots)$ of the scalar h with respect to $\mathbf{u}$, thus (9.9) represents a set of as many simultaneous equations as there are components of $\mathbf{u}$,

$$\frac{\partial h}{\partial u_1} + \sum_j \frac{\partial f}{\partial x_j} \frac{\partial g_j}{\partial u_1} = 0, \qquad \frac{\partial h}{\partial u_2} + \sum_j \frac{\partial f}{\partial x_j} \frac{\partial g_j}{\partial u_2} = 0, \qquad \cdot \quad \cdot \quad \cdot \quad \cdot$$

$$\frac{\partial h}{\partial \mathbf{u}} + \sum_{j=1}^{n} \frac{\partial f}{\partial x_j} \frac{\partial g_j}{\partial \mathbf{u}} = 0. \tag{9.9}$$

The equivalence is demonstrated here by using equations (9.8) and (9.9) to derive the Euler equation for a typical elementary problem in the calculus of variations.

The problem is to find the scalar function $x(t)$ in the interval $t_1 \leqslant t \leqslant t_2$ such that a functional

$$I\big(x(t)\big) = \int_{t_1}^{t_2} h\big(x(t), \dot{x}(t), t\big)\, dt$$

is minimized. The first step towards using the dynamic programming equation is to notice that $x(t)$ can be specified, apart from an arbitrary initial value, by its derivative $\dot{x}(t)$,

$$x(t) = x(t_1) + \int_{t_1}^{t} \dot{x}(t)\, dt,$$

and to reformulate the problem, writing

$$\dot{x}(t) \equiv u(t),$$

as that of finding $u(t)$ in the interval $t_1 \leqslant t \leqslant t_2$ to minimize

$$I = \int_{t_1}^{t_2} h\big(x(t), u(t), t\big)\, dt.$$

In this form the problem is a one-dimensional version of equations (9.1b) and (9.2b) with dynamics

$$g(x, u, t) = u(t),$$

and so equations (9.8) and (9.9) become

$$\frac{\partial f}{\partial t} + h + u \frac{\partial f}{\partial x} = 0,$$

$$\frac{\partial h}{\partial u} + \frac{\partial f}{\partial x} = 0,$$

which can be rewritten, using $\dot{x}$ in place of u,

$$\frac{\partial f}{\partial t} + h + \dot{x} \frac{\partial f}{\partial x} = 0,$$

$$\frac{\partial h}{\partial \dot{x}} + \frac{\partial f}{\partial x} = 0.$$

Now the partial derivative of the first of these equations with respect to x is

$$\frac{\partial^2 f}{\partial x\,\partial t} + \frac{\partial h}{\partial x} + \dot{x}\frac{\partial^2 f}{\partial x^2} = 0,$$

and the total derivative of the second equation with respect to time t is

$$\frac{d}{dt}\left(\frac{\partial h}{\partial \dot{x}}\right) + \dot{x}\frac{\partial^2 f}{\partial x^2} + \frac{\partial^2 f}{\partial x\,\partial t} = 0,$$

and subtracting the first of these equations from the second gives

$$\frac{d}{dt}\left(\frac{\partial h}{\partial \dot{x}}\right) - \frac{\partial h}{\partial x} = 0,$$

which is the Euler equation for this particular problem.†

An example of how the Euler equation specifies an optimal control law is provided by considering a one-dimensional LQP-problem having the above dynamics:

$$\dot{x} = u,$$

and performance criterion

$$I = \int_{t_1}^{t_2} (x^2 + qu^2)\,dt.$$

This problem has already been introduced at Fig. 6.6 in the discussion of fixed configuration design of Section 6.3; it could represent a system where it was required to adjust the torque u applied to a rotating load, having inertia of value unity, so as to regulate the speed of rotation x. It follows from the definitions of the dynamics and the performance criterion here that

$$h = x^2 + q\dot{x}^2,$$

and the Euler equation is

$$\frac{d}{dt}(2q\dot{x}) - 2x = 0$$

or

$$q\ddot{x} - x = 0,$$

a second-order differential equation with general solution

$$x = A\,e^{t/\sqrt{q}} + B\,e^{-t/\sqrt{q}},$$

having two constants of integration, A and B, which must be evaluated using two boundary conditions. In most control problems the initial state $x(t_1)$ is

† Given in textbooks on calculus of variations, for example, Courant and Hilbert (1953) Chapter IV.

given and it is required to find the control $u(t)$ over future time $t_1 \leqslant t \leqslant t_2$; thus one boundary condition is that $x(t_1)$ is fixed so that

$$x(t_1) = A\, e^{t_1/\sqrt{q}} + B\, e^{-t_1/\sqrt{q}}.$$

The second boundary condition is usually specified at the terminal time t_2, either by a fixed terminal state $x(t_2)$ or by equation (9.7).

The general boundary condition of equation (9.7) can be rewritten with the calculus of variations notation:

$$\left(\frac{\partial f}{\partial t} + h\right)\bigg|_{t=t_2} = 0$$

and can be combined with equation (9.8) to give

$$\left(\sum_{j=1}^{n} \frac{\partial f}{\partial x_j} g_j\right)\bigg|_{t=t_2} = 0,$$

and then with equation (9.9) to give

$$\frac{\partial h}{\partial \mathbf{u}}\bigg|_{t=t_2} = 0. \tag{9.10}$$

Equation (9.10) is the general calculus of variations boundary condition, at time t_2, for problems where the terminal state is free; it is sometimes called a transversality condition.

Assuming that the terminal state in the one-dimensional LQP-example is free, equation (9.10) gives

$$\frac{\partial h}{\partial u}\bigg|_{t=t_2} = 2q\dot{x}(t_2) = 0,$$

so that

$$A\sqrt{q}\, e^{t_2/\sqrt{q}} - B\sqrt{q}\, e^{-t_2/\sqrt{q}} = 0.$$

The constants A and B can now be evaluated from the two boundary-value equations, and when they are substituted into the general solution of the second-order differential equation it gives

$$x(t) = \frac{x(t_1)}{e^{(t_2-t_1)/\sqrt{q}} + e^{-(t_2-t_1)/\sqrt{q}}} \left(e^{(t_2-t)/\sqrt{q}} + e^{-(t_2-t)/\sqrt{q}}\right),$$

an equation specifying the optimal trajectory to be followed by the state x. An expression for the optimal control u can be derived from this optimal trajectory using the dynamic equation

$$u = \dot{x}$$

to give

$$u(t) = \frac{(1/\sqrt{q})x(t_1)}{e^{(t_2-t_1)/\sqrt{q}} + e^{-(t_2-t_1)/\sqrt{q}}} \left(e^{-(t_2-t)/\sqrt{q}} - e^{(t_2-t)/\sqrt{q}} \right),$$

an equation specifying the optimal control as a function of time over the interval $t_1 \leqslant t \leqslant t_2$.

To specify the optimal control in this way, as a function of time, requires exact knowledge of the initial state $x(t_1)$, of all parameters of the system, and of the initial and terminal times. In most control systems there is uncertainty about some or all of these factors and this uncertainty is overcome by using feedback control, as discussed in Section 2.1, which requires a control law specifying the control $\mathbf{u}(\mathbf{x})$ as a function of the state $\mathbf{x}$. Dynamic programming leads naturally to optimal control, described by the functions $\mathbf{u}(\mathbf{x})$ defined in Section 8.4, in the form of feedback laws, but the control laws $\mathbf{u}(t)$ resulting from the calculus of variations usually need to be converted into the form of feedback laws.

The analytic expression for the optimal control $u(t)$ in the one-dimensional LQP-example can be converted into a feedback law by regarding current time t as being the initial time for a process having initial state $x(t)$, as in dynamic programming. Then t and t_1 in the expression for $u(t)$ are the same, and the optimal control can be written as a feedback law

$$u(t) = -K(t)x(t),$$

where $K(t)$ is a time-varying feedback gain given by

$$K(t) = \frac{1}{\sqrt{q}} \frac{e^{(t_2-t)/\sqrt{q}} - e^{-(t_2-t)/\sqrt{q}}}{e^{(t_2-t)/\sqrt{q}} + e^{-(t_2-t)/\sqrt{q}}}.$$

The optimal feedback law here has the same structure as in previous LQP-problems. In the limit as terminal time t_2 goes to infinity, the time-varying gain $K(t_2)$ goes to the same value $1/\sqrt{q}$ as was derived in Section 6.3.

Continuous-time, optimal control problems having dynamic equations less elementary than the integrator of the above example can also be formulated as problems in the calculus of variations. For this purpose the dynamic equations (9.1b) of the controlled process are regarded as constraints on the state variables $\mathbf{x}(t)$, and the control function $\mathbf{u}(t)$ is required to minimize a modified integral,

$$\int_{t_1}^{t_2} \left\{ h(\mathbf{x}, \mathbf{u}, t) + \sum_{j=1}^{n} \lambda_j(t) g_j(\mathbf{x}, \mathbf{u}, t) \right\} dt,$$

where the $\lambda_j(t)$ are a set of n Lagrange multipliers and are functions of time. The justification for this standard calculus of variations procedure is derived from the general dynamic programming equation (9.4) in the discussion of Pontriagin's Maximum Principle in Section 9.3.

The calculus of variations could be used to derive all the results of Chapter 8 about continuous-time LQP-problems, but it is not suitable as a general technique for solving non-LQP-problems. Reasons for this unsuitability, illustrated by the above LQP-example, are set out below.

(i) Calculus is only suitable for problems which can be formulated so that optimization is achieved by equations like (9.9) specifying stationary values of a differentiable function.

(ii) The Euler equation leads to a differential equation, generally non-linear and of order $2n$, where n is the number of state variables, specifying optimal trajectories in state space. High-order, non-linear differential equations can rarely be solved analytically.

(iii) The differential equation specifying optimal trajectories is generally a two-point boundary-value equation, having boundary conditions specified at two separate points in time, t_1 and t_2. Such equations cannot be conveniently solved by simulation or by numerical analysis with a digital computer.

(iv) Even if the differential equations can be solved, the resulting information about optimal trajectories in state space requires further conversion before it can be interpreted as a feedback control law.

The relationship between optimal control theory and the calculus of variations is thus of academic, rather than practical, interest.

9.3. The Pontriagin maximum principle

Pontriagin's Maximum Principle generalizes the calculus of variations to include problems where optimization is not achieved by calculus, thus eliminating one of the restrictions ((i) in the above list) on the general usefulness of calculus of variations. It follows directly from the general continuous-time dynamic programming equations when these are rewritten in a different notation. New variables $\mathbf{p}$ are introduced, sometimes called adjoint variables, which are functions of time t taking values equal to partial derivatives of the solution function f,

$$ p_j \equiv \frac{\partial f}{\partial x_j} \quad j = 1, \ldots, n, $$

along the optimal trajectory specified by the equations of optimal control. Also a new function H is introduced, sometimes called the Hamiltonian because it corresponds to the Hamiltonian function of classical mechanics:

$$ H(\mathbf{x}, \mathbf{u}, t) \equiv h(\mathbf{x}, \mathbf{u}, t) + \sum_{j=1}^{n} \frac{\partial f}{\partial x_j} g_j(\mathbf{x}, \mathbf{u}, t); $$

it is the bracketed term { } minimized in the general equation (9.4), and on an optimal trajectory it can be written

$$H = h + \sum_{j=1}^{n} p_j \dot{x}_j. \qquad (9.11)$$

Equation (9.11) shows that the adjoint variables $\mathbf{p}$ correspond to the Lagrange multipliers $\boldsymbol{\lambda}$ of the calculus of variations.

Differential equations for these adjoint variables $\mathbf{p}$ follow from the way they are defined as functions of time only, so that

$$\dot{p}_j \equiv \frac{d}{dt}\left(\frac{\partial f}{\partial x_j}\right) = \frac{\partial^2 f}{\partial t\, \partial x_j} = \frac{\partial}{\partial x_j}\left(\frac{\partial f}{\partial t}\right),$$

and then the Hamilton–Jacobi equation (9.8), together with the definition of the Hamiltonian H, gives

$$\dot{p}_j = -\frac{\partial H}{\partial x_j}. \qquad (9.12)$$

Differential equations (9.1b) for the state variables $\mathbf{x}$ can be rewritten in terms of the Hamiltonian H, using equation (9.11),

$$\dot{x}_j = \frac{\partial H}{\partial p_j}. \qquad (9.13)$$

Equations (9.12) and (9.13) are known as Pontriagin's equations. They form a set of $2n$ simultaneous first-order differential equations specifying values of the Hamiltonian, according to equation (9.11), along an optimal trajectory. The Pontriagin Maximum† Principle is just a statement in words of the minimization in the general dynamic programming equation (9.4), to the effect that the condition for a trajectory to be optimal is that the Hamiltonian H of equation (9.11) be a minimum with respect to the control $\mathbf{u}$ at every point in time.

An example of how the Maximum Principle, together with Pontriagin's equations (9.12) and (9.13) and the Hamiltonian equation (9.11), specifies an optimal control law is provided by reconsidering the one-dimensional LQP-problem of Section 9.2 having dynamic equation

$$\dot{x} = u,$$

and performance criterion

$$I = \int_{t_1}^{t_2} (x^2 + qu^2)\, dt.$$

---

† It would be more rational here to talk about a Minimum Principle, because the optimal control problem has been so formulated that the optimizing condition is that the Hamiltonian be minimized. However, the Principle was originally stated for problems where the Hamiltonian had to be maximized, and the name has stuck.

The Hamiltonian for this problem, given by equation (9.11), is

$$H = x^2 + qu^2 + pu,$$

and the Maximum Principle states that the control u must minimize the Hamiltonian. In this LQP-problem the minimizing condition can be found by calculus; it is an expression

$$u = -p/2q$$

for the optimal control u as a function of the adjoint variable p. This is converted into a control law by eliminating u from the expression for the Hamiltonian H on an optimal trajectory to give

$$H = x^2 - p^2/4q,$$

and then writing the simultaneous first-order differential equations (9.12) and (9.13) for p and x:

$$\dot{p} = -2x,$$

$$\dot{x} = -p/2q,$$

and solving them using appropriate boundary conditions.

Boundary conditions for the state variables $\mathbf{x}(t)$ are usually provided by the given initial state $\mathbf{x}(t_1)$. When the terminal state $\mathbf{x}(t_2)$ is free, boundary conditions for the adjoint variables $\mathbf{p}(t)$ are specified at the terminal time t_2 by equation (9.6) which gives the transversality condition†

$$\left.\frac{\partial f}{\partial \mathbf{x}}\right|_{t=t_2} = \mathbf{p}(t_2) = \mathbf{0}. \tag{9.14}$$

Thus Pontriagin's equations for optimal control have boundary values specified at two separate points in time, and are as intractable as other two-point boundary-value problems.

In the LQP-example the differential equations for p and x are linear and can be combined to give second-order differential equations for the state x,

$$q\ddot{x} - x = 0,$$

and for the adjoint variable p,

$$q\ddot{p} - p = 0,$$

which are similar to the differential equation derived from the Euler equation in Section 9.2. The general solution for the adjoint variable p is,

$$p = A\,e^{t/\sqrt{q}} + B\,e^{-t/\sqrt{q}},$$

---

† When the boundary condition is specified by equation (9.6a) the transversality condition is

$$\mathbf{p}(t_2) = \left.\frac{\partial h_1}{\partial \mathbf{x}}\right|_{t=t_2}. \tag{9.14a}$$

the boundary condition of equation (9.14) gives

$$A\,\mathrm{e}^{t_2/\sqrt{q}} + B\,\mathrm{e}^{-t_2/\sqrt{q}} = 0,$$

the boundary condition given by the initial state $x(t_1)$ is

$$x(t_1) = -\tfrac{1}{2}\dot{p}(t_1)$$

$$= -\frac{1}{2\sqrt{q}}\left(A\,\mathrm{e}^{t_1/\sqrt{q}} - B\,\mathrm{e}^{-t_1/\sqrt{q}}\right),$$

and so the general solution for p is

$$p(t) = \frac{(2\sqrt{q})x(t_1)}{\mathrm{e}^{(t_2-t_1)/\sqrt{q}} + \mathrm{e}^{-(t_2-t_1)/\sqrt{q}}}\left(-\mathrm{e}^{-(t_2-t)/\sqrt{q}} + \mathrm{e}^{(t_2-t)/\sqrt{q}}\right),$$

which converts the calculus condition minimizing the Hamiltonian,

$$u = -p/2q,$$

into the same control law $u(t)$ as was derived in Section 9.2.

The Pontriagin Maximum Principle is unsatisfactory as a general technique for solving non-LQP-problems for most of the reasons ((ii), (iii) and (iv) in the list of Section 9.2) that make the calculus of variations unsatisfactory. However, its ability to handle problems where optimization is not achieved by calculus makes it useful for deriving results about a class of non-linear systems in which the non-linearity takes the form of saturation constraints on the range of the control $\mathbf{u}$.

Problems 9.1–9.6

9.4. Problems with saturating control variables

Constraints on the range of control variables exist in many practical control systems, as has been mentioned in Chapter 6. Such constraints can be represented mathematically in LQP-problems by including a cost-of-control term in the performance criterion, as above, but it is difficult to know what value to assign to the weighting matrix $\mathbf{Q}$. They can alternatively be described by saturation inequalities

$$U_{j1} \leqslant u_j \leqslant U_{j2}, \quad j = 1, \ldots, m \tag{9.15}$$

on the components $u_1, \ldots, u_m$ of the control vector $\mathbf{u}$.

In practice it is not usually difficult to assign values to the saturation levels U_1 and U_2, but inequalities (9.15) are non-linear and so it is not possible to derive general analytic solutions about problems with saturation constraints. Nevertheless the application of Pontriagin's Maximum Principle to problems having linear dynamics with saturation constraints on the controls does lead to useful results.

One such class of problems is where the dynamics are linear,

$$\dot{\mathbf{x}} = \mathbf{Ax} + \mathbf{Bu},$$

where there are constraints (9.15), and where it is required to drive the state $\mathbf{x}$ to a specified target state $\mathbf{x}(t_2)$ as rapidly as possible. These minimum-time problems are described, as discussed in Section 6.1(v), by the performance criterion

$$I = \int_{t_1}^{t_2} 1 \, \mathrm{d}t,$$

where the terminal time t_2 is determined by the optimization. The Hamiltonian, from equation (9.11), here is

$$H = 1 + \mathbf{p'Ax} + \mathbf{p'Bu},$$

and this depends on the control $\mathbf{u}$ only through the last term $\mathbf{p'Bu}$. It is assumed here that all components of the row vector $\mathbf{p'B}$ are non-zero so that the Hamiltonian depends on all the control variables. Problems can be formulated where this assumption is not always true, so that there are trajectories along which the performance criterion is independent of some or all of the control variables. Such trajectories are said to be singular; they represent rather special mathematical cases which are beyond the scope of this book.†
For the general class of non-singular problems the minimization required by the Maximum Principle is a minimization of the term $\mathbf{p'Bu}$ with respect to $\mathbf{u}$. This minimization cannot be achieved by calculus because $\mathbf{p'Bu}$ is linear in $\mathbf{u}$ and is minimized by making each component of $\mathbf{u}$ as negative, or as positive, as possible depending on the sign of its coefficient in $\mathbf{p'B}$. If there were no constraints on $\mathbf{u}$ the optimal control would be infinite, but in the presence of saturation constraints (9.15) the optimal control is to let every control variable take its minimum or maximum allowed value depending on whether the sign of its coefficient in $\mathbf{p'B}$ is positive or negative. This sort of control law, which requires every control variable to take one or other of two limiting values, is called bang–bang control.

An example of a minimum-time problem would be that of driving the rotating inertia of previous examples to a specified target position as rapidly as possible in the presence of constraints on the torque u,

$$-U < u < U.$$

If the inertia has value unity and the state variables x_1, x_2 correspond to its position and velocity so that its dynamic equations are

$$\dot{x}_1 = x_2,$$
$$\dot{x}_2 = u,$$

† See Johnson (1965) or Athans and Falb (1966) Sections 6.21, 6.22 or Hsu and Meyer (1968) Chapter 16 or Bryson and Ho (1969) Chapter 8 or Kirk (1970) Section 5.6.

then the Hamiltonian is

$$H = 1 + p_1 x_2 + p_2 u.$$

This is minimized by the bang–bang optimal control,

$$u = -U \operatorname{sign}(p_2), \tag{9.16}$$

so that on an optimal trajectory the Hamiltonian is

$$H = 1 + p_1 x_2 - U|p_2|,$$

and the Pontriagin equations (9.12) give

$$\dot{p}_1 = 0,$$
$$\dot{p}_2 = -p_1,$$

which combine with the dynamic equations,

$$\dot{x}_1 = x_2,$$
$$\dot{x}_2 = -U \operatorname{sign}(p_2),$$

to specify optimal trajectories. The discontinuous nature of bang–bang control gives rise to the non-linear sign function in the equation for $\dot{x}_2$, which makes it impossible to proceed directly to an analytic solution of this optimal control problem.

The optimal control here can nevertheless be found by considering the form of the solutions to the differential equations for **x** and for **p**,

$$x_2(t) = \pm Ut + A, \tag{9.17a}$$
$$x_1(t) = \pm \tfrac{1}{2}Ut^2 + At + B, \tag{9.17b}$$
$$p_1(t) = C, \tag{9.17c}$$
$$p_2(t) = -Ct + D, \tag{9.17d}$$

where A, B, C, and D are integration constants determined by boundary conditions. The shape of optimal trajectories in state space is found by eliminating time t from equations (9.17a) and (9.17b) to give an expression

$$x_2^2 = \pm 2Ux_1 + A^2 \mp 2UB$$
$$= \pm 2Ux_1 + \text{constant}$$

specifying the two families of parabolas shown in Fig. 9.1; the curvature of trajectories depends on the value of the torque $\pm U$; the position of trajectories depends on the constants of integration and boundary conditions; and the arrows showing direction of increasing time follow from the definition of x_2 as being velocity $\dot{x}_1$. The optimal control, specified in equation (9.16), is to switch from one set of trajectories to the other whenever the adjoint variable p_2 changes sign; equation (9.17d) shows that p_2 is a linear function of time which in any interval $t_1 \leqslant t \leqslant t_2$ can change sign at most once, and it follows that on any optimal trajectory there is at most one switch of torque. When in time, or where in state space, this switching happens depends on the boundary

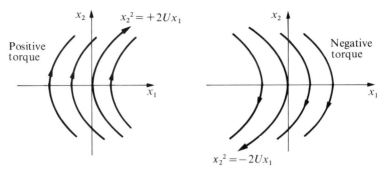

FIG. 9.1. State-space trajectories with positive and negative control U.

conditions, and in particular on the target position to which the system is to be driven. There is no loss of generality in assuming that this desired position is where the state variable x_1 is zero, so that the origin of state space, where velocity x_2 is also zero, represents equilibrium at the desired position and can be regarded as the target state for the optimal control problem. It follows from Fig. 9.1 that the line AOB, shown in Fig. 9.2 and specified by

$$2Ux_1 + x_2|x_2| = 0,$$

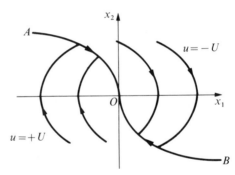

FIG. 9.2. Minimum-time trajectories.

is the set of all points in state space from which it is possible to get to the target state without any switch of torque, and that the remainder of state space consists of points from which it is possible to get to the target state with a single switch of torque on the line AOB as shown in Fig. 9.2.† Thus the optimal control law, using at most one switch of torque on any trajectory, is to make the torque negative or positive according to whether $2Ux_1 + x_2|x_2|$ is positive or negative; this feedback law,

$$u = -U \operatorname{sign} (2Ux_1 + x_2|x_2|),$$

† The argument here can be regarded as a continuous-time version of the discrete-time argument used to introduce controllability in Section 7.2.

giving the optimal control as a function of the state $\mathbf{x}$, completely solves the problem.

The above argument can be generalized to minimum-time problems with any number n of state variables, but does not usually lead to complete solution of high-order problems. What can be proved† is that, when the matrix $\mathbf{A}$ describing system dynamics has all its eigenvalues real, each control variable u_j switches between its limiting values at most $n - 1$ times on an optimal trajectory; but when there is more than one control variable they will not necessarily all switch simultaneously.

Other problems where bang–bang control is optimal include those derived from LQP-problems when the performance criterion is

$$I = \int_{t_1}^{t_2} \mathbf{x}'\mathbf{P}\mathbf{x} \, dt$$

and the cost-of-control term $\mathbf{u}'\mathbf{Q}\mathbf{u}$ is replaced by saturation constraints (9.15). The Hamiltonian for this class of problems is

$$H = \mathbf{x}'\mathbf{P}\mathbf{x} + \mathbf{p}'\mathbf{A}\mathbf{x} + \mathbf{p}'\mathbf{B}\mathbf{u},$$

and so, for the same reasons as for minimum-time problems, the optimal, non-singular control is bang–bang control. If the control $\mathbf{u}$ appears explicitly in the integrand h of the performance criterion I, bang–bang control may not be optimal everywhere in state space; there may be singular trajectories where the Hamiltonian is independent of the control and there may also be regions where the Hamiltonian takes a minimum stationary value with respect to the control.

Problem 9.7

9.5. Distinctive features of dynamic programming

Dynamic programming has been introduced here as the basic mathematical technique of optimal control theory. It is regarded as more basic than the calculus of variations and the Pontriagin Maximum Principle because of the simplicity of its fundamental Principle of Optimality, which leads to all known theoretical results about LQP-control and about bang–bang control, and because its general equations (9.3) and (9.4) are amenable to numerical solution using a digital computer for many problems which cannot be handled at all by other techniques.

Discrete-time problems are most naturally formulated using dynamic programming and the general functional recurrence equation (9.3). The recursive nature of this equation is well-conditioned for numerical solution

† See Pontriagin *et al.*, (1962) Section 17.

using a digital computer, so this is the standard dynamic programming procedure for attacking discrete-time problems. Details of the numerical procedures are beyond the scope of this book,† but certain features can be summarized.

(i) The problem has to be described numerically, and so there is no need for the defining functions **G** and *H* of equations (9.1a) and (9.2a) to be in closed form.

(ii) The minimization of equation (9.3), which establishes the optimal control **u**, can be achieved numerically by a search procedure. Constraints on the range of allowable values of the control **u** or of the state **x** can improve the efficiency of the search by reducing the number of possibilities that need be considered.

(iii) The numerical procedures of dynamic programming share with other numerical procedures an inability to handle problems of high dimensionality where the number of points at which functions must be determined becomes very large. To solve equation (9.3) numerically it is necessary to store values of the solution function $f_N(\mathbf{x})$ and of the optimal control $\mathbf{u}_N(\mathbf{x})$ for every possible state **x**; the number of possible states, and thus the number of words of storage required, increases exponentially with the number *n* of state variables so that, even with coarse quantization on individual variables, the limits of what is physically computable are soon reached. This inability to solve problems of high dimensionality is the main limitation on the general usefulness of equation (9.3).

Continuous-time problems which can be formulated using equation (9.4) can always be reduced to ordinary differential equations, the Pontriagin equations (9.12) and (9.13), but these are two-point boundary-value equations and therefore ill-conditioned for solution using a computer. Dynamic programming offers alternative ways of seeking numerical solutions to continuous-time problems. One way is to approximate the continuous-time process by an equivalent discrete-time process and then to solve the resulting discrete-time equation (9.3) using the standard procedure mentioned above. Another way is to make successive approximations to either the solution function $f(\mathbf{x}, T, t_2)$ or the optimal control function $\mathbf{u}(\mathbf{x}, T, t_2)$ using equation (9.4). These alternatives are limited in their application to problems of low dimensionality, but they do avoid the difficulties of two-point boundary-value problems.

As well as being able to handle problems not specified in closed form, which cannot be handled by other techniques, dynamic programming is well suited for formulating problems where uncertainty is explicitly specified. This feature, which is not shared by the other techniques, is exploited in Chapter 13. It is implicit in the Principle of Optimality and is the reason why

† See Bellman (1957), Bellman and Dreyfus (1962) or Jacobs (1967).

dynamic programming always leads to feedback control laws, even for deterministic problems.

9.6. Bibliography

Applications of dynamic programming to the general class of optimal control problems are discussed in most of the books mentioned in the bibliography of Section 8.6.

The relationship between dynamic programming and the calculus of variations is the subject of Dreyfus (1965). Applications of the calculus of variations to problems of optimal control are in

KALMAN (1960a and b),
Tou (1964) Chapter 5,
NOTON (1965) Chapter 1,
ATHANS and FALB (1966) Sections 5.1–5.10,
McCAUSLAND (1967) Chapter 5,
HSU and MEYER (1968) Chapter 13,
SAGE (1968) Chapter 3,
BRYSON and HO (1969) Chapter 2,
CITRON (1969) Chapters 2, 3, and 7,
PRIME (1969) Sections 4.1–4.4,
KIRK (1970) Chapter 4,
SPEEDY, BROWN, and GOODWIN (1970) Section 7.2.

Pontriagin's Maximum Principle is presented in the book by Pontriagin *et al.* (1962). Its applications to problems of optimal control are discussed extensively in

ATHANS and FALB (1966),
and also in

Tou (1964) Chapter 6,
DORF (1965) Chapter 10,
NOTON (1965) Chapter 4,
LEE and MARKUS (1967),
McCAUSLAND (1967) Chapter 6,
PALLU DE LA BARRIERE (1967) Chapter 18,
FLÜGGE-LOTZ (1968) Chapter 4,
HSU and MEYER (1968) Chapter 14,
SAGE (1968) Chapters 4 and 5,
BRYSON and HO (1969) Chapter 3,
NASLIN (1969) Chapter 6,
PRIME (1969) Chapter 4,
KIRK (1970) Chapter 5, and Section 7.1,
SPEEDY, BROWN, and GOODWIN (1970) Chapter 7,
TAKAHASHI, RABINS, and AUSLANDER (1970) Chapter 14.

Minimum-time control of discrete-time systems having saturating control variables is discussed in

KUO (1970) Chapter 6.

9.7. Problems

9.1. A continuous-time, controlled process has differential equation

$$\dot{x} = -x + u$$

and performance criterion

$$I = \int_{t_1}^{t_2} (x^2 + u^2)\, dt.$$

Write down the Hamiltonian of equation (9.11), show that the optimal value of u is $-p/2$, where p is the adjoint variable, and derive a pair of simultaneous first-order differential equations for x and p. Show that these are satisfied by an optimal feedback-control law in the form

$$u = -Kx$$

provided that

$$\dot{K} = -1 + 2K + K^2,$$

and specify a boundary value for K in this differential equation.

9.2. Write down the Hamiltonian for the general, continuous-time, LQP-problem of equations (8.7b) and (8.8) and show that the optimal control is

$$\mathbf{u} = -\mathbf{Q}^{-1}\mathbf{B}'\mathbf{p},$$

where $\mathbf{p}$ are adjoint variables. Show that the assumption that the adjoint variables are proportional to the state,

$$\mathbf{p} = \mathbf{Vx},$$

leads to the matrix Riccati equation (8.10b). What is the relationship between this assumption and the assumption made in Section 8.3 that the optimal value of the performance criterion is quadratic in $\mathbf{x}$?

9.3. A continuous-time, controlled process has differential equation

$$\dot{x} = -x^3 + u$$

and performance criterion

$$I = \int_{t_1}^{t_2} (x^2 + u^2)\, dt.$$

Write down the Hamiltonian and show that the differential equations for the state and adjoint variables combine to give

$$\ddot{x} - x - 3x^5 = 0.$$

If the initial value $x(t_1)$ is given and the final value $x(t_2)$ of x is free, what are the boundary values on x in this differential equation?

9.4. A continuous-time, controlled process has differential equation

$$\dot{x} = -x + u,$$

'minimum effort' performance criterion

$$I = \int_0^1 u^2 \, dt,$$

and the initial and final states $x(0)$, $x(1)$ are specified. Show that the optimal control can be written in the form

$$u = -Kx,$$

where

$$\dot{K} = 2K + K^2,$$

and specify a boundary value for K in this differential equation.

9.5. A continuous-time, controlled process has differential equation

$$\dot{x} = u,$$

performance criterion

$$I = \int_0^T u^2 \, dt,$$

and the initial and final states are

$$x(0) = X, \qquad x(T) = 0.$$

Show that the optimal control law results in a trajectory

$$x(t) = X(1 - t^2/T^2).$$

Compare this with the trajectories of Problem 8.11.

9.6. A discrete-time version of the non-linear Problem 9.3 has difference equation

$$x(i + 1) = x(i) - \tau x^3(i) + \tau u(i)$$

and quadratic performance criterion

$$I = \tau \sum_{i=1}^{N-1} \left(x^2(i) + u^2(i) \right).$$

Show how dynamic programming could be used to avoid two-point boundary-value problems in computing optimal feedback-control laws for two cases:

(i) when $x(N)$ is unspecified,
(ii) when $x(N)$ must be zero.

9.7. A continuous-time, controlled process has transfer function

$$\frac{Y(s)}{U(s)} = \frac{1}{s(1 + s)}$$

and is to be driven to the origin of state space $(x_1 \equiv y, x_2 \equiv \dot{y})$ in minimum time subject to the constraint

$$|u| \leqslant 1.$$

Show that bang–bang control is optimal and that there is never more than one switch of control on any optimal trajectory. Make a sketch (this can be done by the method of isoclines described in Section 10.1) showing typical trajectories and the switching line.

10. The phase plane

10.1. Introduction

Second-order dynamic systems are particularly well suited to graphical analysis because, with only two state variables, the state space is a plane, which is topologically simpler than higher-order spaces and has trajectories that can be represented by single lines on a two-dimensional sheet of paper. In this context the two-dimensional state space is known as the phase plane.† Phase-plane analysis, like other techniques based on state space, is applicable both to linear and to non-linear systems. It is usually restricted to continuous-time systems and its scope, when applied to control systems, is very limited compared to the general, linear, analytic techniques of Section I or to the fundamental questions about optimal control which are posed, but not generally solved, in Chapter 9. Nevertheless, it can lead to useful results about non-linear problems which have no closed-form solution; Figs 9.1 and 9.2 and the argument used to solve the second-order, minimum-time control problem in Section 9.4 are examples of phase-plane analysis.

The basic technique of phase-plane analysis is to consider continuous-time, stationary, second-order systems under conditions where the control $\mathbf{u}$ is held constant so that the dynamic equations have the form

$$\dot{x}_1 = g_1(x_1, x_2), \tag{10.1}$$

$$\dot{x}_2 = g_2(x_1, x_2).$$

These equations (10.1) give an expression

$$\frac{\mathrm{d}x_2}{\mathrm{d}x_1} = \frac{\dot{x}_2}{\dot{x}_1} = \frac{g_2(x_1, x_2)}{g_1(x_1, x_2)} \tag{10.2}$$

determining the slope, and thus the shape, of trajectories at every point in phase space. An example of how equation (10.2) can be used to sketch trajectories is provided by deriving phase portraits, previously shown in Fig. 7.2, for a linear system with differential equation

$$\ddot{y} + 2\zeta\omega_0\dot{y} + \omega_0^2 y = 0$$

† The techniques of phase-plane analysis were originally developed by Poincaré (1892) who used the word 'phase', as opposed to 'state', variable.

and state variables,

$$x_1 \equiv y, \qquad x_2 \equiv \dot{y}/\omega_0,$$

where x_2 is scaled to eliminate ω_0, so that equation (10.2) gives

$$\frac{\mathrm{d}x_2}{\mathrm{d}x_1} = -\left(2\zeta + \frac{x_1}{x_2}\right).$$

The shape of trajectories depends on the damping ratio ζ as follows.

(i) For zero damping, $\zeta = 0$, the equation for $\mathrm{d}x_2/\mathrm{d}x_1$ is easily integrated to give

$$x_1^2 + x_2^2 = \text{constant},$$

and trajectories are circular, as shown in Fig. 10.1.

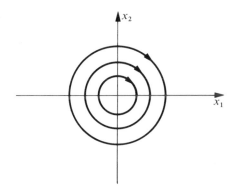

FIG. 10.1. Phase portrait of undamped linear system.

(ii) For damping less than critical, $0 < \zeta < 1$, the equation for $\mathrm{d}x_2/\mathrm{d}x_1$ is not so easily integrated and it is more convenient to sketch the phase portrait by considering loci in the phase plane where all trajectories have the same slope, as shown in Fig. 10.2, which is based on three such loci, where

$$x_1 = 0 \qquad \text{and} \quad \mathrm{d}x_2/\mathrm{d}x_1 = -2\zeta,$$

$$x_2 = 0 \qquad \text{and} \quad \mathrm{d}x_2/\mathrm{d}x_1 = \infty,$$

$$x_1 + 2\zeta x_2 = 0 \quad \text{and} \quad \mathrm{d}x_2/\mathrm{d}x_1 = 0.$$

(iii) For damping ratios greater than critical, $\zeta > 1$, there are, in addition to the loci of Fig. 10.2, straight lines which are crossed by no trajectories because any trajectory on them remains on them. Such lines are described by the equation

$$x_2 = mx_1$$

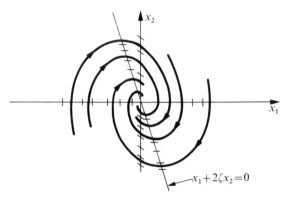

FIG. 10.2. Phase portrait of underdamped linear system.

and have the property that trajectories on them have slope m:

$$\frac{dx_2}{dx_1} = -\left(2\zeta + \frac{1}{m}\right) = m,$$

so that m is a root of the characteristic equation

$$m^2 + 2\zeta m + 1 = 0,$$

which gives

$$m = -\zeta \pm \sqrt{(\zeta^2 - 1)}.$$

Figure 10.3 shows how these lines are used to sketch the phase portrait.

The technique, used in (ii) and (iii) above, of sketching phase portraits by considering loci where all trajectories have the same slope is known as the

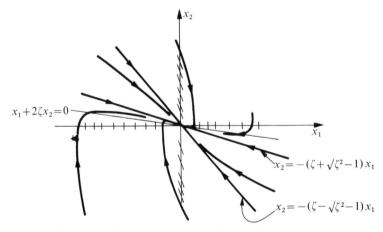

FIG. 10.3. Phase portrait of overdamped linear system.

method of isoclines. It can be used for non-linear systems having differential equations that cannot be solved analytically. Solutions in the form of time-functions can usually be constructed from a phase portrait by taking account of the way state variables are defined. For example, in the linear system where

$$\frac{\mathrm{d}x_1}{\mathrm{d}t} = \omega_0 x_2,$$

the time function $x_1(t)$ can be constructed from an approximate expression

$$\delta t = \delta x_1 / \omega_0 x_2$$

relating small changes δt and δx_1. Figure 10.4 shows how this expression gives the time to travel a small distance from A to B along a trajectory: a complete

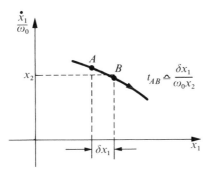

Fig. 10.4. Time along a trajectory.

time-response could be built up by subdividing the trajectory into small portions, like AB where x_2 is approximately constant, and finding the time to travel along each portion.

Problems 10.1, 10.2

10.2. Relay control systems

A relay, used to switch a control signal u between two fixed levels, is a particularly cheap and robust device to use in a practical control system. Although it provides bang–bang control it is more often used for the sake of practical convenience than to assure optimal response, and with a simple, linear, switching law rather than a non-linear optimal control law. The relay itself is a non-linear device and so the analysis of relay control systems is essentially a non-linear analysis; ideas about relay control systems are based on phase-plane analysis of second-order systems, and are introduced here by further discussion of the position-control example.

Suppose that an inertia J is positioned by a motor which is controlled by a relay so that the applied torque takes only the values $\pm U$, and suppose that the desired position x has the constant value X. Convenient state variables for this problem are the error e and its derivative,

$$x_1 \equiv e \equiv X - y,$$

$$x_2 \equiv \dot{x}_1 = -\dot{y},$$

where y is the position that is to be controlled.

The minimum-time, control law, derived in Section 9.4 for this problem, is to switch the control relay according to the sign of the non-linear function

$$2Ux_1/J + x_2|x_2|.$$

This gives good performance, but expensive hardware may be needed to generate the non-linear switching function, so simpler linear switching functions are often used. Figure 10.5 is a block diagram of such a system

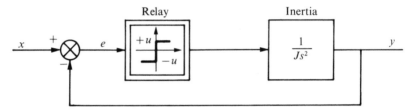

FIG. 10.5. Switching on the sign of the error.

in which the torque is switched according to the sign of the error, so that the inertia is always being driven toward its desired position, and the non-linear differential equation is

$$\dot{x}_2 = -\frac{U}{J}\operatorname{sign}(x_1),$$

and equation (10.2) gives

$$\frac{dx_2}{dx_1} = -\frac{U}{J}\frac{\operatorname{sign}(x_1)}{x_2}.$$

For positive values of error e the torque has the positive value $+U$ and this equation can be integrated to give an equation

$$x_2^2 = -2\frac{U}{J}x_1 + C_p,$$

showing that trajectories are the family of parabolas illustrated in Fig. 10.6(i),† where C_p is a constant of integration depending on initial conditions.

† The relationship between curvature and sign of x_1 here is the reverse of what it was in Fig. 9.1 because the state variable x_1 here is the negative of x_1 there.

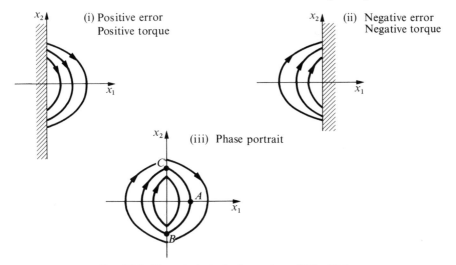

FIG. 10.6. Phase trajectories for system of Fig. 10.5.

Similarly when the error e is negative, trajectories are the family of parabolas

$$x_2^2 = 2\frac{U}{J}x_1 + C_N$$

illustrated in Fig. 10.6(ii). The two partial solutions can be combined to give the phase portrait of Fig. 10.6(iii), which is a complete graphical solution showing how the system behaves when the desired value x is constant. If the system was initially at rest with zero error and the value of x was subject to a step change, the phase-plane point representing the dynamic state would suddenly shift to somewhere like A in Fig. 10.6(iii) and would then move round the closed trajectory $ABCABCA$.... This trajectory represents an oscillatory motion, illustrated by sketching the error $x_1(t)$ as a function of time t in Fig. 10.7, which has no tendency to decay and would be unacceptable

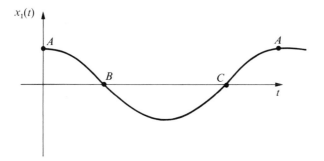

FIG. 10.7. Oscillating error in system of Fig. 10.5.

in a control system. It is not unlike simple harmonic motion, which is described by ellipses in the phase plane, but has discontinuities in acceleration where the torque switches at B and C. If the amplitude of the oscillation in x_1 is A, then the time to travel, at constant acceleration U/J and with zero initial velocity, to the first switching point B is

$$t_{AB} = \sqrt{(2AJ/U)},$$

and the period of the oscillation, which consists of four similar quarters, will be $4t_{AB}$. The phase-plane analysis thus leads to an exact solution describing the behaviour of this non-linear system, as well as showing that it can sustain self-excited oscillations which make it unsatisfactory as a control system.

The relay position-control system can be stabilized, like a linear system, by introducing linear velocity feedback. Figure 10.8 is a block diagram showing

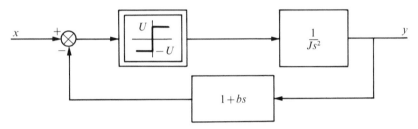

FIG. 10.8. Velocity feedback in the relay control system.

how velocity feedback is introduced so that the differential equation of the control system becomes

$$J\ddot{y} = U \operatorname{sign} (x - y - b\dot{y}),$$

and, with the same state variables as before, equation (10.2) gives

$$\frac{\mathrm{d}x_2}{\mathrm{d}x_1} = -\frac{U}{J} \frac{\operatorname{sign}(x_1 + bx_2)}{x_2}.$$

It follows that trajectories have the same curvature as in Fig. 10.6 but that switching, which depends on sign $(x_1 + bx_2)$, happens on the line

$$x_1 + bx_2 = 0.$$

Figure 10.9 is the phase portrait of the system with this linear velocity feedback; trajectories converge towards the origin and the error $x_1(t)$ decays with time as shown by the corresponding transient responses in Fig. 10.10. Some of these responses are not unlike those of an underdamped linear system but their shape varies with their magnitude, as is common with non-linear systems, becoming oscillatory as the magnitude increases. Only two trajectories, P^* and N^* in Fig. 10.9, actually go to the origin; these are the

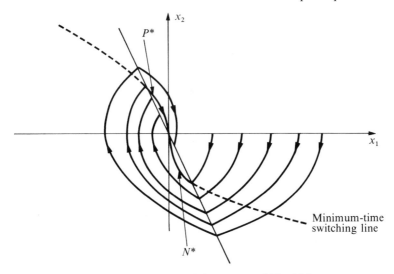

FIG. 10.9. Phase portrait for system of Fig. 10.8.

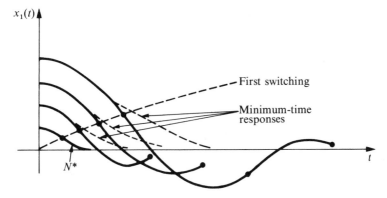

FIG. 10.10. Decaying errors in the system of Fig. 10.8.

trajectories which would form the non-linear switching line for minimum-time control giving the responses shown for comparison in Fig. 10.10. Other trajectories will approach the origin in the way illustrated in Fig. 10.11 which shows the phase portrait in the region of the origin. There is a segment *LL* of the switching line, between tangent curves *P*† and *N*† having the shape of trajectories for positive and negative torque, from which all trajectories lead immediately back to the switching line, so that there appears to be no mechanism for the representative point to move away from any point in *LL*. In practice the relay cannot switch instantaneously, so the switching line is always slightly overshot and the representative point slides along the switching

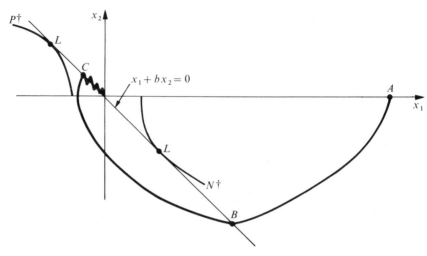

FIG. 10.11. Sliding motion to the origin of the phase plane.

line to the origin, as shown from point C in the figure. This sliding motion is sometimes called chattering, because the rapid oscillation of a real relay can make a chattering sound. During sliding motion the representative point is usually close to the switching line, and so the motion is approximately described by the differential equation

$$x_1 + b\dot{x}_1 = 0,$$

which gives an expression for the error $x_1(t)$ as a function of time,

$$x_1(t) = C_1\, e^{-t/b},$$

where C_1 is a constant of integration, showing that the sliding motion is like the transient response of a first-order linear system with time constant b.

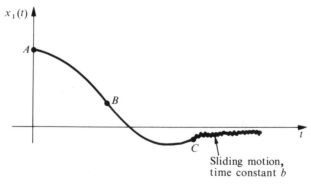

FIG. 10.12. Sliding motion as a function of time.

Figure 10.12 shows the error $x_1(t)$ as a function of time along the trajectory starting from A in Fig. 10.11. As the velocity feedback constant b is increased the error decays to zero more slowly, also the switching line rotates anticlockwise so that the points L in Fig. 10.11, where sliding motion can begin, correspond to increasingly large values of error x_1; thus too much velocity feedback will result in large errors which decay slowly.

The foregoing results about relay control are typical of what can be achieved by phase-plane analysis: quantitative results about stable oscillations and qualitative results about transient responses. They show that a relay control system with a linear switching function can give satisfactory performance, although the amount of damping is related to the amplitude of signals, and there may be sliding motions. The non-linear, minimum-time switching function gives better response, as shown in Fig. 10.10, but may be more expensive to implement; it is sometimes called vee-mod-vee control, describing the non-linear term $x_2|x_2|$ where x_2 is a velocity.

Problem 10.3

10.3. Some other non-linearities

The phase plane can be used to investigate the behaviour of control systems having other non-linearities than the ideal relay discussed in Section 10.2. Examples of some common non-linearities and their phase-plane analyses are presented here by considering their effect on the position-control example with the same state variables as before.

(i) *Saturation*

Saturation non-linearities, in the form of constraints on control variables, have already been mentioned in Section 9.4. Many real systems have saturation characteristics like that shown in Fig. 10.13(a) where an output z saturates

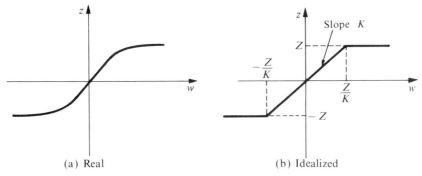

FIG. 10.13. Saturation characteristic.

with respect to an input w; such characteristics are conveniently represented, for purposes of analysis, by the straight line approximation of Fig. 10.13(b).

In position control it is usually the torque which saturates and so systems which are intended to be linear, like that introduced in Section 1.2 and having the block diagram of Fig. 2.7, often include a non-linear torque characteristic

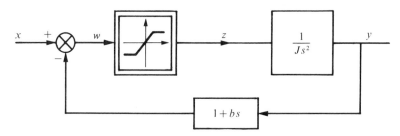

FIG. 10.14. Torque saturation in position-control system.

as shown in the block diagram of Fig. 10.14. When the torque saturation levels are $\pm U$ and the input to the non-linearity is

$$w = X - y - b\dot{y}$$
$$= x_1 + bx_2$$

as in Section 10.2, the phase plane divides into three regions:

(a) where $x_1 + bx_2 > U/K$,
maximum positive torque is applied, and trajectories are

$$x_2^2 = -2\frac{U}{J}x_1 + C_p$$

as in Fig. 10.6(i);

(b) where $x_1 + bx_2 < -U/K$,
maximum negative torque is applied, and trajectories are

$$x_2^2 = 2\frac{U}{J}x_1 + C_N$$

as in Fig. 10.6(ii);

(c) where $-U/K < x_1 + bx_2 < U/K$,
the system satisfies a linear differential equation having natural frequency

$$\omega_0 = \sqrt{(K/J)}$$

and damping ratio

$$\zeta = \frac{b}{2}\sqrt{(K/J)}$$

which determines the shape of trajectories.

Figure 10.15 shows phase portraits of the system for the two cases $\zeta \gtrsim 1$; trajectories in regions (a) and (b) where the torque saturates are similar to those of Fig. 10.9 for a relay system, so large disturbances can cause oscillatory response even when the system is overdamped in the linear region (c).

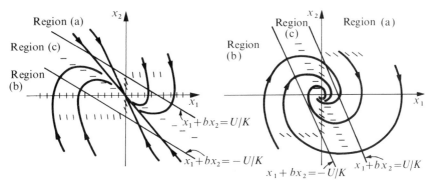

(i) Overdamped linear region (c), $\zeta > 1$ (ii) Underdamped linear region (c), $\zeta < 1$

FIG. 10.15. Phase portraits for system of Fig. 10.14.

(ii) Coulomb friction

A force which is constant in magnitude and opposes any motion is a non-linear phenomenon that can be described in the phase plane. In a position-control system it would be caused by a form of friction known as Coulomb friction and would be related to velocity $\dot{y}$ by the non-linear function

$$\text{Coulomb friction} = -F \operatorname{sign}(\dot{y}),$$

where F is the magnitude of the force.

If Coulomb friction were the only non-linearity in the position-control system of Fig. 2.7, the differential equation of the system would be

$$J\ddot{y} = K(x - y - b\dot{y}) - F \operatorname{sign}(\dot{y})$$

and, with the same state variables as before, phase-plane trajectories would have slopes

$$\frac{dx_2}{dx_1} = -\frac{K}{J}\left(\frac{x_1 + F/K \operatorname{sign}(x_2)}{x_2} + b\right).$$

The phase plane would divide into two regions depending on the sign of the velocity x_2; in each region the system satisfies a linear differential equation having the same natural frequency

$$\omega_0 = \sqrt{(K/J)}$$

and damping ratio

$$\zeta = \frac{b}{2}\sqrt{(K/J)}$$

as above, but the centres to which trajectories converge are offset by an amount $\pm F/K$ as in the phase portraits of Fig. 10.16. These show that the system can come to rest with a finite error x_1 of magnitude up to F/K; when

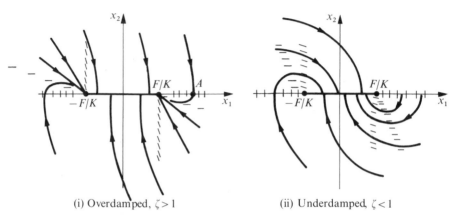

(i) Overdamped, $\zeta > 1$ (ii) Underdamped, $\zeta < 1$

FIG. 10.16. Phase portraits for position-control system with Coulomb friction.

the damping ratio ζ is greater than unity the error in step responses from an initial state like A in portrait (i) never has magnitude less than F/K.

In many position-control systems there is both Coulomb friction and torque saturation. The effect of these two non-linearities can be described in the phase plane by combining the phase portraits of Figs 10.15 and 10.16 to give

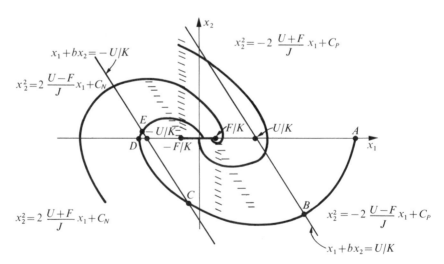

FIG. 10.17. Phase portrait for position-control system with both torque saturation and Coulomb friction.

the phase portrait of Fig. 10.17 for a system satisfying the non-linear differential equation

$$J\ddot{y} = z(x - y - b\dot{y}) - F\operatorname{sign}(\dot{y}),$$

where z is the output of the saturating torque characteristic of Figs 10.13 and 10.14. Figure 10.17 is drawn for the case when the damping ratio ζ in the linear regions is less than unity. It predicts the sort of time responses which are actually observed in real position-control systems; Fig. 10.18 shows a typical transient response of error $x_1(t)$ along the trajectory $ABCDE$.

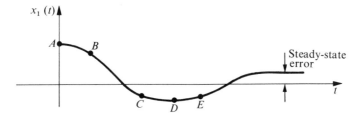

FIG. 10.18. Error response in system with both torque saturation and Coulomb friction.

(iii) *Hysteresis*

Real relays do not always have the ideal characteristic assumed in Section 10.2. There is often a switching threshold so that the relay behaves as though it had the hysteresis characteristic of Fig. 10.19 where the output z switches at different levels $\pm W$ of the input w depending on whether w is increasing or

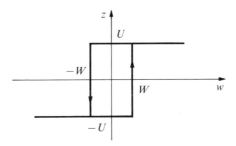

FIG. 10.19. Characteristic of a relay with hysteresis.

decreasing. Figure 10.20 is the phase portrait of the relay control system of Fig. 10.8 when the relay has hysteresis; all trajectories converge on to a single closed trajectory, or limit cycle, which represents an oscillation of fixed amplitude

$$A = \frac{J}{2U}\left(\frac{W}{b}\right)^2$$

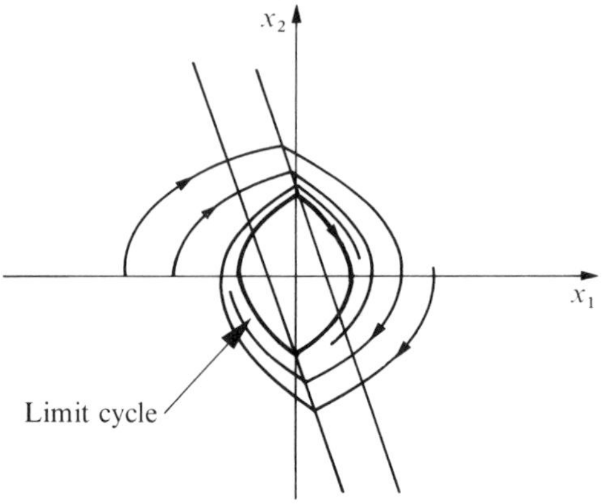

Fig. 10.20. Phase portrait of relay system with hysteresis.

and period, as in Section 10.2,

$$4\sqrt{(2AJ/U)} = 4\frac{JW}{Ub}.$$

This oscillation persists in the steady state, so the system never comes to rest, and errors are not maintained less than the amplitude A.

Problems 10.4–10.7

10.4. Topology of phase portraits

Each trajectory in the phase plane represents a solution to the differential equations (10.1). At ordinary points, where the derivatives $\dot{x}_1$ and $\dot{x}_2$ are not both zero, equation (10.2) specifies a unique slope of trajectory, so that there is only one trajectory through each such point and trajectories do not ordinarily intersect. This topological feature of phase portraits corresponds to an existence theorem stating that solutions to equations (10.1) are uniquely determined by initial values of the state variables x_1 and x_2. Other topological features are associated with the equilibrium behaviour of the equations at singular points and at limit cycles.

At singular points, where the derivatives $\dot{x}_1$ and $\dot{x}_2$ are both zero, the slope of trajectories is not defined. Singular points are points of equilibrium, and the topology of trajectories in the region of a singular point depends on the type of equilibrium. Various types of equilibrium can be demonstrated by linearizing equations (10.1) for small perturbations about a singular point, which is assumed without loss of generality to be the origin. The linearized equations can be written

$$\dot{\mathbf{x}} - \mathbf{A}\mathbf{x},$$

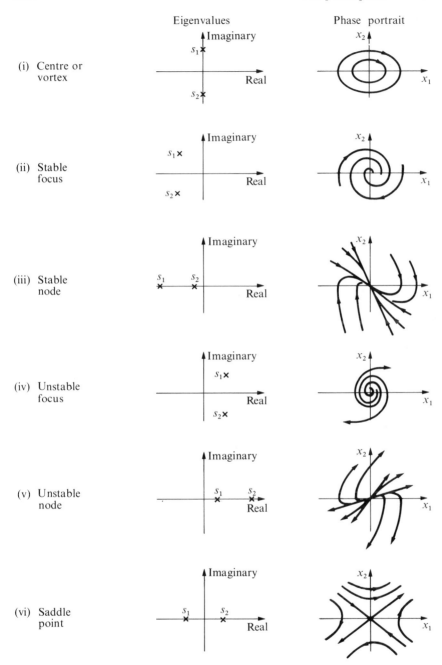

FIG. 10.21. Six types of equilibrium about a singular point at the origin of the phase plane.

where the elements a_{11}, a_{12}, a_{21}, a_{22} of the 2×2 matrix **A** are the partial derivatives $\partial g_1/\partial x_1$, $\partial g_1/\partial x_2$, $\partial g_2/\partial x_1$, $\partial g_2/\partial x_2$ of the functions g_1 and g_2 in equations (10.1), evaluated at the singular point. The type of equilibrium depends on the eigenvalues s_1, s_2 of the matrix **A**; these are roots of the characteristic equation

$$|\mathbf{A} - s\mathbf{I}| = 0,$$

and are given by

$$s_1, s_2 = \tfrac{1}{2}\left\{a_{11} + a_{22} \pm \sqrt{((a_{11} + a_{22})^2 - 4(a_{11}a_{22} - a_{12}a_{21}))}\right\}.$$

The eigenvalues can be real, imaginary, or complex and can have positive or negative real parts depending on the values of the elements a of **A**. Six types of pairs of eigenvalues, corresponding to six types of equilibrium, are illustrated in Fig. 10.21: the first three types (i), (ii), and (iii) correspond to the phase portraits of Figs 10.1, 10.2, and 10.3 and the other three correspond to unstable equilibria. Other types also exist where both eigenvalues are real and one of them is exactly equal to the other or to zero; they can be regarded as special cases of (iii), (v), and (vi) in Fig. 10.21 and are not illustrated here. Figure 10.22 shows how these various types of equilibrium are related to the values of the elements a. The phase portrait of any linear system must have a topology corresponding to one of those mentioned here; there are no other possibilities.

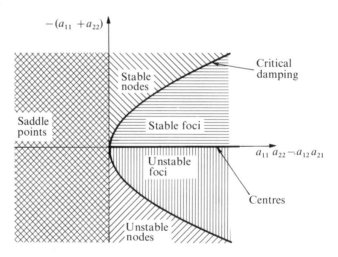

FIG. 10.22. Relationship between type of equilibrium and values of elements a.

In non-linear systems the equilibrium behaviour is not always described by linearization about a singular point. The relay system with hysteresis in Section 10.3(iii) is an example where the functions g_1 and g_2 of equations

(10.1) are discontinuous and linearization using their partial derivatives is impossible; the equilibrium behaviour of this system is a limit cycle. Limit cycles can also be caused by other non-linearities; they are a characteristic of non-linear, as opposed to linear, systems. Two types of limit cycle are possible, stable limit cycles to which trajectories converge as in Fig. 10.20 and unstable limit cycles from which trajectories diverge. The topology of any phase portrait is strongly influenced by its limit cycles, but it is usually not possible to predict where or what type they will be before drawing the phase portrait; indeed one objective of phase-plane analysis is often the investigation of limit cycles in a system. Some guidance as to the distribution of singularities and limit cycles can be derived from general topological considerations that all trajectories must start and finish at a singular point, at a limit cycle, or at infinity, and that trajectories cannot intersect. Mathematical statements of these considerations have been given by Poincaré and by Bendixson but

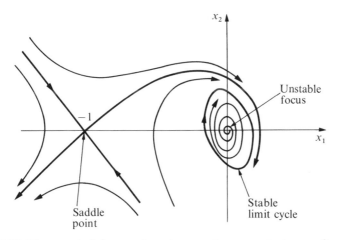

Fig. 10.23. Phase portrait for equations $\dot{x}_1 = x_2$, $\dot{x}_2 = -x_1 + \frac{1}{10}x_2 - x_1^2 - \frac{10}{3}x_2^3$.

they are not discussed here because second-order control systems do not usually have such complicated phase portraits. Figure 10.23 shows the sort of phase portrait that can result from more complicated equations.

Problems 10.8–10.11

10.5. Bibliography

Many books on control engineering have chapters about phase-plane analysis and relay control systems; for example

TRUXAL (1955) Chapter 11,
MURPHY (1959) Chapter 11,
LAUER, LESNICK, and MATSON (1960) Chapter 11,

THALER and BROWN (1960) Chapters 12 and 14,
WEST (1960) Chapters 4–7,
GRAHAM and McRUER (1961) Chapters 7 and 9,
THALER and PASTEL (1962) Chapters 3 and 6,
DOUCE (1963) Sections 2.7 and 7.1,
GIBSON (1963) Chapter 7,
SHINNERS (1964) Sections 7.14–7.18,
LANGILL (1965) Vol. 2, Sections 22.6–22.9,
NASLIN (1965) Chapter 9,
KUO (1967) Sections 11.2–11.5,
HSU and MEYER (1968) Chapter 4,
OGATA (1970) Chapter 12,
TAKAHASHI, RABINS, and AUSLANDER (1970) Section 12.1,
EVELEIGH (1972) Sections 14.3 and 14.4.
Further discussion of topological features of phase-plane analysis is in
MINORSKY (1947) Part I,
ANDRONOW and CHAIKIN (1949) Chapters 5 and 6,
BOGOLIUBOV and MITROPOLSKY (1961) Chapter 2,
HAYASHI (1964) Section 2.2,
STERN (1965) Chapter 7,
AGGARWAL (1972) Chapter 4.
Some discussion on using the phase plane for discrete-time systems is given in
VIDAL (1969) Chapter 3,
KUO (1970) Section 4.13.

10.6. Problems

10.1. Sketch the phase portrait of a system having differential equation

$$\ddot{y} + 2\dot{y} + y = 0.$$

10.2. Sketch the phase portrait of a system having state equations

$$\dot{x}_1 = x_2$$

$$\dot{x}_2 = \begin{cases} -x_2 - (1 + x_2)x_1, & x_1 > 0 \\ -x_2 - (1 - x_2)x_1, & x_1 < 0. \end{cases}$$

10.3. The torque applied to a rotating load of inertia 0.75 kg m^2 can be one of two values, ± 0.6 N m. In order to position the load, the torque is switched whenever a reference voltage changes sign. The reference voltage is the sum of an error voltage of 60 V rad^{-1} of position error and a velocity feedback voltage generated by a tachometer giving 1 V rad^{-1} s^{-1} of load speed.

Sketch the phase portrait of the system.

The system is initially at rest with zero error when the desired position of the load undergoes a step change. If the resulting transient response is that the system settles to its new equilibrium position with exactly two reversals of torque, what was the magnitude of the step in position?

10.4. A position-control servo system has a non-linear clutch in the drive to its output shaft. The load has inertia J and is frictionless. Torque from a motor is transmitted to the load by the clutch which is actuated by a signal $x - y - T\dot{y}$ (where T is the velocity-feedback constant and x and y are input and output shaft positions) and has two operating modes: driving and braking. When the actuating signal exceeds threshold values $\pm a$ the clutch drives with constant torque $\pm U$, otherwise it applies a braking torque with Coulomb friction characteristic of magnitude $\pm C$. The parameter values are:

$$T = 1 \text{ s}, \quad \frac{U}{C} = 2, \quad a = 3 \text{ rads}, \quad \frac{J}{U} = 0.5 \text{ s}^2.$$

Sketch trajectories in the error–error derivative phase plane for various values of input step disturbance. In particular determine the magnitude of input step which results in zero steady-state error.

10.5. In a certain temperature-control system, the controlled object has a thermal capacity Q J K^{-1}. Its temperature is measured by a thermocouple that generates k_1 V K^{-1} and responds to temperature changes like a low-pass filter with time constant T seconds. The control is such that heat is supplied to the controlled object at one of two rates, either zero if the thermocouple voltage is higher than a reference voltage or at a rate A W if the measured temperature is too low. The controlled object loses heat at a rate of k_2 W K^{-1} above a constant ambient temperature.

Find the system differential equation relating the thermocouple voltage to the reference voltage. What are the greatest and the least steady temperature differences between controlled object temperature and ambient temperature that can be achieved by the control system?

If the reference voltage is set to make the controlled object temperature $\frac{2}{3} A/k_2$K above ambient temperature, and if

$$\left(\frac{k_2 T + Q}{\sqrt{(k_2 TQ)}} \right) = 4,$$

sketch the phase portrait of the system in the phase plane of thermocouple voltage and its derivative.

10.6. A controlled process has transfer function

$$\frac{Y(s)}{U(s)} = \frac{1}{(1 + s)^2}.$$

Its output y is made to follow a desired value x by a relay actuated by the error $(x - y)$. The output from the relay is the process input u and has the two values ± 5. There is hysteresis in the relay so that its output u changes value only when the magnitude of its actuating signal increases beyond unity.

Sketch an accurate phase portrait of the system.

10.7. Use the phase portrait of Problem 10.6 to *estimate* the amplitude and period of the steady-state oscillation of the system when the desired value x is constant.

10.8. Sketch the phase portrait of a simple pendulum having differential equation

$$\ddot{y} + \frac{g}{l} \sin y = 0.$$

Show all possible modes of behaviour of the system.

10.9. The growth of a pair of conflicting populations may be described by the equations

$$\dot{x}_1 = k_1 x_1 - k_3 x_1 x_2,$$
$$\dot{x}_2 = k_2 x_2 - k_4 x_1 x_2.$$

Sketch the phase portrait of this system (for positive values only of x_1 and x_2).

10.10. Sketch the phase portrait of a system having differential equations

$$\dot{x}_1 = -2x_1 + x_1 x_2^2.$$
$$\dot{x}_2 = -x_2 + x_2 x_1^2.$$

10.11. Sketch the phase portrait of a system having differential equations

$$\dot{x}_1 = x_2,$$
$$\dot{x}_2 = -x_1 - x_2 - x_1^2.$$

11. Stability

11.1. Introduction

If all controllers were optimal there would be little need to discuss individual aspects of control system behaviour such as stability. However, optimal control theory cannot normally be used to solve specific non-linear problems, so most controllers are not optimal and the question often arises as to whether a non-optimal controller will result in a stable control system.

Non-linear dynamic systems can generally exhibit more complex behaviour than linear systems and require more careful definition of what is meant by 'stability' than was needed in Section I for linear systems. Such complex behaviour has already been introduced in the discussion of the topology of phase portraits of second-order systems in Section 10.4. A non-linear system

$$\dot{\mathbf{x}} = \mathbf{g}(\mathbf{x})$$

may have several equilibrium points with differing types of equilibrium and may have limit cycles, whereas a linear system

$$\dot{\mathbf{x}} = \mathbf{A}\mathbf{x}$$

has only one singular point at the origin of state space (provided $\mathbf{A}$ is non-singular) and the type of equilibrium or stability depends on the eigenvalues of $\mathbf{A}$ which determine the shape of trajectories throughout all state space. Figure 11.1 shows how the state space for the non-linear system of Fig. 10.23 is divided into three regions of which two are unstable and the third can be described as 'stable' in some sense, although not in the sense of Section I that the motion tends to a steady equilibrium value.

The definition of stability to be used here is that a system is *stable* in a region of its state space if its representative point never emerges from the region. This definition includes both the steady equilibrium values of linear systems and the stable region with a stable limit cycle of Fig. 11.1; the two types of stability are distinguished by saying that a system is *asymptotically* stable in a region if all trajectories in the region converge to a single equilibrium point. Asymptotic stability is the stronger form of stability and corresponds to the stability of linear systems in Section I. When the region under discussion is the whole state space, as for linear systems, the stability is said to be *global*, otherwise it is *local*. Thus linear systems which were described in Section I as 'stable' would be described here as being globally, asymptotically stable, and the 'stable region' in Fig. 11.1 is a region of local stability.

Theoretical methods for investigating stability of non-linear control systems are not closely related to methods of controller design as they are for

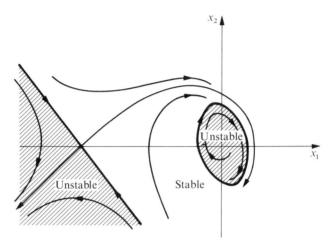

FIG. 11.1. Stable and unstable regions.

linear systems where global behaviour depends only on the roots of a characteristic equation. The general theory, known as Liapunov's direct method, is introduced in Section 11.2. It relates to all dynamic systems and is not specifically concerned with control problems. Although it provides the mathematical foundation for discussions of stability it does not often answer questions about specific non-linear problems nor has it yet provided much new information about control systems. A different sort of method, the describing function, giving approximate results about the size and period of limit cycles in a particular class of feedback systems, is presented in Section 11.3.

11.2. Liapunov's direct method

The objective of Liapunov's direct method is to answer questions about stability of dynamic systems directly without solving the differential equations of motion. It is also known as the 'second method', as opposed to the 'first method' which is to linearize about singularities and then consider eigenvalues of the resulting linear equations, as in Section 10.4. The first method generally can only provide information about local, rather than global, stability, although topological arguments in the phase plane can yield global results for second-order systems.

The second, or direct, method is introduced by considering stability of continuous-time autonomous systems

$$\dot{x} = g(x)$$

having an equilibrium point at the origin.

$$g(0) = 0.$$

Other equilibrium points can be investigated by shifting the origin of coordinates in state space to coincide with the point in question. No control u appears in these equations because they would be describing 'closed-loop' behaviour of a complete system, not the 'open-loop' behaviour of a controlled process as in other Chapters. Extension of the method for discrete-time systems is given at the end of this Section, and generalizations to non-stationary systems are straightforward.

The method is to consider the variation with time of the value of some, as yet unspecified, function $V(x)$ of the state variables, called a Liapunov function. Figure 11.2 is a block diagram of how such a function is generated

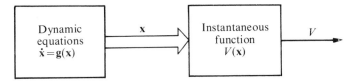

Fig. 11.2. Block diagram of a Liapunov function.

and Fig. 11.3 shows how its value might vary with time along a trajectory in a three-dimensional state space if the function had a form like

$$V(x) = x_1^2 + x_2^2 + x_3^2.$$

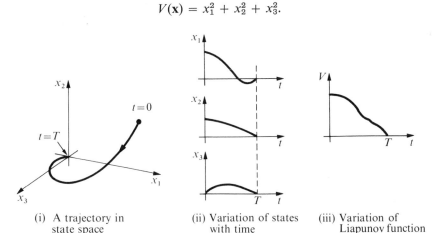

(i) A trajectory in state space

(ii) Variation of states with time

(iii) Variation of Liapunov function with time

Fig. 11.3. Example of how $V(x)$ varies in time.

This form is typical of Liapunov functions in that it has values which can be regarded as some measure of distance in state space from the equilibrium point at the origin.

The test for stability is based on the idea that an equilibrium point will be stable if motion on all trajectories near to it is such that distance from it tends to decrease. This idea is stated mathematically in the following theorem.

The equilibrium point at the origin is globally, asymptotically stable if a positive definite function $V(\mathbf{x})$ exists having time derivative $\dot{V}(\mathbf{x})$ which is negative definite along all trajectories.†

The requirement that the Liapunov function be positive definite assures that there is some sense in which it measures distance from the origin, and the requirement that its derivative also be negative definite assures that this 'distance' decreases to zero. This requirement on $\dot{V}$ can be relaxed to a requirement that $\dot{V}$ be negative semi-definite‡ provided that the set of points in state space where $\dot{V}$ is zero, that is where 'distance' from the origin does not decrease, excludes all trajectories and equilibrium points other than the origin.

The time derivative $\dot{V}$ of the Liapunov function is given by

$$\dot{V} \equiv \frac{\mathrm{d}V}{\mathrm{d}t} = \frac{\partial V}{\partial x_1}\frac{\mathrm{d}x_1}{\mathrm{d}t} + \cdots + \frac{\partial V}{\partial x_n}\frac{\mathrm{d}x_n}{\mathrm{d}t}$$

$$= \frac{\partial V}{\partial x_1}g_1 + \cdots + \frac{\partial V}{\partial x_n}g_n.$$

It can be found directly from a specified set of dynamics $\mathbf{g}(\mathbf{x})$ and Liapunov function $V(\mathbf{x})$ without solving any differential equations. This is the sense in which Liapunov's second method is direct.

An example of the use of the method is provided by applying the theorem to the second-order linear systems with differential equations (as in Section 10.1)

$$\dot{x}_1 = x_2,$$
$$\dot{x}_2 = -x_1 - 2\zeta x_2,$$

which is known (Section 3.2) to be globally, asymptotically stable for all positive values of the damping ratio ζ and unstable for all negative values of ζ. Choice of a suitable Liapunov function is not entirely straightforward; the simple quadratic function

$$V(\mathbf{x}) = x_1^2 + x_2^2$$

† A function $V(\mathbf{x})$ is positive (negative) definite if $V(\mathbf{0}) = 0$ and $V(\mathbf{x}) > 0$ ($V(\mathbf{x}) < 0$) for all $\mathbf{x} \neq 0$.
‡ A function $V(\mathbf{x})$ is positive (negative) semi-definite if $V(\mathbf{0}) = 0$ and $V(\mathbf{x}) \geqslant 0$ ($V(\mathbf{x}) \leqslant 0$).

is positive definite, but its derivative

$$\dot{V} = \frac{\partial V}{\partial x_1} x_2 + \frac{\partial V}{\partial x_2} (-x_1 - 2\zeta x_2)$$

$$= -4\zeta x_2^2$$

is only semi-definite, so this function could be used with the second theorem about stability but cannot be used to prove asymptotic stability. A more useful Liapunov function has the quadratic form

$$V(\mathbf{x}) = ax_1^2 + bx_2^2 + 2cx_1x_2,$$

which is positive definite provided that†

$$a > 0, \qquad b > 0, \qquad ab > c^2.$$

The coefficients a, b, and c can be chosen so that the derivative

$$\dot{V} = -2cx_1^2 + (2a - 2b - 4c\zeta)x_1x_2 + (2c - 4\zeta b)x_2^2$$

has the simple negative definite form

$$\dot{V} = -x_1^2 - x_2^2.$$

Values of the coefficients to achieve this are

$$a = \zeta + \frac{1}{2\zeta}, \qquad b = \frac{1}{2\zeta}, \qquad c = \frac{1}{2};$$

and with these values $V(\mathbf{x})$ is positive definite for all positive ζ and negative definite for all negative ζ. This Liapunov function therefore satisfies the conditions of the theorem about asymptotic stability when ζ is positive and so the method confirms what is already known about this particular linear system.

The general continuous-time linear system

$$\dot{\mathbf{x}} = \mathbf{A}\mathbf{x},$$

which is known to be globally, asymptotically stable if all eigenvalues of $\mathbf{A}$ have negative real parts, can be similarly discussed using a quadratic form of Liapunov function:

$$V(\mathbf{x}) = \mathbf{x}'\mathbf{P}\mathbf{x},$$

with derivative

$$\dot{V} = \mathbf{x}'\mathbf{P}\dot{\mathbf{x}} + \dot{\mathbf{x}}'\mathbf{P}\mathbf{x}$$

$$= \dot{\mathbf{x}}'\mathbf{P}\mathbf{A}\mathbf{x} + \mathbf{x}'\mathbf{A}'\mathbf{P}\mathbf{x}$$

† Conditions for a quadratic form to be positive (or negative) definite are known as Sylvester's conditions; see, for example, Ogata (1967) Section 5.5.

that can be written

$$\dot{V} = \mathbf{x}'\mathbf{Q}\mathbf{x},$$

where

$$\mathbf{P}\mathbf{A} + \mathbf{A}'\mathbf{P} = \mathbf{Q}. \tag{11.1}$$

The condition for asymptotic stability is that $\mathbf{P}$ be positive definite and $\mathbf{Q}$ be negative definite. It can be proved† that this condition is satisfied by equation (11.1) provided that all eigenvalues of $\mathbf{A}$ have negative real parts, so the method again confirms the known result.

The Liapunov theorem about asymptotic stability can be modified to describe other types of stability. For example stability, rather than asymptotic stability, can be investigated using the following theorem.

A system is globally stable if a positive definite function $V(\mathbf{x})$ exists having time derivative $\dot{V}(\mathbf{x})$ which is negative semi-definite and if $V(\mathbf{x})$ goes to infinity as $\mathbf{x}$ goes to infinity.

Here the semi-definite derivative $\dot{V}$ implies that there may be trajectories, such as stable limit cycles, along which 'distance' from the origin does not decrease. The boundary condition that $V(\mathbf{x})$ goes to infinity with $\mathbf{x}$ ensures that trajectories along which $\dot{V}$ is zero do not go to infinity.

Another modification of the theorem about asymptotic stability is

A system is unstable if a positive definite function $V(\mathbf{x})$ exists having time derivative $\dot{V}(\mathbf{x})$ which is positive definite along all trajectories.

The foregoing theorems about types of global stability can be used to investigate local properties in a bounded region surrounding the equilibrium point at the origin. For this purpose it is generally necessary that the boundary of the region investigated should coincide with a surface on which the Liapunov function $V(\mathbf{x})$ takes a constant value.

The main limitation on the usefulness of Liapunov's direct method is that suitable Liapunov functions cannot be found for most non-linear systems. The theorems state what properties the functions should have but give no indication of how to find them. Inability to find a function satisfying a stability theorem is no proof of instability, and in this sense the method can be regarded as giving sufficient but not necessary conditions for discussing stability. No truly general and certain method of finding Liapunov functions for non-linear systems exists; several methods have been suggested which can be used to prove stability, or sometimes instability, over restricted regions of a state space, but discussion of them is beyond the scope of this book.‡

There is however one particular class of non-linear systems where Liapunov

† See for example Willems (1970) Appendix C.
‡ See Schultz (1965), Barnett and Storey (1970) or Willems (1970).

functions can be found. This class includes closed-loop feedback systems with the block diagram of Fig. 11.4(i) where the non-linearity satisfies conditions

$$wz(w) > 0, \quad z \neq 0,$$
$$z(0) = 0,$$

ensuring that it lies in the first and third quadrants as in Fig. 11.4(ii). With these conditions a function of the form

$$\int_0^w z(w)\,dw$$

is positive definite and can be used as a term in the Liapunov function describing the non-linearity. It can be used, for example, to investigate the

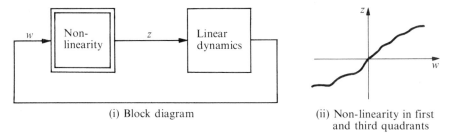

(i) Block diagram

(ii) Non-linearity in first and third quadrants

FIG. 11.4. Feedback system with single non-linearity.

general non-linear position-control system with velocity feedback, shown in Fig. 11.5, where the torque u is related to the actuating signal w by any non-linear (or linear) characteristic lying in the first and third quadrants. The

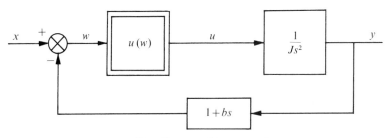

FIG. 11.5. Non-linear position-control system.

error e and its derivative are convenient state variables here (as in Chapter 10),

$$x_1 \equiv e \equiv x - y, \qquad x_2 \equiv \dot{e}$$

and when the desired value x is constant the differential equations of motion are

$$\dot{x}_1 = x_2$$

$$\dot{x}_2 = -\frac{1}{J} u(x_1 + bx_2).$$

A suitable positive-definite Liapunov function is

$$V(x) = \tfrac{1}{2}x_2^2 + \frac{1}{J} \int_0^{x_1 + bx_2} u(w)\, dw\,,$$

and its derivative

$$
\begin{aligned}
\dot{V} &= \frac{\partial V}{\partial x_1} x_2 + \frac{\partial V}{\partial x_2}\left(-\frac{1}{J} u(x_1 + bx_2)\right) \\
&= \frac{1}{J} u(x_1 + bx_2)x_2 - \left(x_2 + \frac{b}{J} u\,(x_1 + bx_2)\right)\left(\frac{1}{J} u(x_1 + bx_2)\right) \\
&= -\frac{b}{J^2}\left(u(x_1 + bx_2)\right)^2
\end{aligned}
$$

is negative semi-definite for all positive b and positive semi-definite for all negative b. The set of points in state space where $\dot{V}$ is zero is the line

$$x_1 + bx_2 = 0$$

which does not coincide with any trajectory. The Liapunov theorems therefore show that the system is asymptotically stable or unstable depending on whether the velocity-feedback coefficient is positive or negative. This result is independent of the exact nature of the non-linearity; it includes cases discussed by phase-plane analysis in Chapter 10 having the various torque characteristics shown in Fig. 11.6.

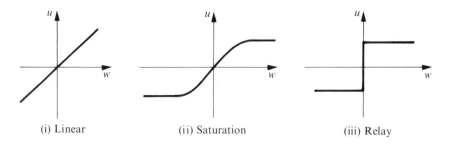

(i) Linear   (ii) Saturation   (iii) Relay

FIG. 11.6. Torque characteristics in first and third quadrants.

Liapunov functions having integral terms can be used to investigate stability of any feedback system in the class of Fig. 11.4 provided that the non-linearity is in the first and third quadrants and the linear dynamics are stable. This is known as Lure's method. The results are however more conveniently expressed in terms of frequency-response functions by Popov's method which translates the sufficient condition for asymptotic stability into a form similar to that of the Nyquist criterion of Section 3.4. Details of Lure's method and the derivation of the Popov criterion are beyond the scope of this book.† The class of feedback systems in Fig. 11.4 is also the subject of analysis using the describing-function method, as discussed in Section 11.3.

For a discrete-time system

$$\mathbf{x}(i + 1) = \mathbf{G}\big(\mathbf{x}(i)\big)$$

having an equilibrium point at the origin

$$\mathbf{G}(0) = \mathbf{0},$$

a positive definite function V will provide, as for continuous-time systems, some measure of distance from the equilibrium point. Questions about whether this distance increases or decreases are answered using the difference

$$\Delta V = V\big(\mathbf{x}(i + 1)\big) - V\big(\mathbf{x}(i)\big)$$

in place of the time-derivative $\dot{V}$ for continuous time. This difference is given, at any point in state space, by

$$\Delta V = V\big(\mathbf{G}(\mathbf{x})\big) - V(\mathbf{x}),$$

and then the Liapunov stability theorems are the same as for continuous time but with ΔV replacing $\dot{V}$ everywhere.

The general discrete-time system

$$\mathbf{x}(i + 1) = \mathbf{A}\mathbf{x}(i)$$

can be discussed using a quadratic Liapunov function, as before,

$$V(\mathbf{x}) = \mathbf{x}'\mathbf{P}\mathbf{x},$$

with difference

$$\Delta V = (\mathbf{A}\mathbf{x})'\mathbf{P}\mathbf{A}\mathbf{x} - \mathbf{x}'\mathbf{P}\mathbf{x}$$

that can be written

$$\Delta V = \mathbf{x}'\mathbf{Q}\mathbf{x},$$

where

$$\mathbf{A}'\mathbf{P}\mathbf{A} - \mathbf{P} = \mathbf{Q}. \qquad (11.2)$$

† For the Lure method see Lasalle and Lefschetz (1961) Chapter 3 or Aggarwal (1972) Section 3.5 where the Popov criterion is also discussed. For the Popov criterion see Hsu and Meyer (1968) Chapter 10.

The condition for asymptotic stability is that **P** be positive definite and **Q** be negative definite; this condition is satisfied by equation (11.2) provided that all eigenvalues of **A** have magnitude less than unity.

General methods of finding Liapunov functions for non-linear, discrete-time systems scarcely exist.

Problems 11.1–11.4

11.3. Describing functions

The describing-function method provides a way of investigating limit cycles in the class of feedback systems, illustrated in Fig. 11.4(i), having a single non-linearity. It differs from the methods of Luré and Popov in that the non-linearity is not restricted to the first and third quadrants, and that the method is concerned only with limit cycles not with any other sort of stability, and that in addition to establishing whether or not a stable limit cycle exists it gives approximate results about the size and period of limit cycles. Similar information can be derived by phase-plane analysis for the special case where dynamics are of second order, but the describing function method is applicable regardless of the order of the system dynamics.

The method is introduced by considering for example the relay control system in Fig. 11.7, which has third-order dynamics. If there were a stable

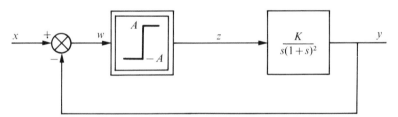

FIG. 11.7. Relay control system with third-order dynamics.

limit cycle oscillation at some frequency ω the signal z at the relay output would be a periodic square wave, and if the desired value x were zero it would be the symmetrical square wave, shown in Fig. 11.8, which can be represented by the Fourier series

$$z(t) = \frac{4A}{\pi}\left(\sin \omega t + \tfrac{1}{3}\sin 3\omega t + \tfrac{1}{5}\sin 5\omega t + \cdots\right).$$

The corresponding signal y at the output of the linear dynamics is the sum of these Fourier components after they have passed through the frequency-response function

$$G(j\omega) = \frac{K}{j\omega(1 + j\omega)^2}.$$

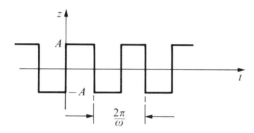

FIG. 11.8. Waveform $z(t)$ in limit cycle oscillation.

The fundamental component y has amplitude Y_1 given by

$$Y_1 = \frac{4A}{\pi} \frac{K}{\omega(1 + \omega^2)},$$

and the next component, the third harmonic, has amplitude Y_3,

$$Y_3 = \frac{4A}{\pi} \frac{K}{9\omega(1 + 9\omega^2)}$$

$$= \frac{Y_1}{9} \frac{1 + \omega^2}{1 + 9\omega^2},$$

which is always less than $Y_1/9$ and if the frequency ω has the value unity would be $Y_1/45$. Thus the linear dynamics tend to filter out high-frequency components of z, and the signal y can be approximated by its fundamental component; the waveforms in the limit cycle would be as shown in Fig. 11.9.

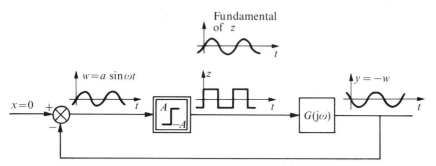

FIG. 11.9. Waveforms in system during limit cycle oscillation.

The signal w at the input to the relay must be in phase with the relay output z and its fundamental component, so w can be written

$$w = a \sin \omega t,$$

and the signal y must then be

$$y = -a \sin \omega t.$$

This analysis shows that the oscillation frequency ω is such that the phase shift $\underline{/G(j\omega)}$ due to the linear dynamics is $-\pi$, which implies that

$$-\left(\frac{\pi}{2} + 2\tan^{-1}\omega\right) = -\pi,$$

so

$$\tan^{-1}\omega = \pi/4$$

and

$$\omega = 1.$$

Having found the frequency ω of the oscillation, and noting that its value unity is such that the third harmonic of y has amplitude one forty-fifth of the fundamental so that the approximation of y by its fundamental is justifiable, the amplitude a of the oscillation in y is approximated by the amplitude Y_1 of its fundamental component at frequency ω:

$$a \simeq \frac{2AK}{\pi}.$$

Thus it has been shown that there is a stable limit cycle in the system of Fig. 11.7 and that the resulting waveform in y is

$$y \simeq \frac{2AK}{\pi}\sin t.$$

The above approximate analysis can be generalized for any system in the class illustrated in Fig. 11.4(i). The essential feature of the method is the assumption that the signal z at the output of the non-linearity can be represented by its fundamental component only. This assumption is justified by the tendency of most real linear dynamical systems to filter out high-frequency components. It leads to estimates of frequencies and amplitudes which are usually accurate to within 10% of their true values.

The method always assumes that the input w and the output z of the non-linearity are periodic, as they must be in a limit cycle oscillation. Under these conditions the effect of the non-linearity on the fundamental component is represented by a *describing function* F which is a complex number, like a frequency-response function, having modulus M and argument ϕ,

$$F = M\,e^{j\phi}$$

defined by

$$M \equiv \frac{\text{amplitude of fundamental component of } z}{\text{amplitude of fundamental component of } w},$$

$$\phi \equiv \text{phase angle by which fundamental of } z \text{ leads}$$
$$\text{fundamental of } w,$$

such that when

$$\text{fundamental of } w = a \sin \omega t$$

then the

$$\text{fundamental of } z = Ma \sin (\omega t + \phi).$$

The non-linearity is assumed to exclude any dynamic effects, so the describing function does not depend on the frequency ω as does a linear frequency-response function $G(j\omega)$. Instead the describing function usually depends on the amplitude a of the signal w at the input to the non-linearity; this dependence can be indicated by the notation

$$F(a) = M(a)\,e^{j\phi(a)}.$$

The describing function of the ideal relay in the above example is

$$F(a) = \frac{4A}{\pi a},$$

a real number with zero argument $\phi(a)$ because the ideal relay introduces no phase lag.

The general describing-function method is to assume that the desired value x is zero, and to replace the non-linearity by its describing function $F(a)$, as

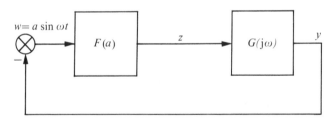

Fig. 11.10. Replacing the non-linearity by its describing function.

in Fig. 11.10, and then to recognize that the condition for sustained oscillations in the resulting system is that the signal

$$w = a \sin \omega t$$

must propagate around the system without distortion so that

$$y = -a \sin \omega t.$$

This condition will be satisfied if the amplitude a and frequency ω of the oscillation are such that

$$F(a)G(j\omega) = -1$$

or

$$G(j\omega) = -1/F(a). \qquad (11.3)$$

Equation (11.3) can, if necessary, be solved graphically using a Nyquist diagram. Figure 11.11 illustrates the graphical method for the relay system of

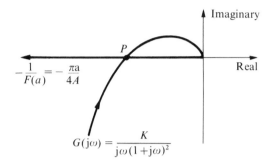

FIG. 11.11. Graphical solution of equation (11.3) for relay system.

the foregoing example. The point P where the loci $G(j\omega)$ and $-1/F(a)$ intersect must correspond to a frequency ω and an amplitude a such that

$$\underline{/G(j\omega)} = -\pi, \qquad |G(j\omega)| = \frac{\pi a}{4A}$$

as before.

To find the describing function of a given non-linearity $z(w)$ it is necessary to find the fundamental component in the Fourier series representing the periodic signal $z(a \sin \omega t)$ at the non-linearity's output when its input is sinusoidal

$$w = a \sin \omega t.$$

For symmetrical† non-linearities the Fourier series is

$$z = \sum_{k=1}^{\infty} (a_k \sin k\omega t + b_k \cos k\omega t),$$

and the coefficients a_k and b_k are given by

$$a_k = \frac{1}{\pi} \int_0^{2\pi} z(a \sin \omega t) \sin k\omega t \, \mathrm{d}(\omega t),$$

$$b_k = \frac{1}{\pi} \int_0^{2\pi} z(a \sin \omega t) \cos k\omega t \, \mathrm{d}(\omega t).$$

† Systems with asymmetric non-linearities, when there is a non-zero constant b_0 in the Fourier series, can be investigated using the dual-input describing function. See West (1960) Chapters 10 and 11, Gibson (1963) Sections 9.13, 9.14, Gelb and Vander Velde (1968) Chapter 6, or Hsu and Meyer (1968) Chapter 7.

The fundamental component is determined by the coefficients a_1 and b_1; the describing function then has modulus

$$M(a) = \sqrt{(a_1^2 + b_1^2)}/a$$

and argument

$$\phi(a) = \tan^{-1} b_1/a_1.$$

Using this method it follows, for example, that the relay with hysteresis shown in Fig. 11.12(i) has describing function

$$F(a) = \frac{4A}{\pi a}\, e^{j\phi(a)}$$

with

$$\phi(a) = -\sin^{-1} B/a$$

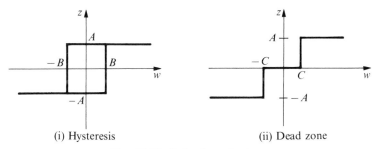

(i) Hysteresis   (ii) Dead zone

FIG. 11.12. Relay imperfections.

such that the locus $-1/F(a)$ on the Nyquist diagram is the straight line shown in Fig. 11.13. Similarly the relay with dead zone shown in Fig. 11.12(ii) has describing function

$$F(a) = 0, \qquad\qquad a \leqslant C$$

$$F(a) = \frac{4A}{\pi a^2}\, \sqrt{(a^2 - C^2)}, \quad a > C$$

which is always real and takes values as illustrated in Fig. 11.14.

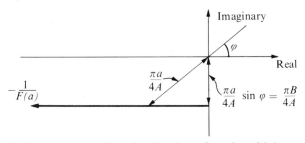

FIG. 11.13. Inverse describing function locus for relay with hysteresis.

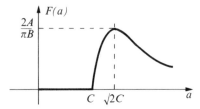

FIG. 11.14. Describing function for relay with dead zone.

When a describing function is not single valued, as the dead-zone describing function of Fig. 11.14 is not, the equation (11.3) specifying amplitude and frequency of oscillations can have more than one solution. For example, if the relay in the system of Fig. 11.7 had a dead zone the Nyquist diagram for solving equation (11.3) would be as shown in Fig. 11.15. The two points P_1

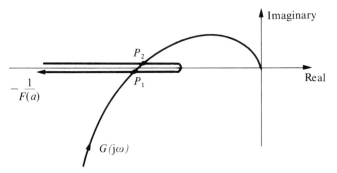

FIG. 11.15. Nyquist diagram for relay system with dead zone.

and P_2 where the loci $G(j\omega)$ and $-1/F(a)$ intersect both correspond to the same frequency as before, determined by the condition

$$\underline{/G(j\omega)} = -\pi.$$

But two solutions for the amplitudes follow from the condition

$$|G(j\omega)| = \frac{\pi a}{4A\sqrt{(a^2 - C^2)}},$$

which for this particular system leads to a quadratic equation in a^2,

$$\pi^2 a^4 - 4A^2 K^2 a^2 + 4A^2 K^2 C^2 = 0,$$

having the solutions

$$a_1, a_2 = \frac{\sqrt{(2)}\,AK}{\pi} \left(1 \pm \sqrt{(1 - \pi^2 C^2/A^2 K^2)}\right)^{1/2}.$$

One of these solutions corresponds to a stable limit cycle and the other to an unstable limit cycle. The stability here usually depends on the sense of intersection of the $G(j\omega)$ and $-1/F(a)$ loci. The stable intersection is that having the same sense as the intersection specifying the stable limit cycle in Fig. 11.11; namely P_1 which corresponds to the higher of the two amplitudes.

It is possible that the loci shown in Fig. 11.15 might not intersect at all, as

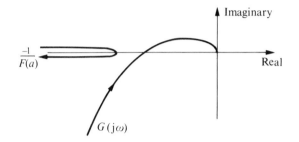

Fig. 11.16. No oscillation in relay system with dead zone.

shown in Fig. 11.16. The condition for the loci to intersect, giving real solutions to the equation for a_1 and a_2, is

$$K > 2\pi C/A.$$

This is the limiting value of gain in the system below which there will be no oscillations because they cannot build up beyond the dead zone of the relay.

The above methods of describing-function analysis can be used to investigate limit cycles in any system having the form of Fig. 11.4(i). Describing functions for some common, practical non-linearities are tabulated in Fig. 11.17.

Describing functions for other non-linearities which can be expressed as a weighted sum of two component non-linearities,

$$z(w) = k_1 z_1(w) + k_2 z_2(w),$$

can be shown† to be given by the weighted sum of the describing functions of their components:

$$F(a) = k_1 F_1(a) + k_2 F_2(a).$$

Problems 11.5–11.11

† See, for example, Gelb and Vander Velde (1968) Section 2.5.

Non-linearity	Describing function $F(a)$
Ideal relay	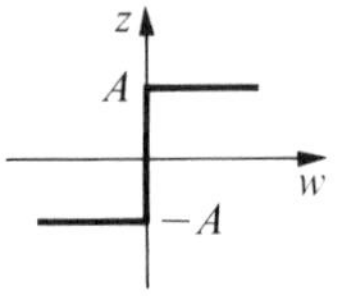 $\dfrac{4A}{\pi a}$
Relay with hysteresis $(B<a)$	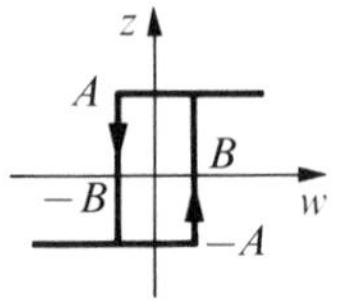 $\dfrac{4A}{\pi a}\left\{\sqrt{\left(1-(B/a)^2\right)}-\mathrm{j}\dfrac{B}{a}\right\}$
Relay with dead zone $(C<a)$	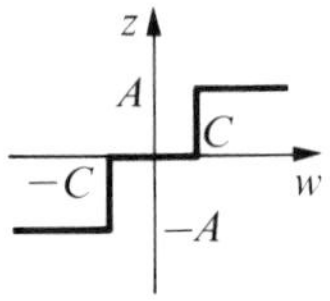 $\dfrac{4A}{\pi a}\sqrt{\left(1-(C/a)^2\right)}$
Relay with hysteresis and dead zone $(C<a)$	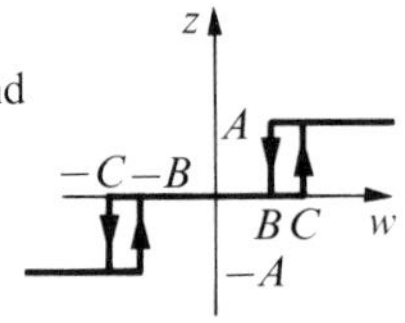 $\dfrac{2A}{\pi a^2}\left\{\sqrt{(a^2-B^2)}+\sqrt{(a^2-C^2)}+\right.$ $\left.+\mathrm{j}(C-B)\right\}$
Saturation $(D<a)$	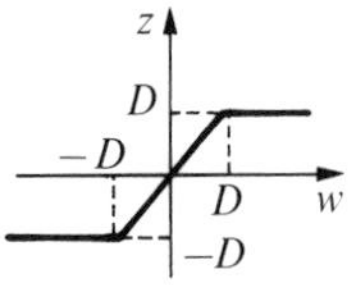 $\dfrac{2}{\pi}\left\{\sin^{-1}\dfrac{D}{a}+\dfrac{D}{a}\sqrt{\left(1-(D/a)^2\right)}\right\}$
Dead zone $(C<a)$	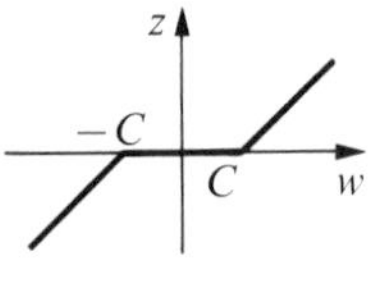 $\dfrac{2}{\pi}\left\{\cos^{-1}\dfrac{C}{a}-\dfrac{C}{a}\sqrt{\left(1-(C/a)^2\right)}\right\}$
Backlash $(B<a)$	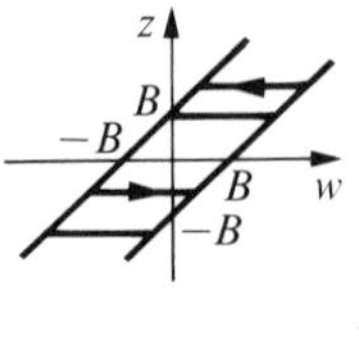 $\dfrac{1}{\pi}\left\{\cos^{-1}\left(\dfrac{2B}{a}-1\right)+\dfrac{2B}{a}\left(\dfrac{2B}{a}-1\right)\times\right.$ $\left.\times\sqrt{\left(\dfrac{a}{B}-1\right)}\right\}-\mathrm{j}\dfrac{4B}{\pi a}\left(1-\dfrac{B}{a}\right)$

FIG. 11.17. Describing functions of some common non-linearities.

11.4. Bibliography

Liapunov's own account of his methods for investigating stability is available in a French translation of the Russian original:

LIAPUNOV (1892).

More recent English-language discussion of the direct method together with descriptions of methods which have been proposed for finding Liapunov functions are in

KALMAN and BERTRAM (1960),
LA SALLE and LEFSCHETZ (1961),
THALER and PASTEL (1962) Chapter 11,
GIBSON (1963) Chapter 8,
DORF (1965) Chapter 4,
SCHULTZ (1965),
DE RUSSO, ROY, and CLOSE (1967) Chapter 7,
OGATA (1967) Chapter 8; (1970) Chapter 15,
SCHULTZ and MELSA (1967) Chapter 5,
HSU and MEYER (1968) Chapter 9,
BARNETT and STOREY (1970),
WILLEMS (1970),
AGGARWAL (1972) Chapter 3.

Applications of Liapunov's direct method to discrete-time systems are mentioned in

KALMAN and BERTRAM (1960) Part II,
MONROE (1962) Chapters 23 and 24,
DORF (1965) Section 6.7,
FREEMAN (1965) Sections 7.8–7.12,
LINDORF (1965) Chapter 8,
BROMBERG (1967) Chapter 3,
OGATA (1967) Section 8.7,
VIDAL (1969) Section 6.2,
KUO (1970) Section 5.6,
TAKAHASHI, RABINS, and AUSLANDER (1970) Section 4.4.

Describing-function analysis is the subject of

GELB and VANDER VELDE (1968),
and is discussed in many books on automatic control, for example,

TRUXAL (1955) Chapter 10,
MURPHY (1959) Chapter 10,
WEST (1960) Chapters 8 and 9,
GRAHAM and McRUER (1961) Chapters 3 and 4,
THALER and PASTEL (1962) Chapter 4,
GIBSON (1963) Chapter 9,
SHINNERS (1964) Sections 7.7–7.9,
KUO (1967) Sections 11.6–11.8,
HSU and MEYER (1968) Chapter 6,
TAKAHASHI, RABINS, and AUSLANDER (1970) Section 12.2.

11.5. Problems

11.1. Use a quadratic Liapunov function based on the unit matrix $(\mathbf{P} = \mathbf{I})$ to prove that a linear system represented by

$$\dot{\mathbf{x}} = \mathbf{Sx},$$

where $\mathbf{S}$ is a diagonal matrix of distinct real eigenvalues, is asymptotically stable if and only if all the eigenvalues are negative.

11.2. Use a Liapunov function having time derivative $-\mathbf{x}'\mathbf{x}$ to establish conditions for asymptotic stability of the general, second-order linear system

$$\dot{x}_1 = a_{11}x_1 + a_{12}x_2,$$
$$\dot{x}_2 = a_{21}x_1 + a_{22}x_2.$$

11.3. Use the Liapunov function

$$V = x_1^2 + x_2^2$$

(i) to prove that a system having differential equations

$$\dot{x}_1 = -2x_1 + x_1x_2,$$
$$\dot{x}_2 = -x_2 + x_1x_2,$$

is asymptotically stable when $x_1 < 1$ and $x_2 < 2$. Sketch a phase portrait of the system,

(ii) to establish a region of asymptotic stability for a system having differential equations

$$\dot{x}_1 = -2x_1 + x_1x_2^2,$$
$$\dot{x}_2 = -x_2 + x_2x_1^2.$$

Compare this result with the phase portrait of Problem 10.10.

11.4. Use the Liapunov function

$$V = x_1^2 + x_2^2,$$

to investigate stability of a system having differential equations

$$\dot{x}_1 = x_2 - ax_1(x_1^2 + x_2^2),$$
$$\dot{x}_2 = -x_1 - ax_2(x_1^2 + x_2^2).$$

The result of this analysis differs from the result of approximate analysis using equations linearized about the equilibrium point; which result is stronger?

11.5. Use the describing-function method to estimate the frequency and amplitude of oscillations of a closed-loop system in which the output of a perfect relay has magnitude $\pm A$ and drives a linear controlled object with transfer function:

$$G(s) = \frac{K}{s(1 + sT_1)(1 + sT_2)}.$$

The input to the relay is the difference between controlled-object output and a reference signal that can be assumed to be zero.

11.6. Derive the describing function for an idealized saturating characteristic $z = f(w)$ defined by

$$\begin{cases} z = Kw, & -\dfrac{A}{K} \leqslant w \leqslant \dfrac{A}{K} \\[2mm] z = -A, & w < -\dfrac{A}{K} \\[2mm] z = A, & w > \dfrac{A}{K}. \end{cases}$$

11.7. In the control system shown in Fig. 11.18 the controller is intended to be a

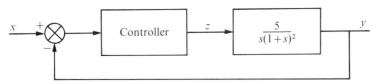

FIG. 11.18.

linear error amplifier such that

$$z = K(x - y),$$

where

$$K = 0\cdot04.$$

The system is inadvertently constructed with the value of K 30 times greater than intended. It is then found that for zero input ($x = 0$) an oscillation builds up until the controller output saturates at $z = \pm 60$. Show that an estimate of the amplitude a of the oscillation in the system output y satisfies the equation

$$6 \sin^{-1} \frac{50}{a} + \frac{300}{a} \sqrt{\left(1 - \frac{2500}{a^2}\right)} = \pi.$$

11.8. Derive an expression for the describing function $F(a)$ of a relay having both hysteresis and dead zone as shown in Fig. 11.19. Make a drawing using polar

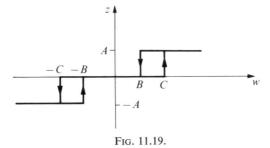

FIG. 11.19.

coordinates to show how the complex number $-1/F(a)$ varies in the Argand diagram when values of the relay parameters are

$$A = \pi/2, \quad B = 1, \quad C = 2.$$

(Calculate $F(a)$ for $a = 2, 2·1, 2·5, 3, 4, ...$)

11.9 The relay of Problem 11.8 is driven by the error in a feedback-control system where the transfer function $G(s)$ of the controlled process is specified by the following frequency-response test data:

ω	1	2	3	5	10	20
$|G(j\omega)|$	2·2	1·95	1·85	1·82	1·8	1·8
$\underline{/G(j\omega)}$	$-128°$	$-144°$	$-156°$	$-166°$	$-174°$	$-182°$

Estimate the amplitude and frequency of the steady-state oscillation of the system.

11.10. The rolling motion of a space rocket is controlled by a relay system that fires one of two control jets as shown in Fig. 11.20. The moment of inertia of the

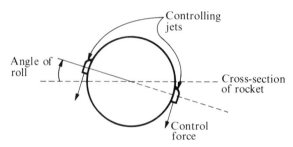

FIG. 11.20.

rocket about its roll axis is $0·2$ kg m². Each jet is at a radial distance of $0·6$ m from the roll axis and can develop a force of 14 N. The firing of the jets is controlled by a hydraulic relay actuated by a hydraulic pressure signal. This signal is composed of a roll-angle feedback signal and a roll-rate feedback signal, which are summed and amplified by a hydraulic amplifier that introduces two lags having time constants $0·05$ s and $0·01$ s. The feedback gain from roll angle to hydraulic signal is 420 kN m^{-2} rad^{-1}, and from roll rate to hydraulic signal is 42 kN m^{-2} rad^{-1} s. The hydraulic relay has hysteresis of ± 5 kN m^{-2}.

Estimate the amplitude and frequency of the steady-state oscillation in the system.

(This problem gives rise to a quintic equation which need be solved only approximately.)

11.11. Prove that the describing function of a non-linearity $z(w)$ which can be expressed as a weighted sum of two component nonlinearities

$$z(w) = k_1 z_1(w) + k_2 z_2(w)$$

is the weighted sum of the describing functions of the components.

Find the describing function of the symmetric non-linearity in Fig. 11.21.

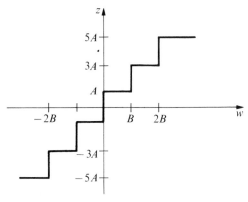

FIG. 11.21.

Part III
Systems with uncertainty

12. Probability and random processes

12.1. Introduction

The outstanding feature of real control systems, that they must be feedback systems, does not follow from the deterministic theory of previous chapters. The reason for using feedback is to ameliorate the effects of uncertainty in real systems, so a control theory which satisfactorily describes real control systems must take account of uncertainty. Another well-known feature of real controllers, that they often include integrators as discussed in Chapters 2 to 5 although the deterministic optimal control theory of Chapters 8 and 9 implies that the control should be a function of the error and its time derivatives, is also explained by taking account of uncertainty.

Uncertainty can arise in two different forms:

(i) uncertainty about the value of unknown constants such as the thermal capacity of a temperature-controlled oven and its contents;

(ii) uncertainty associated with unpredictable time-varying quantities, called *stochastic* variables, such as the position of a moving target for a steerable aerial, or the torque on the aerial due to the wind, or measurement noise which is usually an unavoidable side-effect of feedback.

Uncertainty is described for purposes of mathematical analysis by probability theory; unknown constants by well-known probability models which are summarized in Sections 12.2 and 12.3, and stochastic variables by more complex mathematical models known as random processes which are presented in Sections 12.4 to 12.6. These standard descriptions of uncertainty provide the mathematical language used in Chapters 13 and 15 to introduce ideas about uncertainty into control theory. They also provide the mathematical framework for interpreting measurements (statistics) relating to uncertain quantities, that is for processing feedback signals, as discussed in Chapter 14.

12.2. Probability

Mathematical descriptions of uncertainty are based on the concept of *probability* which is used as a measure of uncertainty; an event which is certain to occur is said to have probability of value one and an event which is certain not to occur is said to have probability of value zero. An uncertain event, for example the event that it will snow on Christmas Day or the event that a controlled variable is within one per cent of its desired value, has probability greater than zero and less than one. The assignment of a specific value for the probability of any given uncertain event must be preceded by a formal definition of the concept of probability.

The question of how to define probability has been notoriously controversial and it would seem to be impossible to give a satisfactory practical definition in terms of a physical model or a real experiment. Two of the most plausible attempts at a practical definition are the following:

(i) *Classical or* a priori *probability*
If m out of n equally likely events have an attribute E, the probability of the occurrence of an event with attribute E is $P(E) = m/n$.

(ii) *Statistical or empirical or relative-frequency probability*
If an experiment is repeated n times and if the result has attribute E on m occasions, the probability of the result of an experiment having attribute E is
$$P(E) = \lim_{n \to \infty} (m/n).$$

Objections to these definitions are well known. Classical probability depends on the phrase 'equally likely', which can only be defined as 'having equal probabilities' so that the definition is circular. Statistical probability depends on unjustifiable assumptions that the ratio m/n tends to a limit as n goes to infinity and that the limit is the same for all sets of n repetitions of the experiment.

The reason why probability is so difficult to define is because it is a concept that appears to have meaning both in the real world of observable experimental results and as a mathematical term used in purely mathematical models. The difficulty in defining probability reflects the difficulty in defining the exact relationship, if any, between pure mathematics and the real world: it is a problem in philosophy or metaphysics and beyond the scope of this book.

In the absence of any definition of the relationship between pure mathematics and the real world, it is customary to work on the assumption that real practical problems can in some sense be represented by mathematical models, as for example when the dynamics of a real controlled process are represented by a linear differential equation. The justification for this assumption is that it leads to usable design procedures. When the real practical problem contains uncertainty the mathematical model uses a purely mathematical definition of probability.

(iii) *Mathematical or axiomatic probability*
If there is a set of m possible outcomes $\{E_1, E_2, \ldots, E_m\}$ of some experiment S, and if each outcome E_i has a number $P(E_i)$ associated with it, subject to the restrictions

$$P(E_i) \geqslant 0 \quad i = 1, \ldots, m$$

$$\sum_{i=1}^{m} P(E_i) = 1,$$

then the set of numbers $\{P(E_i)\}$ are the mathematical probabilities of the events $\{E_i\}$.

This definition, although physically meaningless, is mathematically consistent and provide a basis for useful mathematical models of uncertainty in real practical problems. The assignment of a specific value for the probability $P(E_i)$ of any real uncertain event E_i will depend on what assumptions are made about the relationship between the real practical problem and the mathematical model.

Mathematical probability is thus defined with reference to the set of all possible outcomes of an experiment. This set, sometimes called the sample space, must be defined as the first step in applying probability theory to any real problem; some examples are given below.

(i) Tossing a coin once has two possible outcomes, and so the sample space is a set with two members which might be labelled H and T, with probabilities which would be written $P(H)$ and $P(T)$.

(ii) Drawing one card from a standard pack leads to a set with 52 members.

(iii) Tossing a coin twice leads to a set with four members, HH, HT, TH, and TT.

(iv) Measuring the value of a continuous variable has an infinite number of possible outcomes. Probability theory is still applicable when the sample space is an infinite set, but it becomes necessary to use ideas of integration and differentiation which are not needed for finite sets.

It is sometimes convenient to discuss the probability of a compound event, for example the result of tossing a coin and simultaneously throwing a six-sided die. The relevant probability is called *joint probability* and the notation used is:

If A is one of the set of possible outcomes of an experiment S_1 and B is one of the set of possible outcomes of an experiment S_2, the joint probability that A and B result from a joint experiment S_1 and S_2 is written $P(A, B)$.

The sample space for joint probability is the set of all possible combinations of the outcomes of the elementary experiments S_1 and S_2; in the example of tossing a coin and throwing a die the joint sample space has twelve members

$(H \times 6) + (T \times 6)$. When the experiments S_1 and S_2 do not affect each other the joint probability is the product of the elementary probabilities:

$$P(A, B) = P(A)P(B), \tag{12.1}$$

and the experiments are said to be *statistically independent*. If the coin and the die are both unbiased and independent the probability of each possible combination should be $1/12$.

When two experiments are not independent and the result of one is influenced by the result of another it is convenient to discuss *conditional probability*, using the notation:

If the outcome B of an experiment S_2 is influenced by the outcome A of another experiment S_1, the conditional probability of B, given A, is written $P(B \mid A)$.

An example of conditional probability is given by considering the probability that two cards dealt at random from a standard pack are both aces. If A is the event that the first card is an ace, the probability of this event should be

$$P(A) = \frac{4}{52}.$$

If the first card is an ace there will then be only three aces left in the pack and so the conditional probability of the event B that the second card is an ace should be

$$P(B \mid A) = \frac{3}{51},$$

and the probability of the joint event that both cards are aces is

$$P(A, B) = \frac{4 \times 3}{52 \times 51}.$$

This is an example of a general result relating joint probability to conditional probability,

$$P(A, B) = P(A)P(B \mid A) = P(B)P(A \mid B). \tag{12.2}$$

Another general result gives the probability that B will be the outcome of experiment S_2 as the sum of all the conditional probabilities $P(B \mid E_i)$ for all the possible outcomes E_i of experiment S_1:

$$P(B) = \sum_{i=1}^{m} P(B \mid E_i)P(E_i).$$

This is sometimes written, using a slightly less rigorous notation,

$$P(B) = \sum_A P(B \mid A)P(A) \qquad (12.3a)$$

and there is a corresponding expression for $P(A)$, which can be written

$$P(A) = \sum_B P(A \mid B)P(B). \qquad (12.3b)$$

An important application of conditional probability is in the interpretation of a noisy measurement Y of a variable X; equation (12.2) can be rewritten to give an expression for the conditional probability of X given the measurement Y

$$P(X \mid Y) = \frac{P(Y \mid X)P(X)}{P(Y)} = \frac{P(Y \mid X)P(X)}{\sum_X P(Y \mid X)P(X)}. \qquad (12.4)$$

Equation (12.4) is known as *Bayes's rule*; it is the starting point for the statistical methods discussed in Chapter 14.

Problems 12.1–12.3

12.3. Probability distributions

Consider an idealized experiment which consists of n independent tosses of a biased coin having probabilities

$$P(H) = p, \qquad P(T) = 1 - p.$$

There are 2^n possible outcomes of this experiment, so the sample space has 2^n members, and the number of outcomes containing a combination of x heads and $n - x$ tails is

$$^nC_x = n!/x!(n - x)!$$

The probability of each such outcome (given by repeated application of equation (12.1)) is $p^x(1 - p)^{n-x}$ and so the probability of getting x heads in n tosses is

$$P(x) = {}^nC_x p^x(1 - p)^{n-x}. \qquad (12.5)$$

Equation (12.5) is an example of a *probability distribution*; it is a mathematical function which relates numerical values of probability to numerical values of a random variable x. The expression for the probability in equation (12.5) is a typical term of the binomial expansion for $\big(p + (1 - p)\big)^n$ and so this particular probability distribution is known as the *Binomial distribution* (also known as the Bernoulli distribution). Figure 12.1 illustrates the binomial distribution for the case when n is ten and p is one half.

Adding together the binomial probabilities for all possible values of the random variable x gives

$$\sum_{x=0}^{n} P(x) = \big(p + (1 - p)\big)^n = 1.$$

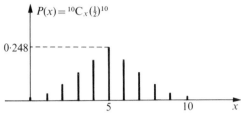

FIG. 12.1. Example of binomial distribution.

Every probability distribution must sum to unity, or normalize, in this way which is sometimes written

$$\sum_x P(x) = 1. \tag{12.6}$$

A random variable is any numerical function whose value is determined for each member of a sample space. It is not necessarily in one-to-one correspondence with members of the sample space (in the coin tossing example there are 2^n members but only $n + 1$ possible values of the random variable x); nor should it be confused with the probabilities of the members, those being a special set of numbers associated with the members. Random variables can be discrete-valued, like the number of heads in n tosses of a coin, or they can be continuous-valued like the measured value of a physical variable. Numerical values of probability for either type of random variable are described by *probability distribution functions*.

A probability distribution function is a mathematical function giving the probability that the value of a random variable x will be less than or equal to some specified number x^1. For discrete-valued random variables the probability distribution function is given by

$$F(x^1) = \sum_{x = -\infty}^{x^1} P(x). \tag{12.7}$$

Figure 12.2 shows the probability distribution function corresponding to the probability distribution of Fig. 12.1. Continuous-valued, random variables

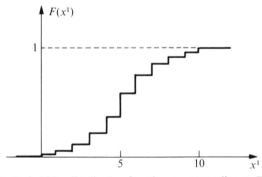

FIG. 12.2. Probability distribution function corresponding to Fig. 12.1.

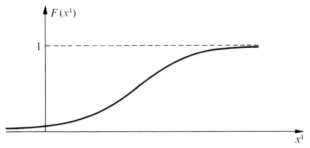

FIG. 12.3. Probability distribution function for a continuous-valued variable.

have smooth probability distribution functions like that illustrated in Fig. 12.3. Certain common properties of all probability distribution functions follow from the way they are defined:

(i) $F(x^1)$ increases (or remains constant) as x^1 increases.
(ii) $F(-\infty) = 0$, $F(\infty) = 1$.   (12.8)
(iii) $P(a < x \leqslant b) = F(b) - F(a)$.

For continuous-valued random variables, equation (12.8(iii)) leads to an expression for the probability that the value x of the random variable is close to a specified value x^1:

$$P(x^1 < x \leqslant x^1 + \delta x) = F(x^1 + \delta x) - F(x^1)$$
$$\simeq \delta x \left(\frac{\mathrm{d}F}{\mathrm{d}x}\right)_{x=x^1}. \qquad (12.9)$$

The derivative in equation (12.9) is written

$$\left(\frac{\mathrm{d}F}{\mathrm{d}x}\right)_{x=x^1} \equiv p(x^1),$$

and is known as the probability density of the continuous-valued random variable; it corresponds to the probability distribution P of a discrete-valued variable. Probability densities satisfy normalization and summation conditions in the form of integrals corresponding to the summations of equations (12.6), (12.7), and (12.8) for probability distributions:

$$\int_{-\infty}^{\infty} p(x)\,\mathrm{d}x = 1, \qquad (12.10)$$

$$\int_{-\infty}^{x^1} p(x)\,\mathrm{d}x = F(x^1), \qquad (12.11)$$

$$\int_{a}^{a} p(x)\,\mathrm{d}x = P(a < x \leqslant b). \qquad (12.12)$$

A particularly important probability density is the *normal distribution* (also known as the Gaussian distribution) defined by

$$p(x) = \frac{1}{\sigma\sqrt{(2\pi)}} \exp\left(-(x - m)^2/2\sigma^2\right) \tag{12.13}$$

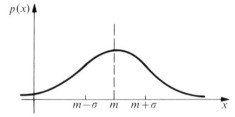

FIG. 12.4. Normal probability density.

and illustrated in Fig. 12.4; the significance of the two parameters m and σ is discussed below. The normal distribution is important because, like a linear dynamic equation, it is amenable to mathematical analysis and also because of the *Central Limit Theorem* which states that:

The probability distribution of the sum of a large number of independent random variables approximates to the normal distribution, regardless of the distributions of the individual random variables.

Confirmation of the Central Limit Theorem is provided in practical situations, ranging from IQ testing to observation of thermal noise voltages, where uncertainty is caused by a large number of independent factors and where random variables are observed to have a bell-shaped distribution like that in Fig. 12.4. Mathematical proofs of the theorem also exist; for example it can be shown that the sum of two or more normally distributed random variables is itself normally distributed and that the binomial distribution tends to a normal shape as n, the number of independent trials (coin tosses), becomes large.†

It is often convenient to summarize the information contained in a probability distribution by the *moments* of the distribution. The Nth moment is the average, or expected value, of x^N given by

$$E(x^N) \equiv \sum_x x^N P(x), \quad \text{for discrete } x. \tag{12.14a}$$

$$E(x^N) \equiv \int_{-\infty}^{\infty} x^N p(x)\,\mathrm{d}x, \quad \text{for continuous } x. \tag{12.14b}$$

These average values are sometimes represented by the notation

$$E(x^N) = \overline{x^N}.$$

† See for example Cramer (1955) Sections 6.4, 7.3, and 7.4 or Feller (1957) Chapter 10.

The first moment, sometimes called the mean value, gives an indication of the likely value of the random variable; the second moment of the distribution about the mean value, $E((x - \bar{x})^2)$, gives an indication of the spread of values about the average; it is called the *variance*. It can be shown that, for any distribution, the variance is given by

$$E((x - \bar{x})^2) = E(x^2) - (\bar{x})^2. \tag{12.15}$$

The square root of the variance also has a common name, it is known as the *standard deviation*. The mean value and the variance for the binomial distribution of equation (12.5) and for the normal distribution of equation (12.13) are

$$Binomial \quad \bar{x} = pn, \quad \text{variance} = p(1 - p)n.$$

$$Normal \quad \bar{x} = m, \quad \text{variance} = \sigma^2.$$

The parameters m and σ which characterize the normal distribution are thus the mean value and the standard deviation of the distribution. Parameters which completely characterize a probability distribution, as m and σ do for the normal distribution of equation (12.13) and p and n do for the binomial distribution of equation (12.5), are known as *sufficient statistics*.

Probability distributions have been introduced above as a way of describing the distribution of probability over a single random variable x. They can also be used to describe the distribution of joint probability over a set of n random variables $x_1, \ldots, x_n$ (written $\mathbf{x}$). The most important joint probability distribution is the multidimensional normal distribution

$$p(\mathbf{x}) = \frac{1}{(2\pi)^{n/2}\sqrt{|\boldsymbol{\Sigma}|}} \exp(-\tfrac{1}{2}(\mathbf{x} - \mathbf{m})'\boldsymbol{\Sigma}^{-1}(\mathbf{x} - \mathbf{m})), \tag{12.16}$$

where $\mathbf{m}$ is the set of mean values and $\boldsymbol{\Sigma}$ is the *covariance matrix* defined by

$$\boldsymbol{\Sigma} \equiv E((\mathbf{x} - \mathbf{m})(\mathbf{x} - \mathbf{m})').$$

The typical element σ_{ij} of a covariance matrix is

$$\sigma_{ij} = E((x_i - m_i)(x_j - m_j))$$

and so covariance matrices are always symmetric.

Some of the foregoing ideas are illustrated by deriving expressions for the mean value m and variance σ^2 of a random variable x that is the sum of two independent random variables x_1, x_2 having probability densities $p_1(x_1)$, $p_2(x_2)$ with mean values m_1, m_2 and variances σ_1^2, σ_2^2. When

$$x = x_1 + x_2$$

the conditional probability density for x, given a value x_1^1 of x_1 and a corresponding value

$$x_2^1 = x - x_1^1$$

of x_2, is

$$p(x \mid x_1^1) = p_2(x_2^1).$$

So, by analogy with equations (12.3), the probability density for x is

$$p(x) = \int_{-\infty}^{\infty} p(x \mid x_1^1) p_1(x_1^1) \, dx_1^1$$

$$= \int_{-\infty}^{\infty} p_2(x_2^1) p_1(x_1^1) \, dx_1^1.$$

The mean value m of x is then given by equation (12.14b), and noting that dx here can be replaced by dx_2^1,

$$m = \int_{-\infty}^{\infty} \int_{-\infty}^{\infty} (x_1^1 + x_2^1) p_1(x_1^1) p_2(x_2^1) \, dx_1^1 \, dx_2^1$$

$$= m_1 + m_2. \tag{12.17}$$

This result about the mean value m holds whether or not x_1 and x_2 are independent of each other. The corresponding result for the variance σ^2, using equation (12.15) to give

$$\sigma^2 + m^2 = \int_{-\infty}^{\infty} \int_{-\infty}^{\infty} (x_1^1 + x_2^1)^2 p_1(x_1^1) p_2(x_2^1) \, dx_1^1 \, dx_2^1$$

is simplified if x_1 and x_2 are independent so that

$$\overline{x_1 x_2} = \bar{x}_1 \bar{x}_2 = m_1 m_2;$$

the variance then becomes

$$\sigma^2 = \sigma_1^2 + \sigma_2^2. \tag{12.18}$$

Equations (12.17) and (12.18) can be generalized to give the mean and variance of the sum of any number of weighted variables $x_1, \ldots, x_n$. When

$$x = \sum_{i=1}^{n} a_i x_i$$

the mean value of x is

$$m = \sum_{i=1}^{n} a_i m_i \tag{12.19}$$

and, if the random variables are independent of each other, the variance of x is

$$\sigma^2 = \sum_{i=1}^{n} a_i^2 \sigma_i^2. \tag{12.20}$$

A quantity which arises in connection with control systems having quadratic performance criteria is the mean of a quadratic form such as $\mathbf{x}'\mathbf{P}\mathbf{x}$ when $\mathbf{x}$ is a set of n random variables having mean $\mathbf{m}$ and covariance matrix $\boldsymbol{\Sigma}$, and $\mathbf{P}$ is an $n \times n$-matrix. It can be written

$$E(\mathbf{x}'\mathbf{P}\mathbf{x}) = E\big((\mathbf{x} - \mathbf{m})'\mathbf{P}(\mathbf{x} - \mathbf{m}) + \mathbf{m}'\mathbf{P}\mathbf{x} + \mathbf{x}'\mathbf{P}\mathbf{m} - \mathbf{m}'\mathbf{P}\mathbf{m}\big),$$

and because $\mathbf{m}$ is $E(\mathbf{x})$ this becomes

$$E(\mathbf{x}'\mathbf{P}\mathbf{x}) = E\big((\mathbf{x} - \mathbf{m})'\mathbf{P}(\mathbf{x} - \mathbf{m})\big) + \mathbf{m}'\mathbf{P}\mathbf{m}.$$

The quadratic term on the right of this equation is rewritten as the trace† of a matrix

$$(\mathbf{x} - \mathbf{m})'\mathbf{P}(\mathbf{x} - \mathbf{m}) = \mathrm{tr}\big(\mathbf{P}(\mathbf{x} - \mathbf{m})(\mathbf{x} - \mathbf{m})'\big)$$

so that the mean value can be expressed in terms of the covariance matrix $\boldsymbol{\Sigma}$ as

$$E(\mathbf{x}'\mathbf{P}\mathbf{x}) = \mathrm{tr}\,(\mathbf{P}\boldsymbol{\Sigma}) + \mathbf{m}'\mathbf{P}\mathbf{m}. \tag{12.21}$$

Problems 12.4–12.13

12.4. Discrete-time random processes

A useful mathematical model for a discrete-time stochastic variable $x(i)$ assumes that it is generated by a sequence of independent random variables $\xi(i - 1),\ \xi(i),\ \xi(i + 1), \ldots$ according to a linear difference equation

$$x(i) + a_{n-1}x(i - 1) + \cdots + a_0 x(i - n) = b_n\xi(i) + \cdots + b_0\xi(i - n), \tag{12.22}$$

which has z-transform

$$X(z) = \frac{B(z^{-1})}{A(z^{-1})}\,\Xi(z) = g(z^{-1})\Xi(z) = G(z)\Xi(z). \tag{12.23}$$

Here $g(z^{-1})$ is the impulse response of the difference equation and results from writing the z-transfer function $G(z)$ as a polynomial in z^{-1}, as discussed in Section 1.5. This mathematical model regards the stochastic variable x as the output of a linear dynamic system driven by noise ξ, as shown in Fig. 12.5.

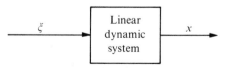

FIG. 12.5. Stochastic variable x as output of linear dynamic system driven by noise ξ.

† The trace of a matrix $\mathbf{A}$ is the sum of its diagonal elements
$$\mathrm{tr}\,\mathbf{A} = a_{11} + a_{22} + a_{33} + \ldots.$$

It can be represented as a set of n simultaneous first-order difference equations by using the transformation of equation (7.11b) to give

$$\mathbf{x}(i + 1) = \mathbf{A}\mathbf{x}(i) + \mathbf{B}\xi(i), \tag{12.24}$$

where the matrices $\mathbf{A}$ and $\mathbf{B}$ are as in equations (7.12). Equation (12.24) is an example of the type of random process known as a Markov process.†

If the random variables $\xi(i)$ are all drawn from a common probability distribution and the coefficients of the difference equation (12.22) are constant and such that the system of Fig. 12.5 is asymptotically stable, the process is *stationary*. The Central Limit Theorem ensures that the probability distribution of $x(i)$ tends to be normal regardless of whether or not the common distribution of the $\xi(i)$ is normal. Stationary processes describe common practical situations, and it is usual to assume that the mean value of the probability distribution for ξ is zero so that the mean value of the stochastic variable x, given by combining equations (12.19) and (12.23), is also zero. The variance σ_ξ^2 of the distribution for ξ is reflected in a special function of the stochastic variable x, called the *autocorrelation function*, which summarizes the variability of x both in amplitude and in time.

The autocorrelation function of a stochastic variable x is defined to be the average or expected value of the product $x(i)x(i + k)$. It is written

$$R_x(k) \equiv E\big(x(i)x(i + k)\big) \tag{12.25}$$

with argument k, showing that it is a function of the time interval between its two components, and with suffix x to identify it with the variable x. The autocorrelation function $R_\xi(k)$ of the sequence of independent, zero-mean, random variables ξ in the mathematical model of equation (12.22) follows immediately from the above definition:

$$\left.\begin{aligned}
R_\xi(0) &\equiv E\big(\xi^2(i)\big) = \sigma_\xi^2 \\
R_\xi(k) &\equiv E\big(\xi(i)\xi(i + k)\big) = 0, \ k \neq 0
\end{aligned}\right\} \tag{12.26}$$

and is illustrated in Fig. 12.6. This particular autocorrelation function is widely used in mathematical models for stochastic variables.

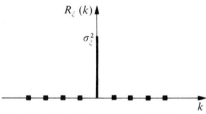

FIG. 12.6. Autocorrelation function of a sequence of independent random variables.

† The characteristic feature of a Markov process is that its future depends on its present but not on its past. This can be written

$$p(\mathbf{x}(i + 1) \mid \mathbf{x}(i), \mathbf{x}(i - 1), \dots,) = p(\mathbf{x}(i + 1) \mid \mathbf{x}(i)).$$

See Feller (1957) Chapter 15.

The autocorrelation function $R_x(k)$ of the stochastic variable x in the mathematical model of equation (12.22) can be related to the sequence of random variables ξ by replacing x in equation (12.25) by the impulse response $g(z^{-1})$ from equation (12.23) to give

$$
\begin{aligned}
R_x(k) &= E\{(g_0\xi(i) + g_1\xi(i-1) + \cdots)(g_0\xi(i+k) + g_1\xi(i+k-1) + \cdots)\} \\
&= E\{g_0^2\xi(i)\xi(i+k) + g_0g_1\xi(i)\xi(i+k-1) + \cdots + g_0g_k\xi^2(i) + \cdots \\
&\quad + g_1g_{k+1}\xi^2(i-1) + \cdots\}.
\end{aligned}
$$

In this equation the right-hand side consists of terms which have products of different values of ξ and terms which have squared values of ξ. It follows from the definition of ξ as a sequence of independent random variables having the autocorrelation function of equation (12.26) that the expected value of the product terms is zero and the expected value of the squared terms is σ_ξ^2 so that the equation for $R_x(k)$ here becomes

$$
R_x(k) = \sigma_\xi^2 \sum_{j=0}^{\infty} g_j g_{j+k}. \tag{12.27}
$$

The autocorrelation function of a stochastic variable gives a measure of the correlation between values of the variable separated by time intervals k. For example there might be high correlation between positions of a target separated by an interval of one second, but low correlation between positions separated by an interval of one hour; the rate at which correlation decreases as the interval increases would be related to the target's ability to take evasive action. For stationary processes such as the mathematical model of equation (12.22) the autocorrelation function is independent of the absolute value i of time and has the following properties.

(i) When k is zero

$$
R_x(0) \equiv E(x^2(i))
$$

the autocorrelation function gives the second moment, or variance σ_x^2 when the mean is zero, of x. Using equation (12.27) this is

$$
\sigma_x^2 = \sigma_\xi^2 \sum_{j=0}^{\infty} g_j^2, \tag{12.28}
$$

which can be evaluated from the transfer function $G(z)$ of equation (12.23) using the methods of Section 6.2:

$$
\sigma_x^2 = \frac{\sigma_\xi^2}{2\pi j} \oint G(z)G(z^{-1}) \frac{\mathrm{d}z}{z}. \tag{12.29}
$$

(ii) The autocorrelation function is symmetric in k:

$$
R(k) = R(-k)
$$

because, for a stationary process,

$$E\big(x(i)x(i + k)\big) = E\big(x(i - k)x(i)\big).$$

(iii) The value of the autocorrelation function never exceeds the value when k is zero:

$$R(k) \leqslant R(0) \quad \text{for all } k$$

because no value of x can correlate with any other value more strongly than it does with itself. Figure 12.7 is a sketch of a typical discrete-time auto-correlation function.

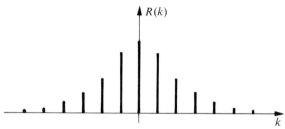

FIG. 12.7. Typical discrete-time autocorrelation function.

Equations (12.27)–(12.29) are convenient for analysis of linear systems. For example, if the stochastic variable x were the input to a system with z-transfer function $D(z)$ and impulse response $d(z^{-1})$, the z-transform of the output y from the system would be

$$Y(z) = D(z)X(z) = d(z^{-1})g(z^{-1})\,\Xi(z)$$

and the variance of y would be given by the sum of the squares of the co-efficients of an impulse response h, defined by

$$h(z^{-1}) \equiv d(z^{-1})g(z^{-1}),$$

in the form

$$\sigma_y^2 = \sigma_\xi^2 \sum_{j=0}^{\infty} h_j^2 , \tag{12.30}$$

or by using the corresponding z-transform $H(z)$:

$$H(z) \equiv D(z)G(z)$$

and evaluating

$$\sigma_y^2 = \frac{\sigma_\xi^2}{2\pi j} \oint H(z)H(z^{-1})\,\frac{\mathrm{d}z}{z}. \tag{12.31}$$

Problems 12.14–12.16

12.5. Mathematical models for continuous-time stochastic variables

Mathematical models for continuous-time stochastic variables are not as simple as for discrete-time because there is no simple continuous-time equivalent to the discrete-time sequence of independent random variables $\xi(i - 1)$, $\xi(i)$, $\xi(i + 1), \ldots$. However, the idea of the autocorrelation function does carry over; it need not be restricted to stationary processes and the general definition of the autocorrelation function for continuous-time processes is†

$$R_x(t, t + \tau) \equiv E\big(x(t)x(t + \tau)\big). \tag{12.32}$$

For a stationary process the autocorrelation function would be independent of time t and could be written $R_x(\tau)$. Continuous-time autocorrelation functions have the same properties as those listed as (i), (ii), (iii) in Section 12.4 for discrete-time.

The continuous-time equivalent to the autocorrelation function, equation (12.26), of the discrete-time noise ξ is

$$R_\xi(\tau) = N^2\delta(\tau), \tag{12.33}$$

where $\delta(\tau)$ is the unit impulse function introduced in Chapter 1 and N^2 is a constant specifying the magnitude of ξ, but is not the variance. It follows from the way that the unit impulse function and the autocorrelation function are defined that the variance of the continuous-time noise ξ would be infinite:

$$\sigma_\xi^2 = R_\xi(0) = N^2\,\delta(0) = \infty.$$

Real signals do not have infinite variance, so the autocorrelation function of equation (12.33) is not the autocorrelation function of any real signal. Nevertheless, the idea of a stochastic variable having this autocorrelation function and known as *white noise*‡ plays the same role in mathematical models for continuous-time variables as does the sequence $\xi(i - 1)$, $\xi(i)$, $\xi(i + 1), \ldots$ for discrete time.

The most common mathematical model for a continuous-time stochastic variable $x(t)$ assumes that it is the output of a linear dynamic system driven by white noise $\xi(t)$ as illustrated in Fig. 12.5. The Central Limit Theorem ensures that the probability distribution of x here tends to be normal regardless of whether or not the white noise ξ, with its infinite variance, can be said to have a normal distribution.

There are technical, mathematical difficulties in writing continuous-time differential equations corresponding to the difference equations (12.22) and (12.24) when $\xi(t)$ is white noise. The continuous-time equivalent of equation (12.22) would be

† The meaning of the expectation operation $E(\cdot)$ in this equation (12.32) is discussed in Section 12.6.
‡ The name white noise arises because the continuous-time variable ξ can be represented in the frequency domain by a power spectral density function which is uniform, like the spectrum of white light. See Jacobs (1969).

$$\frac{d^n x}{dt^n} + \cdots + a_1 \frac{dx}{dt} + a_0 x = b_n \frac{d^n \xi}{dt^n} + \cdots + b_1 \frac{d\xi}{dt} + b_0 \xi.$$

One difficulty is that the derivatives $d\xi/dt$, $d^2\xi/dt^2$, ... of the white noise ξ do not exist because neighbouring values $\xi(t)$ and $\xi(t + \Delta)$ are by definition independent even for infinitesimal Δ, so that the usual definition of a derivative:

$$\frac{d\xi}{dt} = \lim_{\Delta \to 0} \left(\frac{\xi(t + \Delta) - \xi(t)}{\Delta} \right)$$

has no meaning. The continuous-time equivalent of equation (12.24), which results from transforming the above differential equation according to equation (7.11a), would be

$$\dot{\mathbf{x}} = \mathbf{A}\mathbf{x} + \mathbf{B}\xi,$$

but the vector $\mathbf{x}$ here again includes non-existent derivatives of ξ.

Derivatives of ξ can be avoided by considering state space models having the form

$$\dot{\mathbf{x}} = \mathbf{A}\mathbf{x} + \mathbf{B}\boldsymbol{\xi},$$

where $\boldsymbol{\xi}$ is a vector with m components ($m \leqslant n$) representing up to m mutually independent white noises. Here $\mathbf{A}$ and $\mathbf{B}$ are matrices of appropriate dimension and not necessarily the same matrices as in the previous paragraph or equation (12.24); $\mathbf{B}$ is, without loss of generality, assumed to have only diagonal terms so that each noise $\boldsymbol{\xi}$ affects only one of the variables $\mathbf{x}$. Although derivatives of $\boldsymbol{\xi}$ are now avoided, there is a further mathematical difficulty due to the infinite variance of white noise because this implies that $\boldsymbol{\xi}$ can take infinite values, and hence that the derivative $\dot{\mathbf{x}}$ can take infinite values and so proper solutions to the differential equations of the state-space model do not exist. This difficulty is avoided by rewriting the differential equations in the form

$$\begin{aligned} d\mathbf{x} &= \mathbf{A}\mathbf{x}\, dt + \mathbf{B}\boldsymbol{\xi}\, dt \\ &= \mathbf{A}\mathbf{x}\, dt + \mathbf{B}\, d\mathbf{w}, \end{aligned} \qquad (12.34)$$

known as a stochastic differential equation, where $\mathbf{w}$ is a vector with m components, each of which is the time-integral of one of the original white noises. Integrated white noise,

$$w = \int_{t_1}^{t_1 + \tau} \xi\, dt,$$

known by the names Wiener process and Brownian motion, can be shown† to have finite variance proportional to the interval of integration τ. Thus, equation (12.34) is a proper mathematical equation relating the increment $d\mathbf{x}$ to the current value of $\mathbf{x}$ and to an independent stochastic variable $d\mathbf{w}$ of finite variance.

Further developments and applications of stochastic differential equations requires mathematical techniques of the Ito calculus which are beyond the scope of this book.† For this reason the analysis of problems of optimal stochastic control is presented here mostly in terms of discrete-time problems; the results about discrete-time stochastic control frequently have obvious continuous-time equivalents. There is, however, one class of stochastic continuous-time problems, corresponding to the deterministic LQP-design problems of Chapter 6, which can be conveniently solved by continuous-time analysis and is discussed in Section 13.2.

12.6. Continuous-time random processes

Every mathematical model of uncertainty is based on the fundamental concepts of sample space and probability introduced in Section 12.2. For unknown constants the sample space is the set of possible values of the constant and the probability distribution is as discussed in Section 12.3. For discrete-time random processes the set of all possible values of the sequence of variables $\xi(i-1), \xi(i), \xi(i+1), \ldots$ provides a multidimensional sample space having a joint probability distribution which, according to equation (12.1), is the product of the probability distributions of the independent variables $\xi(i-1), \xi(i), \xi(i+1), \ldots$. For continuous-time stochastic variables the relationship between the mathematical models discussed in Section 12.5 and the fundamental concepts of Section 12.2 is complicated by the mathematical difficulties mentioned in Section 12.5.

The difficulties are avoided by taking as sample space the ensemble of all possible future realizations of the stochastic variable $x(t)$. Figure 12.8 shows some typical members, labelled $\Omega_1, \Omega_2, \Omega_3$, of such an ensemble. The number of ensemble members is usually infinite, but generally some will be more likely to occur than others and it is possible to imagine a probability distribution $p(\Omega)$ defined over the ensemble. Here Ω is used as a parameter labelling ensemble members so that the complete ensemble is represented by a function $x(t, \Omega)$. The combination of ensemble function $x(t, \Omega)$ and probability distribution $p(\Omega)$, known as a continuous-time random process, provides the fundamental basis for mathematical models of continuous-time, stochastic variables.

Such random processes are appropriate for describing uncertain future realizations of a stochastic variable, they are inappropriate for past realizations where a record exists and there is no uncertainty. There is no loss of generality in choosing the time origin to be such that the boundary between the uncertain future and the known past is where t is zero. Thus random processes will be assumed here to exist for positive values of time t only; as

† See Åström (1970) Chapter 3 or Kushner (1971) Chapter 10 or Melsa and Sage (1973) Chapter 8. When the white noise has the autocorrelation function $N^2\delta(\tau)$ of equation (12.33) the variance of its integral w is $N^2\tau$.

time evolves the ensemble of possible future realizations becomes transformed into a single record of past history.

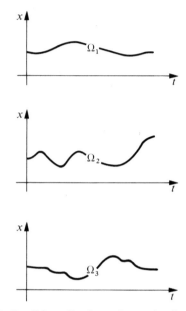

FIG. 12.8. Possible realizations of a stochastic variable $x(t)$.

An example of a pair of functions defining a random process is the ensemble function

$$x(t, \Omega) = x(0) \cos \Omega t$$

together with the normal probability distribution

$$p(\Omega) = \frac{1}{\sqrt{(2\pi)}} \exp(-\Omega^2/2).$$

Figure 12.9 shows some members of this ensemble together with its associated probability distribution. This example of a cosine wave with random frequency is not typical of the sort of random process associated with control systems where stochastic variables are usually such that the functions $x(t, \Omega)$ and $p(\Omega)$ are both complicated and unknown. Stochastic control theory does not require exact specification of random processes. The idea of a random process is introduced here in order to establish a relationship with the fundamental probability concepts of Section 12.2 and to give meaning to operations such as the expectation used to define the continuous-time autocorrelation function in equation (12.32).

The labelling parameter Ω may have to take its value in some high-dimensional or non-Euclidean space depending on what determines the set of all

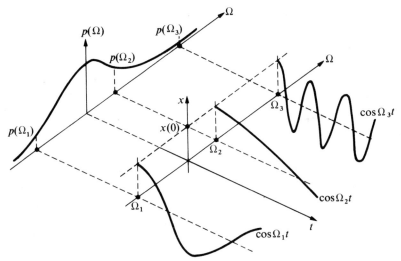

FIG. 12.9. Example of a random process.

possible realizations of $x(t)$. Nevertheless it is sometimes convenient to discuss Ω as though it were a single variable taking values on a line, as it does in
the example of Fig. 12.9. The expectation operation can then be defined, for
any function $f(x)$ of the stochastic variable at future time t, by writing

$$E\big(f(x)\big)(t) \equiv \int_{\Omega} f\big(x(t,\,\Omega)\big)p(\Omega)\,\mathrm{d}\Omega \qquad (12.35)$$

so that the autocorrelation function of equation (12.32) is

$$R(t,\,t+\tau) = \int_{\Omega} x(t,\,\Omega)x(t+\tau,\,\Omega)p(\Omega)\,\mathrm{d}\Omega.$$

Expected values are defined by equation (12.35) at a specific instant of time t
and may in general be time-varying, as indicated by the notation on the left-
hand side of the equation. If the expected values are independent of time t the
process is stationary.

The random process illustrated in Fig. 12.9 is non-stationary because all
ensemble members take the value $x(0)$ when t is zero even though their future
values are uncertain. The mean and variance can be evaluated (using tabulated integrals) and are

$$E(x)(t) = \frac{x(0)}{\sqrt{(2\pi)}} \int_{-\infty}^{\infty} \cos \Omega t\; \mathrm{e}^{-\Omega^2/2}\,\mathrm{d}\Omega = x(0)\,\mathrm{e}^{-t^2/2},$$

$$E\{(x - E(x))^2\} = \frac{x^2(0)}{\sqrt{(2\pi)}} \int_{-\infty}^{\infty} (\cos \Omega t - \mathrm{e}^{-t^2/2})^2\, \mathrm{e}^{-\Omega^2/2}\,\mathrm{d}\Omega$$

$$= x^2(0)(\tfrac{1}{2} - \mathrm{e}^{-t^2} + \tfrac{1}{2}\mathrm{e}^{-2t^2}).$$

Figure 12.10 shows how these expected values change with time.

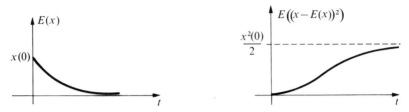

FIG. 12.10. Mean and variance of the random process of Fig. 12.9.

Figure 12.11 illustrates another type of non-stationarity which can arise when changes are expected in the physical mechanism generating a stochastic variable. This type of non-stationary process is appropriate when there is prior information about the times t_1, t_2, t_3, etc. when changes will occur. In the absence of such prior information a non-stationary process is not usually appropriate. The question whether or not this type of non-stationary process should be used as the mathematical model for a given practical problem must be settled by assumptions based on prior information. It is not possible to

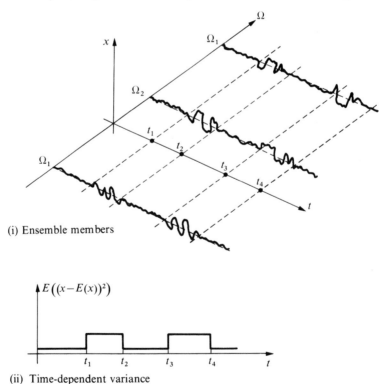

(i) Ensemble members

(ii) Time-dependent variance

FIG. 12.11. Example of non-stationary random process.

deduce from a record of past history such as Fig. 12.12(i) whether it was a member of a non-stationary ensemble like that of Fig. 12.11(i), which is re-drawn as Fig. 12.12(ii), or of an ensemble where the times of change were uncertain so that the ensemble becomes stationary again as in Fig. 12.12(iii).

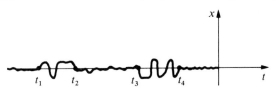

(i) Record of past history

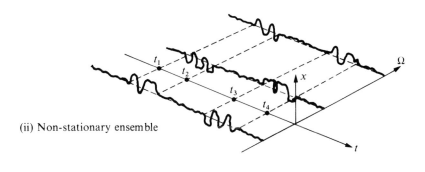

(ii) Non-stationary ensemble

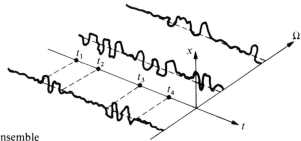

(iii) Stationary ensemble

FIG. 12.12. Record (i) could have come from either of ensembles (ii) or (iii).

The expected values defined by equation (12.35) can be regarded as averages with respect to the probability distribution $p(\Omega)$ over all members of the ensemble at a fixed time t. Another class of averages are the time averages defined by averaging over all future time for a given ensemble member Ω:

$$\overline{f(x)}\,(\Omega) \equiv \lim_{T \to \infty} \frac{1}{T} \int_0^T f(x(t, \Omega))\mathrm{d}t. \tag{12.36}$$

These time averages may in general depend on the particular ensemble member chosen as indicated by the notation on the left-hand side of the equation. Figure 12.13 shows the difference between time averages and ensemble averages or expected values.

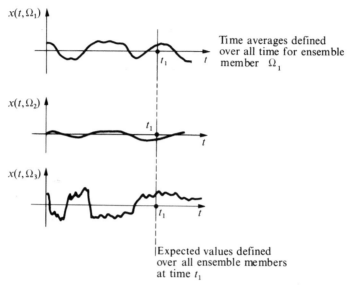

FIG. 12.13. The distinction between expected values and time averages.

There is a time average corresponding to every expected value, for example:

$\bar{x}(\Omega)$ corresponds to $E(x)(t)$,

$\overline{x(t)x(t + \tau)}(\Omega)$ corresponds to the autocorrelation function $R(t, t + \tau)$.

The time averages are independent of the uncertainty associated with the probability distribution $p(\Omega)$ of the random process, but they correspond to the sort of measurement that can be made in practice from a single record of the past history of $x(t)$. In this respect time averages are the complement of expected values which depend on the probability distribution $p(\Omega)$ but cannot usually be measured in practice.

Under certain special conditions time averages measured from a single record of past history can be used as estimates of corresponding expected values of a stochastic variable in the future. One condition is that the random process be stationary, so that ensemble averages are independent of time. A second condition is that the process be such that time averages are independent of ensemble member. If the time averages are independent of ensemble member, and thus of the labelling parameter Ω, the process is said to be *homogeneous*. It can be shown† that if a process is both stationary and

† Jacobs (1969).

homogeneous the time average of any function $f(x)$ is the same as the ensemble average:

$$\overline{f(x)} = E(f(x)),$$

and such processes are said to be *ergodic*.

Erdogic random processes are used wherever possible as mathematical models for stochastic variables so that future expected values can be estimated by measuring past time averages. The expected value of any function $f(x)$ is estimated by

$$E(f(x)) = \overline{f(x)} \simeq \frac{1}{T} \int_{-T}^{0} f(x)\, dt, \qquad (12.37)$$

and in particular the autocorrelation function is estimated by

$$R_x(\tau) \equiv \overline{x(t)x(t+\tau)} \simeq \frac{1}{T} \int_{-T}^{0} x(t-\tau)x(t)\, dt.$$

The accuracy of estimation improves with the measuring time T: as a general rule accuracy tends to be proportional to the square root of T, and when T goes to infinity the estimation becomes an exact measurement, equation (12.36). In many practical situations it is impossible to derive useful results from the mathematical model unless it is ergodic. To assume that an ergodic random process is an appropriate model requires sufficient prior information to justify assumptions of stationarity and homogeneity; although such information is frequently not available the arbitrary assumption that an ergodic process is appropriate, sometimes called the 'ergodic hypothesis', is often justified by the usefulness of results which follow from it.

The ergodic hypothesis is used here to derive an expression for the auto-correlation function $R_x(\tau)$ of the output $x(t)$ of a continuous-time system like that in Fig. 12.5 having input $\xi(t)$. For ergodic processes the auto-correlation function, defined by equation (12.32) as an ensemble average, can be written as a time average:

$$R_x(\tau) = \overline{x(t)x(t+\tau)},$$

which can be expressed in terms of the input $\xi(t)$ by using the convolution integral, equation (1.16b),

$$x(t) = \int_{0}^{\infty} g(u)\xi(t-u)\, du,$$

where g is the impulse response of the linear dynamic system, to give

$$R_x(\tau) = \overline{\int_{0}^{\infty}\int_{0}^{\infty} g(u_1)\xi(t-u_1)g(u_2)\xi(t+\tau-u_2)\, du_1\, du_2},$$

where the suffices on u_1 and u_2 distinguish between the variables of integration in the convolution integrals for $x(t)$ and for $x(t + \tau)$. In the above expression $g(u)$ is deterministic and

$$\overline{\xi(t - u_1)\xi(t + \tau - u_2)} = R_\xi(\tau + u_1 - u_2),$$

so that

$$R_x(\tau) = \int_0^\infty \int_0^\infty g(u_1)g(u_2)R_\xi(\tau + u_1 - u_2) \, du_1 \, du_2. \tag{12.38}$$

If $\xi(t)$ is white noise having the autocorrelation function of equation (12.33), equation (12.38) gives

$$R_x(\tau) = N^2 \int_0^\infty g(u)g(u + \tau) \, du. \tag{12.39}$$

The variance of x is given by $R_x(0)$:

$$\sigma_x^2 = N^2 \int_0^\infty g^2(u) \, du, \tag{12.40}$$

and can be evaluated from the transfer function $G(s)$ of the dynamic system using Parseval's theorem, as discussed in Section 6.2:

$$\sigma_x^2 = \frac{N^2}{2\pi j} \int_{-j\infty}^{j\infty} G(s)G(-s) \, ds. \tag{12.41}$$

Equations (12.39)–(12.41) are convenient for analysis of linear systems. For example, if the stochastic variable x were the input to a system with transfer function $D(s)$ the Laplace transform of the output y from the system would be

$$Y(s) = D(s)X(s) = D(s)G(s)\Xi(s) = H(s)\Xi(s),$$

where

$$H(s) \equiv D(s)G(s),$$

and the variance of y would be given by the integral of the square of the corresponding impulse response h:

$$\sigma_y^2 = N^2 \int_0^\infty h^2(u) \, du, \tag{12.42}$$

or could be evaluated using Parseval's theorem

$$\sigma_y^2 = \frac{N^2}{2\pi j} \int_{-j\infty}^{j\infty} H(s)H(-s) \, ds \tag{12.43}$$

as at equation (12.31).†

Problems 12.17–12.23

12.7. Bibliography

The theory of probability and random processes is a well-documented branch of mathematics. The introduction to probability theory given here follows that in
CRAMER (1955) Parts I and II,
FELLER (1957) Chapters 1–10,
PARZEN (1960),
GNEDENKO and KHINCHIN (1961),
PAPOULIS (1965) Part I,
BREIPHOL (1970) Chapters 1–6,
DAVENPORT (1970) Chapters 1–8,
MELSA and SAGE (1973) Chapters 1–4.
The discussion of random processes follows that in
LANING and BATTIN (1956),
BENDAT (1958),
DAVENPORT and ROOT (1958),
PARZEN (1962),
COX and MILLER (1965),
PAPOULIS (1965) Part II,
JACOBS (1969),
MEDITCH (1969) Chapter 4,
ÅSTRÖM (1970) Chapters 2–4,
BREIPHOL (1970) Chapters 11 and 12,
DAVENPORT (1970) Chapters 9–14,
MELSA and SAGE (1973) Chapters 5–8.

12.8. Problems

12.1. Specify the set of all possible outcomes (the sample space) in each of the following experiments (random events).

(i) A single throw of a six-sided die.

(ii) Two successive throws of a six-sided die.

(iii) A number generated by summing the results of each of two successive throws of a six-sided die.

---

† Equations (12.41) and (12.43) can be written in the form

$$\sigma_x^2 = \int_0^\infty \phi_x(\omega) \, d\omega, \qquad \sigma_y^2 = \int_0^\infty \phi_y(\omega) \, d\omega,$$

where

$$\phi_x(\omega) \equiv \frac{N^2}{\pi} |G(j\omega)|^2, \qquad \phi_y(\omega) \equiv \frac{N^2}{\pi} |H(j\omega)|^2$$

are known as the power spectral density functions of x and of y. The power spectral density function of an ergodic random process can be shown to be the Fourier transform of the autocorrelation function; its argument ω has the significance of angular frequency and it translates the mathematical model of a stochastic variable into the language of frequency responses. See the references to random processes given in Section 12.7.

(iv) The number of people inside the next aeroplane to fly overhead.

(v) The average age of the people in (iv) above.

(vi) The time when the aeroplane in (iv) above will fly overhead.

(vii) The number of aeroplanes which will fly overhead in the next hour.

(viii) The value at some arbitrarily chosen moment in time of the speed of an aeroplane.

(ix) The result of a physical measurement of the speed of an aeroplane at a moment when it is flying at speed V.

(x) The value at some arbitrarily chosen moment in time of the velocity (speed and direction) of an aeroplane in flight.

(xi) Values at some arbitrarily chosen moment in time of the angular position and speed of the rotating inertia in the position-control system of Fig. 1.7.

(xii) The position of a pin-point placed at random on a sheet of paper.

(xiii) Values at some arbitrarily chosen moment of time of the states x of either of the dynamic systems specified by equations (7.1) or (7.2).

(xiv) A prediction before the aeroplane in (iv) appears of its speed.

(xv) An estimate after the aeroplane in (iv) has passed overhead of its speed.

(xvi) A record of the speed of the aeroplane in (iv) during the hour after it has passed.

12.2. In a population composed of 55% women and 45% men, 15% are left-handed although only 10% of the women are left-handed. What is the probability that a person chosen at random from this population will be

(i) a right-handed woman?

(ii) a right-handed man?

12.3. Motorists could be classified according to their probability of making an insurance claim in any one year as follows.

Class A having 10% probability of claiming and consisting of 60% of motorists.
Class B having 20% probability of claiming and consisting of 30% of motorists.
Class C having 60% probability of claiming and consisting of 10% of motorists.

(i) What is the probability that a motorist chosen at random will make a claim within one year?

(ii) If he does not make a claim within the year what is the conditional probability that he belongs to class A? to class B? to class C?

(iii) If he does make a claim within the year what can be deduced about his classification?

12.4. A procedure to generate a sequence of random digits is as follows:

(a) open a book at a random page and note the page number;

(b) add the penultimate digit of the number in (a) to the previous digit of the sequence (assume the first number to be zero);

(c) let the last digit of the sum in (b) be the next digit of the sequence.

(i) What are the sample spaces for the random events of (a) and (b)?

(ii) Suggest a probability distribution for the random numbers of (c).

(iii) Use the above procedure to generate some random numbers and sketch the distribution of their relative frequencies.

12.5. Sketch typical probability distribution functions for the random variables in Sections (iii)–(ix) of Problem 12.1.

12.6. Figure 12.4 shows the triangular probability density function of a continuous-valued random variable x.

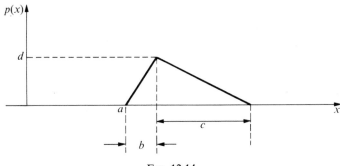

FIG. 12.14.

Prove that $d = 2/(b + c)$.

Sketch the probability distribution function of x.

12.7. If x is a random variable having probability-density function $p_x(x)$ and $y = f(x)$ is a function of x having derivative $dy/dx = f'(x)$ and unique inverse $f^{-1}(x)$, prove that the probability density function of y is

$$p_y(y) = \frac{p_x(f^{-1}(y))}{f'(f^{-1}(y))}.$$

Find $p_y(y)$ when

(i) x is uniformly distributed between 0 and 1 and $y = e^x$.

(ii) x is uniformly distributed between 0 and 1 and $y = x^2$.

(iii) y is generated by the mechanism shown in Fig. 12.15 where the angle θ of the line AB is uniformly distributed between $-\tau/2$ and $\pi/2$.

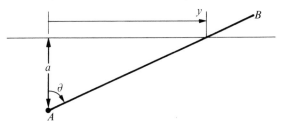

FIG. 12.15.

Check that each of the above three probability-density functions satisfies the normalizing condition of equation (12.10).

12.8. Derive equation (12.15).

12.9. Calculate the mean value and the variance of the following probability distributions.

(i) $p(x) = 1/a$, $0 \leqslant x \leqslant a$ (uniform density).

(ii) $P(x) = \frac{1}{10}$, $x = 0, 1, \ldots, 9$ (uniform distribution corresponding to that of Problem 12.4).

(iii) The triangular density of Problem 12.6 when $a = -b$ and $c = b$.

(iv) The binomial distribution of equation (12.5).

(v) The normal distribution of equation (12.13).

(vi) $p(x) = \dfrac{a}{\pi(a^2 + x^2)}$, the Cauchy distribution (of Problem 12.7(iii)).

12.10. A random variable y is the sum of two independent normally distributed random variables having means m_1, m_2 and variances σ_1^2, σ_2^2, respectively. Show that the probability density function of y is

$$p(y) = \frac{1}{2\pi\sigma_1\sigma_2} \int_{-\infty}^{\infty} \exp\left\{ -\frac{(y - x - m_1)^2}{2\sigma_1^2} - \frac{(x - m_2)^2}{2\sigma_2^2} \right\} dx$$

and confirm that this is normal with mean $m_1 + m_2$ and variance $\sigma_1^2 + \sigma_2^2$.

12.11. Confirm that it follows from the definition of covariance matrix that the covariance matrix of the joint probability distribution of any number of independent random variables is diagonal.

12.12. Two random variables have a normal joint distribution with means $m_1 = m_2 = 0$ and covariance matrix

$$\Sigma = \begin{bmatrix} \sigma_{11} & \sigma_{12} \\ \sigma_{12} & \sigma_{22} \end{bmatrix}$$

and one of them, say x_1, is observed to take the value X. Use equation (12.2) to show that the conditional probability density for the other has the normal form

$$p(x_2 \mid X) = \text{constant} \times \exp\left\{ -\frac{\left(x_2 - \dfrac{\sigma_{12}}{\sigma_{11}} X \right)^2}{2\sigma_{22}\left(1 - \dfrac{\sigma_{12}^2}{\sigma_{11}\sigma_{22}} \right)} \right\}.$$

12.13. Three random variables x_1, x_2, and x_3 have a normal joint distribution with means $m_1 = -1$, $m_2 = 0$, $m_3 = 1$ and covariance matrix

$$\Sigma = \begin{bmatrix} 1 & 0\cdot1 & 0\cdot5 \\ 0\cdot1 & 2 & 0\cdot25 \\ 0\cdot5 & 0\cdot25 & 3 \end{bmatrix}.$$

Evaluate the expected value of the random variable

$$y = 3x_1^2 + 2x_2^2 + x_3^2 + 2x_1x_2 + 4x_1x_3 + 8x_2x_3.$$

12.14. Generate realizations of discrete-time stochastic variables $x(i)$ over N stages, $i = 1, 2, \ldots, N$ ($N > 10$), as follows:

(a) use the procedure of Problem 12.4 to generate a sequence of random numbers $\xi_1(0)$, $\xi_1(1)$, $\xi_1(2)$, ...;

(b) form a related, zero-mean sequence

$$\xi(i) = \xi_1(i) - 4\tfrac{1}{2}, \quad i = 0, 1, 2, \ldots;$$

(c) use these random variables $\xi(i)$ to drive the difference equations:

(i) $x(i) - 0.5x(i - 1) = \xi(i)$

(ii) $x(i) + x(i - 1) = \xi(i)$

(iii) $x(i) + x(i - 1) + \tfrac{1}{2}x(i - 2) = \xi(i) - \tfrac{1}{2}\xi(i - 1)$.

(Assume all initial values $x(i)$, $i \leqslant 0$, to be zero.)

12.15. Assuming that the random variables $\xi(i)$ of Problem 12.14(b) are independent and uniformly distributed,

(i) calculate the variance σ_x^2 and sketch the autocorrelation function of the stochastic variable $x(i)$ in Problem 12.14(i);

(ii) explain why the variance of the stochastic variable $x(i)$ in Problem 12.14(ii) is infinite;

(iii) use equation (12.29) to calculate the variance of the stochastic variable $x(i)$ in Problem 12.14(iii).

12.16. A system with difference equation

$$y(i) = x(i) - \alpha x(i - 1)$$

has as its input x the stochastic variable of Problem 12.14(i). Use equation (12.31), together with the assumption of Problem 12.15 about the distribution of $\xi(i)$, to find the variance of y.

12.17. Sketch some typical members of the sample space of each of the following stochastic variables.

(i) The discrete-time variable of Problem 12.14(i).

(ii) The set of midday temperatures at a particular place every day next week.

(iii) The discrete-time variable of Problem 12.14(ii).

(iv). A record of the speed of the next aeroplane to fly overhead (cf., Problem 12.1) during the hour after it will have passed.

12.18. Sketch some typical ensemble members (members of the sample space) of the continuous-time random processes specified by the following functions.

(i) $x(t, \Omega) = x(0)\, e^{\Omega t}$, $p(\Omega)$ normal with zero mean and unit variance.

(ii) $x(t, \Omega) = x(0)\, e^{\Omega t}$, $p(\Omega)$ uniformly distributed over the range $-1 < \Omega < 1$.

(iii) $x(t, \Omega) = x(0)\, e^{\Omega t}$, $p(\Omega)$ uniformly distributed over the range $-3 < \Omega < -1$.

(iv) $x(t, \Omega) = x(0) + \Omega \sin t$, $p(\Omega)$ normal with zero mean and unit variance.

(v) $x(t, \Omega) = x(0) + \sin \Omega t$, Ω one of the integers -3, -1, 1, 3 with probabilities $P(-3) = P(-1) = P(1) = P(3) = \tfrac{1}{4}$.

(vi) $x(t, \Omega) = \sin (t + \Omega)$, $p(\Omega)$ uniformly distributed over the range $0 \leqslant \Omega < 2\pi$.

(vii) $x(t, \Omega_1, \Omega_2) = \Omega_1 \sin (t + \Omega_2)$, $p(\Omega_1)$ normal with zero mean and unit variance, $p(\Omega_2)$ uniformly distributed over the range $0 \leqslant \Omega < 2\pi$, Ω_1 independent of Ω_2.

12.19. Calculate the ensemble average $E(x)(t)$, the time average $\bar{x}(\Omega)$, the auto-correlation function $E(x(t)x(t + \tau))$, the time average $x(t)x(t + \tau)(\Omega)$, and find the variance of each of the random processes (ii), (iv), (vi), (vii) of Problem 12.18.

12.20. Which of the random processes of Problem 12.18 are

(i) stationary?
(ii) homogeneous?
(iii) ergodic?

12.21. Use equation (12.39) to calculate the autocorrelation function of the continuous-time output x of a system like that shown in Fig. 12.5 when its transfer function is

$$G(s) = \frac{K}{1 + sT}$$

and its input ξ is white noise having the autocorrelation function of equation (12.33).

What is the variance of x? Confirm this result using Parseval's theorem (equation (12.41) and the table of Fig. 6.5).

12.22. A system with differential equation

$$\ddot{y} + 2\zeta\dot{y} + y = x$$

has as its input x the stochastic variable of Problem 12.21. Use equation (12.43) to find the variance of y.

12.23. Explain the apparent discrepancy between the statement in Section 12.5 that integrated white noise has finite variance and the result of substituting the transfer function of an integrator

$$G(s) = \frac{1}{s}$$

in equation (12.41).

13. Optimal stochastic LQP-control

13.1. Introduction

UNCERTAINTY in control systems may be due to unknown constants such as parameter values of the controlled process and its environment or it may be due to stochastic variables such as target positions, unpredictable disturbances, or measurement noises. However it arises, the effect of uncertainty in a dynamic system is generally that time-dependent variables characterizing the system, such as the state $\mathbf{x}$ or the output $\mathbf{y}$, become stochastic variables. The theory of control in the presence of uncertainty is therefore known as stochastic control theory. It is a combination of the mathematical models of uncertainty which were introduced in Chapter 12 and the analytic techniques which were introduced for deterministic control problems in Chapters 6 and 8.

The analytic design procedure of Chapter 6 is easily extended for fixed-configuration stationary LQP-systems with uncertainty in the form of stochastic variables. This is presented in Section 13.2 to provide an introduction to stochastic control theory in the same way that Chapter 6 provides an introduction to Chapter 8.

The semi-free problem of optimal stochastic control, where a controlled process, a performance criterion, and some uncertainty are specified and it is required to find the control law to optimize performance, is perhaps the central problem in control theory. It asks the fundamental question of what is the optimal control law and takes account of the fundamental reason for using feedback, namely uncertainty. It is presented here mainly in terms of discrete-time systems so as to avoid the mathematical difficulties associated with continuous-time stochastic variables mentioned in Section 12.5: the principal results about feedback control laws have obvious continuous-time equivalents.

Uncertainty in the dynamics of discrete-time controlled processes can be represented by generalizing the deterministic dynamic equation (8.1b) to give

$$\mathbf{x}(i + 1) = \mathbf{G}_1\big(\mathbf{x}(i), \mathbf{u}(i), \boldsymbol{\xi}_1(i), i\big), \qquad (13.1)$$

where $\boldsymbol{\xi}_1(i)$ is a set of independent zero-mean random variables driving dynamic equations such as equations (12.24) included in the set $\mathbf{G}_1$ to account for stochastic variables. A linear version of this controlled process,

$$x(i + 1) = \mathbf{A}x(i) + \mathbf{B}u(i) + \boldsymbol{\xi}_1(i), \tag{13.2}$$

is considered in Sections 13.3 and 13.4 on optimal LQP-problems.

Unknown constants must be regarded as components of the state $\mathbf{x}$ which do not change with time and are not available for observation. For example, the first-order version of the linear equation (13.2) is

$$x(i + 1) = ax(i) + bu(i) + \xi(i),$$

and if the coefficients a and b were unknown constants they would be represented by state variables x_2 and x_3 in a set of three dynamic equations:

$$x_1(i + 1) = x_2(i)x_1(i) + x_3(i)u(i) + \xi(i),$$
$$x_2(i + 1) = x_2(i),$$
$$x_3(i + 1) = x_3(i),$$

describing a system having three state variables of which only the state x_1 is available for observation. This system is non-linear because of the non-linearity in the first dynamic equation. Unknown constants often lead to such non-linearities and to problems which are not LQP but which belong to the more general class of stochastic control problems discussed in Chapter 15.

When there are states which are not available for observation and when there is measurement noise, uncertainty exists about the present value of the state $\mathbf{x}$. Such uncertainty can be distinguished from the uncertainty about future states represented by equation (13.1) and is represented by assuming that there are measurable outputs $\mathbf{y}$ related to the state $\mathbf{x}$, which cannot otherwise be observed, by functions of the general form

$$\mathbf{y}(i) = \mathbf{G}_2(\mathbf{x}(i), \boldsymbol{\xi}_2(i), i), \tag{13.3}$$

where $\boldsymbol{\xi}_2(i)$ are independent, zero-mean random variables representing measurement noise. A linear version of this output equation,

$$\mathbf{y}(i) = \mathbf{C}\mathbf{x}(i) + \boldsymbol{\xi}_2(i), \tag{13.4}$$

is introduced into the discussion of LQP-problems in Section 13.4. The number of outputs $\mathbf{y}$ may be less than the number of states $\mathbf{x}$ so that, as in Section 7.4, the matrix $\mathbf{C}$ is not necessarily square.

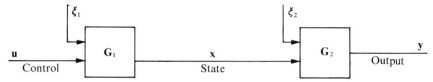

FIG. 13.1. Controlled process with uncertainty.

Figure 13.1 illustrates the general class of controlled processes, defined by equations (13.1) and (13.3), with uncertainty about both the future and the present state. To specify the mathematical model for a particular process within this class, the functions G_1 and G_2 must be specified, as by equations (13.2) and (13.4), and also probability distributions must be specified for any uncertain variables x, ξ_1, ξ_2. Normal probability distributions are convenient for analysis in the same way as are linear dynamics and quadratic perform-ance criteria and so are assumed whenever possible. The dynamic program-ming methods introduced in Chapter 8 for deterministic semi-free problems carry over to stochastic problems with controlled processes of the general class illustrated in Fig. 13.1.

In stochastic problems the scalar function $H(x(i), u(i), i)$ which specifies the deterministic performance criterion in equation (8.2b), repeated here as equation (13.5)

$$ I = \sum_{i=i_1}^{i_2} H\big(x(i), u(i), i\big), \tag{13.5} $$

is stochastic, so the sum I will be a random number and therefore unsuitable as a basis for design calculations. Deterministic performance criteria for stochastic problems are provided by taking expected values with respect to current probability distributions for the uncertain variables x and ξ. The expected value J_1 of the sum I evaluated at the initial time i_1, and written

$$ J_1 = E(I \mid i_1), \tag{13.6a} $$

is one such deterministic performance criterion. Another version of this criterion is given by dividing by the operating time $i_2 - i_1 + 1$ and letting i_2 go to infinity to give J_2:

$$ J_2 = E(\bar{H} \mid i_1) \tag{13.6b} $$

the expected value of the time average $\bar{H}$ of H. If the stochastic variable H were known to be ergodic, the time average H would be the same as the expected value $E(H(i) \mid i_1)$ of $H(i)$ at any time $(i \geqslant i_1)$ and this would provide a third version J_3 of the criterion:

$$ J_3 = E\big(H\big(x(i + 1), u(i)\big)\big). \tag{13.6c} $$

In equation (13.6c) the explicit dependence of H on i and the conditional dependence of $E(H)$ on i_1 is suppressed because the ergodic hypothesis is that the system is stationary with no such dependences. The state $x(i + 1)$ rather than $x(i)$ appears as the argument of H because in discrete-time systems having dynamic equations (13.1) $x(i + 1)$ is the state affected by $u(i)$. This distinction was unimportant with summation criteria such as J_1 which account for all values of $x(i)$ and $u(i)$, including the initial state $x(i_1)$ which is beyond control and the final control $u(i_2)$ which can have no useful effect; however, the single-stage criterion J_3 must be based on the state $x(i + 1)$ which depends on $u(i)$.

Problems 13.1, 13.2

13.2. Fixed-configuration problems

The analytic design procedure introduced in Chapter 6 for deterministic fixed-configuration problems carries over to problems with stochastic variables.

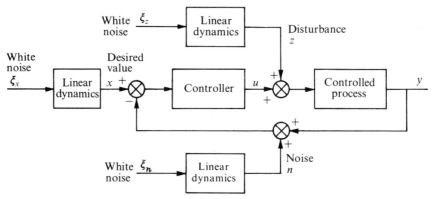

FIG. 13.2. Fixed-configuration control system.

Figure 13.2 is a block diagram of the class of single-variable systems considered. Transfer functions of the controlled process and of the controller are constant and specified except for values of certain controller parameters. All three types of stochastic variable, desired value x, disturbance z, and measurement noise n can be included. They are assumed to be generated by independent white noises ξ_x, ξ_z, ξ_n of specified magnitudes driving linear dynamics having specified constant transfer functions. The stochastic variables are thus assumed to be ergodic with specified autocorrelation functions. It is required to find values of the unspecified controller parameters so as to optimize a quadratic performance criterion.

In this stochastic system the error e:

$$e \equiv x - y$$

and the control u are unpredictable, ergodic stochastic variables. Because they are ergodic the time averages of the squares e^2 and u^2 which were used to define the quadratic performance criteria in Chapter 6 are the same as the ensemble averages or variances

$$\overline{e^2} = E(e^2) = \sigma_e^2,$$
$$\overline{u^2} = E(u^2) = \sigma_u^2.$$

These variances are completely determined by the above problem specification and can therefore be used to construct a quadratic performance criterion for the stochastic problem in the form J_3 of equation 13.6c. The typical criterion is

$$J = \sigma_e^2 + q\sigma_u^2.$$

Provided that the noises ξ_x, ξ_y, ξ_n are mutually independent, J can be evaluated using the principle of superposition, as in Section 2.4, by summing three terms:

$$J = J_x + J_y + J_n$$

each of which is due to one of the three external variables in the absence of the other two. These terms can be evaluated from equations (12.31) or (12.43), depending on whether the system is discrete-time or continuous-time, using Parseval's theorem and the methods of Section 6.2. The fixed-configuration design procedure is then the same as in Chapter 6.

A simple continuous-time example is the system shown in Fig. 13.3 where

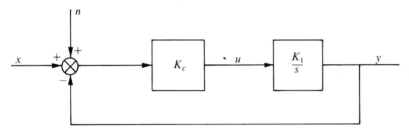

FIG. 13.3. A simple stochastic system.

the desired value x and the noise n in measuring error are assumed to be generated by independent white noises ξ_x and ξ_n, as in Fig. 13.2, each having unit magnitude and driving linear dynamic systems with transfer functions

$$G_x(s) = \frac{X}{1 + sT}, \quad G_n(s) = N.$$

Autocorrelation functions of these stochastic variables are, using equation (12.39),

$$R_x(\tau) = \frac{X^2}{2T} e^{-|\tau|/T}, \quad R_n(\tau) = N^2 \delta(\tau).$$

There are no disturbances z and it is required to choose a value for the gain K_c so as to minimize

$$J = \sigma_e^2 + q\sigma_u^2.$$

This is a stochastic version of the deterministic problem of Fig. 6.6 for which the choice of an optimal value of overall gain,

$$K \equiv K_c K_1,$$

is discussed in Section 6.3.

The variance σ_e^2 of the error is evaluated by regarding the error as the sum of two terms:

(i) $$e_x \equiv x - y_x,$$

due to imperfect following of the desired value x in the absence of noise. This
is related to x by the transfer function

$$\frac{E_x(s)}{X(s)} = \frac{1}{1 + K/s} = D_x(s);$$

(ii) $$e_n \equiv y_n,$$

due to the noise n. This is related to n by the transfer function

$$\frac{E_n(s)}{N(s)} = \frac{K/s}{1 + K/s} = D_n(s).$$

Provided that x and n are uncorrelated, the variance σ_e^2 is the sum of two
terms, one due to each component. These can be evaluated by substituting
the above transfer functions $G(s)$, $D(s)$ into equation (12.43) and using the
table of Fig. 6.5 which gives

$$\sigma_e^2 = \frac{X^2}{2T(1 + KT)} + \frac{KN^2}{2}.$$

The variance σ_u^2 of the control can similarly be regarded as the sum of two
terms, but the term due to the noise n is infinite. The significance of this is
that the simple controller with transfer function K_c will cause the control
signal u in this particular problem to take infinite values. The fixed-configuration design here is therefore restricted to minimizing performance
criteria in which the cost of control, measured by σ_u^2, is neglected:

$$J(q = 0) = \sigma_e^2.$$

When the cost of control is important the original performance criterion J
can be minimized using a dynamic controller having a transfer function which
is derived by the semi-free design procedure of Section 13.5.

For the restricted, fixed-configuration problem the minimizing value of K,
and of the controller gain K_c, is found by setting $\partial \sigma_e^2 / \partial K$ to zero which gives

$$K_c^* = \frac{1}{K_1 T}\left(\frac{X}{N} - 1\right).$$

This optimal value depends on the signal-to-noise ratio X/N. When X/N is
unity the optimal gain is zero, which implies that when the noise n is as large
as the desired-value signal x control cannot be usefully exercised and the best
control is no control. When the noise is not so large and does not swamp the
signal, the minimum value of error variance, achieved by using the above gain
K_c^* in the controller, is

$$(\sigma_e^2)^* = \frac{N^2}{T}\left(\frac{X}{N} - \frac{1}{2}\right).$$

Problems 13.3–13.5

13.3. Semi-free problems with uncertainty only about future states

When there is uncertainty about the state $\mathbf{x}$, its current value is described by a conditional probability density which evolves according to Bayes's rule, equation (12.4), as discussed in subsequent sections. In the present Section 13.3 some fundamental results of stochastic control theory are derived by considering the restricted class of discrete-time stochastic problems having uncertainty about future values of the state but not about present values.

Consider now this class of problems having uncertainty about future states as specified in equation (13.1),

$$\mathbf{x}(i+1) = \mathbf{G}\big(\mathbf{x}(i),\, \mathbf{u}(i),\, \xi(i),\, i\big)$$

and where it is required to find the series of controls $\mathbf{u}(i_1)\ldots\mathbf{u}(i_2)$ so as to minimize the performance criterion of equation (13.6a) with the mean value taken with respect to known probability densities for the set of independent random variables $\xi(i)$. Dynamic programming can be used here with solution functions, corresponding to those in the general deterministic discrete-time equation (8.13a), defined by

$f_N(\mathbf{x}, i_2) \equiv$ Expected value of I using optimal control over N stages, finishing at time i_2 and starting from state
$$\mathbf{x}(i_2 - N + 1) = \mathbf{x};$$
the expectation or mean value being with respect to known probability densities for
$$\xi(i_2 - N + 1), \ldots, \xi(i_2 - 1).$$

$\mathbf{u}_N(\mathbf{x}, i_2) \equiv$ Optimal control $\mathbf{u}^*(i_2 - N + 1)$ for a process finishing at time i_2 and starting from state
$$\mathbf{x}(i_2 - N + 1) = \mathbf{x}.$$

In a process finishing at time i_2 and starting from state
$$\mathbf{x}(i_2 - N + 1) = \mathbf{x}$$
when there are N stages-to-go, the state resulting from a first control
$$\mathbf{u}(i_2 - N + 1) = \mathbf{u},$$
will be
$$\mathbf{x}(i_2 - N + 2) = \mathbf{G}(\mathbf{x}, \mathbf{u}, \xi, i_2 - N + 1),$$
where ξ is the value taken by the set of random variables $\xi(i_2 - N + 1)$. The Principle of Optimality implies that when the optimal control is used the expected contribution to the sum I from the remaining $N - 1$ stages will be
$$f_{N-1}\big(\mathbf{G}(\mathbf{x}, \mathbf{u}, \xi, i_2 - N + 1), i_2\big).$$

The contribution to I from the first stage is

$$H(\mathbf{x}, \mathbf{u}, i_2 - N + 1),$$

and so the expected value of the performance criterion, provided that the random variables $\xi(i_2 - N + 1)$ take the value ξ, is

$$I = H(\mathbf{x}, \mathbf{u}, i_2 - N + 1) + f_{N-1}\big(G(\mathbf{x}, \mathbf{u}, \xi, i_2 - N + 1), i_2\big).$$

When the known probability density $p_{i_2-N+1}(\xi)$ for the random variables $\xi(i_2 - N + 1)$ is taken into account the expected value of I is

$$J_1 = \int_\zeta I\, p_{i_2-N+1}(\xi)\, \mathrm{d}\xi$$

$$= H(\mathbf{x}, \mathbf{u}, i_2 - N + 1) + \int_\zeta f_{N-1}\big(G(\mathbf{x}, \mathbf{u}, \xi, i_2 - N + 1), i_2\big) p_{i_2-N+1}(\xi)\, \mathrm{d}\xi.$$

When the optimal control is used this value of the performance criterion J_1 is minimized with respect to all controls, including the value $\mathbf{u}$ of the first control $\mathbf{u}(i_2 - N + 1)$, so the optimal value of J_1 over the N stages is

$$f_N(\mathbf{x}, i_2) = \min_{\mathbf{u}} \big\{ H(\mathbf{x}, \mathbf{u}, i_2 - N + 1) +$$

$$+ \int_\zeta f_{N-1}\big(G(\mathbf{x}, \mathbf{u}, \xi, i_2 - N + 1), i_2\big) p_{i_2-N+1}(\xi)\, \mathrm{d}\xi \big\}. \qquad (13.7)$$

Equation (13.7) is the functional recurrence equation specifying the solution function $f_N(\mathbf{x}, i_2)$ and the optimal control $\mathbf{u}_N(\mathbf{x}, i_2)$. An initial solution is provided by the single-stage process, $N = 1$, where the definitions give

$$f_1(\mathbf{x}, i_2) = \min_{\mathbf{u}} \{ H(\mathbf{x}, \mathbf{u}, i_2) \}. \qquad (13.8)$$

Equations (13.7) and (13.8) are the generalization of the deterministic dynamic programming equations of Chapter 8 to the class of stochastic problems considered here.

For stationary LQP-problems with linear dynamic equations (13.2),

$$\mathbf{x}(i + 1) = \mathbf{A}\mathbf{x}(i) + \mathbf{B}\mathbf{u}(i) + \xi(i)$$

and the quadratic scalar function H of equation (8.8),

$$H = \mathbf{x}'\mathbf{P}\mathbf{x} + \mathbf{u}'\mathbf{Q}\mathbf{u}, \qquad (13.9)$$

the solution functions f to the dynamic programming equations (13.7) and (13.8) are found to have the quadratic form

$$f_N(\mathbf{x}) = \mathbf{x}'\mathbf{V}_N\mathbf{x} + w_N,$$

where w_N is independent of $\mathbf{x}$. The validity of this solution is established by substituting it into the functional recurrence equation (13.7) to give

$$\mathbf{x}'\mathbf{V}_N\mathbf{x} + w_N = \min_{\mathbf{u}} \{\mathbf{x}'\mathbf{P}\mathbf{x} + \mathbf{u}'\mathbf{Q}\mathbf{u} +$$
$$+ E\big((\mathbf{A}\mathbf{x} + \mathbf{B}\mathbf{u} + \boldsymbol{\xi})'\mathbf{V}_{N-1}(\mathbf{A}\mathbf{x} + \mathbf{B}\mathbf{u} + \boldsymbol{\xi}) + w_{N-1}\big)\},$$

where the notation $E(\cdot)$ used for the expected value is as defined in equation (12.14). This expected value inside the bracketed term { } can be expressed, according to equation (12.21), in terms of the mean and the covariance matrix $\boldsymbol{\Sigma}_1$ of the random variables $\boldsymbol{\xi}$; when the mean of $\boldsymbol{\xi}$ is zero the functional recurrence equation becomes

$$\mathbf{x}'\mathbf{V}_N\mathbf{x} + w_N = \min_{\mathbf{u}} \{\mathbf{x}'\mathbf{P}\mathbf{x} + \mathbf{u}'\mathbf{Q}\mathbf{u} + (\mathbf{A}\mathbf{x} + \mathbf{B}\mathbf{u})'\mathbf{V}_{N-1}(\mathbf{A}\mathbf{x} + \mathbf{B}\mathbf{u})$$
$$+ \operatorname{tr}\mathbf{V}_{N-1}\boldsymbol{\Sigma}_1 + w_{N-1}\}.$$

The minimizing value of $\mathbf{u}$ here is the same as for the corresponding deterministic problem in Section 8.3 and is given by equations (8.9a) which are repeated here as equations (13.10):

where

$$\left.\begin{array}{c} \mathbf{u}_N(\mathbf{x}) = -\mathbf{K}_N\mathbf{x}, \\[2mm] \mathbf{K}_N = (\mathbf{Q} + \mathbf{B}'\mathbf{V}_{N-1}\mathbf{B})^{-1}\mathbf{B}'\mathbf{V}_{N-1}\mathbf{A}. \end{array}\right\} \qquad (13.10)$$

Substituting this minimizing value of $\mathbf{u}$ into the functional recurrence equation leads to the same difference equations (8.10a), repeated here as (13.11), as in Section 8.3 for $\mathbf{V}_N$:

$$\mathbf{V}_N = \mathbf{P} + \mathbf{A}'\mathbf{V}_{N-1}\mathbf{A} - \mathbf{A}'\mathbf{V}_{N-1}\mathbf{B}(\mathbf{Q} + \mathbf{B}'\mathbf{V}_{N-1}\mathbf{B})^{-1}\mathbf{B}'\mathbf{V}_{N-1}\mathbf{A} \qquad (13.11)$$

and to a difference equation for the extra term w_N:

$$w_N = w_{N-1} + \operatorname{tr}\mathbf{V}_{N-1}\boldsymbol{\Sigma}_1. \qquad (13.12)$$

Boundary conditions for the solution of these equations (13.11) and (13.12) are provided by equation (13.8) which gives

$$\mathbf{u}_1(\mathbf{x}) = \mathbf{0}, \qquad \mathbf{K}_1 = \mathbf{0}$$
$$\mathbf{V}_1 = \mathbf{P}, \qquad w_1 = 0. \qquad (13.13)$$

The boundary condition (13.13) on $\mathbf{V}_1$ is the same as equation (8.11a) for the deterministic problem and it follows that the matrices $\mathbf{V}_N$ and gains $\mathbf{K}_N$ specified by equations (13.11) and (13.10) are also the same as for the deterministic problem.

When the operating time and the number N of stages-to-go are infinite, equations (13.10) and (13.11) may have solutions which are independent of time as discussed in Section 8.3. If such steady-state solutions for $\mathbf{K}$ and $\mathbf{V}$ exist, the corresponding value of the term w_∞ given by equation (13.12) is infinite and the resulting value of the performance criterion J_1 is therefore also infinite. This is illustrated in Fig. 13.4 which shows how the stochastic scalar function H used in the performance criterion of equations (13.5) and (13.6)

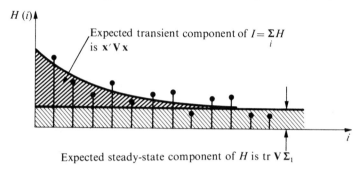

FIG. 13.4. Typical realization of $H(i)$ and composition of expected value of $I=\sum_i H(i)$.

might behave during a typical control sequence. The sum

$$I = \sum_{i=1}^{\infty} H(i)$$

of equation (13.5) can be regarded as being composed of two terms, a transient term having expected value $\mathbf{x}'\mathbf{V}\mathbf{x}$ and a steady-state term which increases at every stage by an increment having expected value tr $\mathbf{V}\boldsymbol{\Sigma}_1$ and so becomes infinite. The optimal control minimizes the infinite expected value J_1 of the sum I; it also minimizes the finite expected value J_2 of the time average of I and when the system has settled to an ergodic steady state it minimizes the corresponding finite expected value J_3 of H which becomes tr $\mathbf{V}\boldsymbol{\Sigma}_1$.

Equations (13.10), (13.11), (13.13) giving the optimal control are the same as equations (8.9a), (8.10a), (8.11a) for the corresponding deterministic system without random variables $\boldsymbol{\xi}$, and they have the same solution, which is the feedback control illustrated in Fig. 13.5. There is, however, an essential

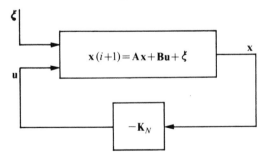

FIG. 13.5. Optimal feedback control for the LQP-problem.

difference between solutions to deterministic problems and solutions to stochastic problems. It is that although the dynamic programming analysis gives the optimal control in both cases as a feedback control law, that is as a function $-\mathbf{K}_N\mathbf{x}$ of the current state $\mathbf{x}$, the deterministic control need not be expressed in this form whereas the stochastic control must be. In deterministic problems, an initial state and a series of controls completely determine a series of states and so a control law can be expressed as a function of initial state and of time, as mentioned in Chapter 9. In stochastic problems with uncertainty about future states the only way to find the value of state needed to determine the optimal control of equation (13.10) is by measurement, or feedback, of the current state.

The conclusion that feedback control is essential in stochastic problems is not specific to the LQP-problem considered here. It is implicit in the assumption that solution functions giving optimal controls have the general form $\mathbf{u}_N(\mathbf{x}, i_2)$ of a function of the current state $\mathbf{x}$ and it is justified by the existence of solutions to the resulting general functional recurrence equation (13.7). It is one of the fundamental results of stochastic control theory and has already been quoted in previous chapters.

A conclusion which is specific to LQP-problems is that the feedback control law illustrated in Fig. 13.5 which is optimal for the stochastic problem is also optimal for the corresponding deterministic problem. In stochastic control a control law having the same feedback equations as the optimal control for a corresponding deterministic problem is called the *certainty-equivalent* control. Stochastic problems like this LQP-problem where the certainty-equivalent control is the optimal stochastic control are said to be certainty equivalent. Certainty equivalence is discussed more fully in Chapter 15.

Problems 13.6–13.10

13.4. Semi-free problems with uncertainty about present states

The fundamental result derived in Section 13.3, that the optimal control for a stochastic problem is a feedback control, applies to all stochastic control problems including those with uncertainty about the present state represented by an output equation in the general form of equation (13.3). In the presence of such uncertainty the current value of the state is described by a probability density. This is a conditional probability density in which information provided by the known sequences of past controls $\mathbf{u}$ and observed outputs $\mathbf{y}$ is combined with prior information about the initial value of the state $\mathbf{x}(i_1)$ according to the equations (13.1) and (13.3) governing the process. It evolves according to Bayes's rule, in a way which is discussed more fully in Chapter 14, as control is exercised and outputs are observed. In general the feedback law must specify the control $\mathbf{u}$ as a function of this conditional probability density.

The controller for a controlled process with uncertainty about present states therefore has two separate functions to perform:

(i) combine knowledge of previous control $\mathbf{u}(i-1)$ and resulting current output $\mathbf{y}(i)$ with previous information to give updated conditional probability density for current state $\mathbf{x}(i)$;

(ii) generate control $\mathbf{u}(i)$ as a feedback function of resulting conditional probability density for $\mathbf{x}(i)$.

Figure 13.6 illustrates the separation of these two functions in a controller for the general class of systems of Fig. 13.1. The design of optimal feedback laws to perform function (ii) is discussed here; the other function (i) of updating conditional probability densities is discussed in Chapter 14.

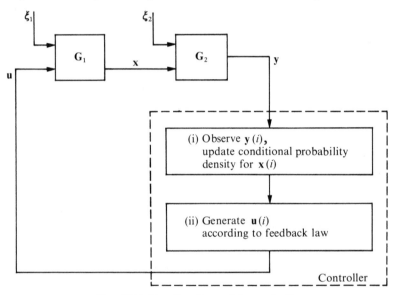

FIG. 13.6. Two functions of the controller.

The analysis of feedback laws in the presence of uncertainty about the present state has to take account of the conditional probability density for the state and of known probability densities for the random variables ξ_1 and ξ_2. It is thus more complex than previous analyses and so it is convenient to introduce some new notation to simplify the algebra. The sequences of controls $\mathbf{u}$ and outputs $\mathbf{y}$ which contribute knowledge about the current state $\mathbf{x}(i)$ are represented by y^i:

$$y^i \equiv \{\mathbf{u}(i_1), \ldots, \mathbf{u}(i-1), \mathbf{y}(i_1+1), \ldots, \mathbf{y}(i)\},$$

where the sequence of controls stops at $\mathbf{u}(i-1)$ because the control $\mathbf{u}(i)$ has no effect on $\mathbf{x}(i)$ in equation (13.1), and the sequence of outputs starts at

$\mathbf{y}(i_1 + 1)$ because it is assumed that the initial output $\mathbf{y}(i_1)$ has already been used to establish a prior probability density for the initial state $\mathbf{x}(i_1)$. The sequences of random variables $\boldsymbol{\xi}_1$ and $\boldsymbol{\xi}_2$ which will contribute future uncertainty affecting the sum I in the performance criterion of equation (13.6a) are represented by ξ^i:

$$\xi^i \equiv \{\boldsymbol{\xi}_1(i), \ldots, \boldsymbol{\xi}_1(i_2 - 1), \boldsymbol{\xi}_2(i + 1), \ldots, \boldsymbol{\xi}_2(i_2 - 1)\},$$

where both sequences stop at time $i_2 - 1$ because the value of each at the final time i_2 has no effect on the sum I defined by equation (13.5), and the sequence of noises $\boldsymbol{\xi}_2$ starts at $\boldsymbol{\xi}_2(i + 1)$ because the current noise $\boldsymbol{\xi}_2(i)$ will have already been accounted for in the conditional probability density for the current state $\mathbf{x}(i)$.

The conditional probability density is written

$$p(\mathbf{x}(i) \mid y^i_\cdot)$$

and in this context the notation $p(\bullet)$ is not supposed to represent a specific function of the argument $\bullet$ but is shorthand for 'the (appropriate) probability density for $\bullet$' where what is appropriate is specified by the argument $\bullet$. Thus

$p(\mathbf{x}(i_1))$ is shorthand for 'the prior probability density of $\mathbf{x}(i_1)$',
$p(\mathbf{x}(i) \mid y^i)$ is shorthand for 'the probability density of $\mathbf{x}(i)$ conditional on y^i',

and there is no implication that these density functions are identical.

With this notation the solution functions for a general dynamic programming analysis to minimize the performance criterion of equation (13.6a) become

$f_N(p(\mathbf{x} \mid y), i_2) \equiv$ Expected value of I using optimal control over N stages, finishing at time i_2 and starting from uncertain state $\mathbf{x}(i_2 - N + i)$ described by

$$p(\mathbf{x}(i_2 - N + 1) \mid y^{i_2 - N + 1}) = p(\mathbf{x} \mid y),$$

the expectation being with respect to $p(\mathbf{x} \mid y)$ and to known probability densities for $\xi^{i_2 - N + 1}$;

$\mathbf{u}_N(p(\mathbf{x} \mid y), i_2) \equiv$ Optimal control $\mathbf{u}^*(i_2 - N + 1)$ for a process finishing at time i_2 and starting from uncertain state $\mathbf{x}(i_2 - N + 1)$ described by

$$p(\mathbf{x}(i_2 - N + 1) \mid y^{i_2 - N + 1}) = p(\mathbf{x} \mid y).$$

These are now functionals rather than functions; that is the argument of $f_N(\ )$ and of $\mathbf{u}_N(\ )$ includes the probability density function $p(\mathbf{x} \mid y)$ which is itself a function of the state variables $\mathbf{x}$. General dynamic programming recurrence equations specifying optimal controls can be written in terms of such functionals but they are intractable and little is known about their solutions.

Attention is therefore concentrated on stochastic problems where the conditional probability density function $p(\mathbf{x} \mid y)$ is completely specified by values of a finite number of sufficient statistics such as the mean m and variance σ^2 of the normal density function of equation (12.13). LQP-problems have probability densities specified by sufficient statistics when their random variables are all normally distributed. It is shown in Chapter 14 that for a controlled process with

(i) linear dynamic equation (13.2),
(ii) linear output equation (13.4),
(iii) normally-distributed independent random variables ξ_1 and ξ_2 having zero means and covariance matrices Σ_1 and Σ_2,
(iv) normal prior probability density for initial state $\mathbf{x}(1)$ having mean $\mathbf{m}(1)$ and covariance matrix $\Sigma(1)$,

the conditional probability density $p(\mathbf{x}(i) \mid y^i)$ is normal with mean $\mathbf{m}(i)$ and covariance matrix $\Sigma(i)$, which are sufficient statistics, specified by the difference equations†

$$\mathbf{m}(i) = \mathbf{A}\mathbf{m}(i - 1) + \mathbf{B}\mathbf{u}(i - 1) +$$

$$+ \ \Sigma(i)\mathbf{C}'\Sigma_2^{-1}\{\mathbf{y}(i) - \mathbf{C}(\mathbf{A}\mathbf{m}(i - 1) + \mathbf{B}\mathbf{u}(i - 1))\} \quad (13.14a)$$

$$\Sigma(i) = \{(\Sigma_1 + \mathbf{A}\Sigma(i - 1)\mathbf{A}')^{-1} + \mathbf{C}'\Sigma_2^{-1}\mathbf{C}\}^{-1}. \quad (13.14b)$$

The significance of various terms in these equations (13.14) is discussed in Chapter 14; here they are quoted as examples of sufficient statistics and as part of the solution to LQP-control problems. The function of updating conditional probability densities according to Bayes's rule, shown as (i) in Fig. 13.6, is in general complicated and non-linear and so it is not usual to find sufficient statistics in problems having non-linear equations (13.1) or (13.3) or non-normal random variables.

For stationary LQP-problems with the sufficient statistics of equations (13.14), solution functions to the dynamic programming equations are

$f_N(\mathbf{m}, \Sigma) \equiv$ Expected value of I using optimal control over N stages starting from uncertain state $\mathbf{x}(i)$ described by conditional probability density $p(\mathbf{x}(i) \mid y^i)$ having mean $\mathbf{m}(i) = \mathbf{m}$ and variance $\Sigma(i) = \Sigma$; the expectation being with respect to this probability density and to known probability densities for ξ^i;

$\mathbf{u}_N(\mathbf{m}, \Sigma) \equiv$ Optimal control $\mathbf{u}^*(i)$ for a process with N stages-to-go and starting from uncertain state described by the above conditional probability density.

The Principle of Optimality can be used here, as previously, to derive equations specifying the optimal control in terms of the sufficient statistics $\mathbf{m}$ and

† These equations are an example of the class of estimators known as Kalman filters.

Σ. Derivation of the equations is more complicated than previously because they have to be expressed in terms of the information available to the controller at time i when the control $\mathbf{u}(i)$ is implemented.

If at time i the control $\mathbf{u}(i)$ were to take the value $\mathbf{u}$, the uncertain value of the next state $\mathbf{x}(i + 1)$ would be described by a conditional probability density

$$p\big(\mathbf{x}(i + 1) \,|\, y^i, \mathbf{u}(i) = \mathbf{u}\big),$$

having expected value, given by equations (12.19) and (13.2),

$$E\big(\mathbf{x}(i + 1) \,|\, y^i, \mathbf{u}(i) = \mathbf{u}\big), = \mathbf{Am} + \mathbf{Bu} \tag{13.15a}$$

and covariance matrix, given by a multivariable version of equation (12.18) together with equation (13.2) and a result† which follows from the definition in Section 12.3 of covariance matrices,

$$\text{Covariance}\,\big(\mathbf{x}(i + 1) \,|\, y^i, \mathbf{u}(i) = \mathbf{u}\big) = \mathbf{A}\Sigma\mathbf{A}' + \Sigma_1. \tag{13.15b}$$

This conditional probability density affects the term

$$f_{N-1}\big(\mathbf{m}(i + 1), \Sigma(i + 1)\big)$$

which will be used in the dynamic programming analysis to account for the expected contribution to the sum I from the $N - 1$ stages following stage i. The value of $\mathbf{m}(i + 1)$ here, given by equation (13.14a),

$$\mathbf{m}(i + 1) = \mathbf{Am} + \mathbf{Bu} +$$
$$+ \Sigma(i + 1)\mathbf{C}'\Sigma_2^{-1}\big(\mathbf{Cx}(i + 1) + \xi_2(i + 1) - \mathbf{C}(\mathbf{Am} + \mathbf{Bu})\big), \tag{13.16}$$

is random because of the random terms $\mathbf{Cx}(i + 1) + \xi_2(i + 1)$ which appear, according to the output equation (13.4), in place of the measurement $\mathbf{y}(i + 1)$ which is not available at stage i. It is described by a conditional probability density

$$p\big(\mathbf{m}(i + 1) \,|\, y^i, \mathbf{u}(i) = \mathbf{u}\big),$$

having expected value, given by equations (12.19), (13.15a), and (13.16),

$$E\big(\mathbf{m}(i + 1) \,|\, y^i, \mathbf{u}(i) = \mathbf{u}\big) = \mathbf{Am} + \mathbf{Bu} \tag{13.17a}$$

and covariance matrix, given by the result used above together with equations (13.15b) and (13.16):

$$\text{Covariance}\,\big(\mathbf{m}(i + 1) \,|\, y^i, \mathbf{u}(i) = \mathbf{u}\big) =$$
$$\Sigma(i + 1)\mathbf{C}'\Sigma_2^{-1}\big(\mathbf{C}(\mathbf{A}\Sigma\mathbf{A}' + \Sigma_1)\mathbf{C}' + \Sigma_2\big)\Sigma_2^{-1}\mathbf{C}\Sigma(i + 1). \tag{13.17b}$$

The value of the covariance matrix $\Sigma(i + 1)$ in the above equations is not random but is given by writing $i + 1$ in place of i in equation (13.14b). It is

---

† The result used here is that the covariance matrix of a linear sum $\mathbf{Ax}$ of random variables $\mathbf{x}$ having covariance matrix Σ is $\mathbf{A}\Sigma\mathbf{A}'$.

independent of the sequences of controls and outputs y^i and so is completely determined in advance.

The Principle of Optimality implies that when the optimal control is used the expected contribution to the sum I from the $N - 1$ stages following stage i will be $f_{N-1}(\mathbf{m}(i + 1), \Sigma(i + 1))$. The contribution to I from the first stage is

$$\mathbf{x}'(i)\mathbf{P}\mathbf{x}(i) + \mathbf{u}'\mathbf{Q}\mathbf{u}.$$

The expected value of I is then

$$J = E\{(\mathbf{x}'(i)\mathbf{P}\mathbf{x}(i) + \mathbf{u}'\mathbf{Q}\mathbf{u} + f_{N-1}(\mathbf{m}(i + 1), \Sigma(i + 1)))| y^i, \mathbf{u}(i) = \mathbf{u}\}$$

and when the optimal control is used the performance criterion J is minimized with respect to all controls, including the value $\mathbf{u}$ of the first control $\mathbf{u}(i)$, so the optimal value of J over the N stages is

$$f_N(\mathbf{m}, \Sigma) = \min_{\mathbf{u}} \left[E\{(\mathbf{x}'(i)\mathbf{P}\mathbf{x}(i) + \mathbf{u}'\mathbf{Q}\mathbf{u} + \right.$$
$$\left. + f_{N-1}(\mathbf{m}(i + 1), \Sigma(i + 1)))| y^i, \mathbf{u}(i) = \mathbf{u} \}\right]. \quad (13.18)$$

Equation (13.18) is the functional recurrence equation specifying solutions to this LQP-problem. The solution functions f are found to be quadratic in $\mathbf{m}$:

$$f_N(\mathbf{m}, \Sigma) = \mathbf{m}'\mathbf{V}_N\mathbf{m} + w_N$$

as in Section 13.3. The validity of this solution is established by substituting it into the functional recurrence equation (13.18); then using the result of equation (12.21) to evaluate mean values with respect to the conditional probability densities of equation (13.14) and (13.17) equation (13.18) becomes

$$\mathbf{m}'\mathbf{V}_N\mathbf{m} + w_N = \min_{\mathbf{u}} \{\mathbf{m}'\mathbf{P}\mathbf{m} + \operatorname{tr} \mathbf{P}\Sigma + \mathbf{u}'\mathbf{Q}\mathbf{u} +$$
$$+ (\mathbf{A}\mathbf{m} + \mathbf{B}\mathbf{u})'\mathbf{V}_{N-1}(\mathbf{A}\mathbf{m} + \mathbf{B}\mathbf{u}) +$$
$$+ \operatorname{tr} \mathbf{V}_{N-1}\left(\Sigma(i + 1)\mathbf{C}'\Sigma_2^{-1}(\mathbf{C}(\mathbf{A}\Sigma\mathbf{A}' + \Sigma_1)\mathbf{C}' + \Sigma_2)\Sigma_2^{-1}\mathbf{C}\Sigma(i + 1)\right) +$$
$$+ w_{N-1}\}. \quad (13.19)$$

The minimizing value of $\mathbf{u}$ here is given by expressions similar to those for the deterministic problem in Section 8.3 and for the stochastic problem in Section 13.3:

$$\mathbf{u}_N(\mathbf{m}, \Sigma) = -\mathbf{K}_N\mathbf{m},$$

where

$$\mathbf{K}_N = (\mathbf{Q} + \mathbf{B}'\mathbf{V}_{N-1}\mathbf{B})^{-1}\mathbf{B}'\mathbf{V}_{N-1}\mathbf{A}. \quad \left.\right\} \quad (13.20)$$

Substituting this minimizing value of $\mathbf{u}$ into the recurrence equation (13.19) leads to the same difference equation for $\mathbf{V}_N$ as equation (8.10a) or (13.11) and to an equation for the extra term w_N,

$$w_N = w_{N-1} +$$
$$+ \text{ tr } \mathbf{V}_{N-1}\left(\boldsymbol{\Sigma}(i+1)\mathbf{C}'\boldsymbol{\Sigma}_2^{-1}(\mathbf{C}(\mathbf{A}\boldsymbol{\Sigma}\mathbf{A}' + \boldsymbol{\Sigma}_1)\mathbf{C}' + \boldsymbol{\Sigma}_2)\boldsymbol{\Sigma}_2^{-1}\mathbf{C}\boldsymbol{\Sigma}(i+1)\right) +$$
$$+ \text{ tr } \mathbf{P}\boldsymbol{\Sigma}, \tag{13.21}$$

which is a generalization of equation (13.12). Boundary conditions for the solution of these difference equations are derived from the single-stage process, $N = 1$, where the definitions give

$$\mathbf{V}_1 = \mathbf{P}, \qquad w_1 = \text{ tr } \mathbf{P}\boldsymbol{\Sigma},$$
$$\tag{13.22}$$
$$\mathbf{u}_1(\mathbf{m}, \boldsymbol{\Sigma}) = \mathbf{0}, \qquad \mathbf{K}_1 = \mathbf{0}.$$

Equations (13.20)–(13.22) specify the optimal feedback control law for the function (ii) of generating the control $\mathbf{u}(i)$ in the controller of Fig. 13.6. Together with equations (13.14) which specify the function (i) of updating the normal conditional probability densities they give the solution to the stochastic LQP-problem of equations (13.2), (13.4), and (13.9). Figure 13.7 shows the optimal controller performing these two separate functions. The

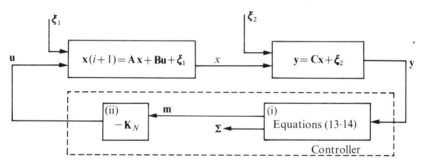

FIG. 13.7. Optimal stochastic LQP-control.

feedback-control law here is certainty equivalent because equations (13.20) specifying the optimal control for the stochastic problem are identical to the equations (8.9a) of the optimal feedback-control law for the corresponding deterministic problem having no uncertainty about the state $\mathbf{x}$. In the stochastic problem the certainty-equivalent control is generated by using the conditional mean $\mathbf{m}$ in place of the actual value of the state. The covariance matrix $\boldsymbol{\Sigma}$ has no effect on the control function (ii), although it is needed as a sufficient statistic for the other controller function (i) of updating the conditional probability density according to equations (13.14). Stochastic problems such as this one in which the argument of the optimal control law is the conditional mean $\mathbf{m}$ and not any other feature of the conditional probability density $p(\mathbf{x} \mid y^i)$, such as for example the variance $\boldsymbol{\Sigma}$, are sometimes described as being *separable*. This technical meaning of the word 'separable' is more specific than common usage which would be to describe the general stochastic controller of Fig. 13.6 as separable regardless of what features of the conditional probability density $p(\mathbf{x} \mid y^i)$ were needed to generate the feedback control. Separability is discussed more fully in Chapter 15.

Problems 13.11–13.15

13.5. Generalizations

The results in Sections 13.3 and 13.4 about optimal stochastic LQP-control can be generalized in the same way that deterministic LQP-results were generalized in Section 8.4. Stochastic desired-value variables, such as target positions, can be included by augmenting the state vector $\mathbf{x}$ to include state variables of a dynamic system like that in Fig. 12.5 describing the stochastic variables. Non-stationary problems, including terminal-control problems, can also be solved.

Corresponding results exist for continuous-time systems. The continuous-time version of the problem in Section 13.4 has dynamics described by a stochastic differential equation† like equation (12.34),

$$dx = \mathbf{A}\mathbf{x} \, dt + \mathbf{B}\mathbf{u} \, dt + d\mathbf{w}_1,$$

and noisy output, also described by a stochastic differential equation,†

$$\mathbf{y} \, dt = \mathbf{C}\mathbf{x} \, dt + d\mathbf{w}_2,$$

where $\mathbf{w}_1$ and $\mathbf{w}_2$ are independent Wiener processes having zero means and covariance matrices, proportional to time, $\mathbf{\Sigma}_1 \int dt$ and $\mathbf{\Sigma}_2 \int dt$ respectively. The quadratic performance criterion is

$$J = E\left(\int_0^T (\mathbf{x}'\mathbf{P}\mathbf{x} + \mathbf{u}'\mathbf{Q}\mathbf{u}) \, dt \right),$$

where the expected value is with respect to probability distributions for all the uncertain variables, states $\mathbf{x}$ and independent random variables $\mathbf{w}$. When these are all normal, sufficient statistics $\mathbf{m}$ and $\mathbf{\Sigma}$ of the conditional probability density for the state $\mathbf{x}$ are given, as discussed in Chapter 14, by the differential equations‡

$$\frac{d\mathbf{m}}{dt} = \mathbf{A}\mathbf{m} + \mathbf{B}\mathbf{u} + \mathbf{\Sigma}\mathbf{C}'\mathbf{\Sigma}_2^{-1}(\mathbf{y} - \mathbf{C}\mathbf{m}), \qquad (13.23a)$$

$$\frac{d\mathbf{\Sigma}}{dt} = \mathbf{\Sigma}_1 + \mathbf{A}\mathbf{\Sigma} + \mathbf{\Sigma}\mathbf{A}' - \mathbf{\Sigma}\mathbf{C}'\mathbf{\Sigma}_2^{-1}\mathbf{C}\mathbf{\Sigma}. \qquad (13.23b)$$

† It is sometimes convenient to write these equations in the mathematically improper form

$$\dot{\mathbf{x}} = \mathbf{A}\mathbf{x} + \mathbf{B}\mathbf{u} + \mathbf{\xi}_1,$$
$$\mathbf{y} = \mathbf{C}\mathbf{x} + \mathbf{\xi}_2,$$

where $\mathbf{\xi}_1, \mathbf{\xi}_2$ are sets of white noises having autocorrelation functions $\mathbf{\Sigma}_1 \, \delta(\tau), \mathbf{\Sigma}_2 \, \delta(\tau)$.

‡ These are the equations of a continuous-time Kalman filter.

The optimal value of the performance criterion J is found to have the quadratic form

$$f(\mathbf{m}, \boldsymbol{\Sigma}, T) = \mathbf{m}'\mathbf{V}(T)\mathbf{m} + w(T),$$

where $\mathbf{V}(T)$ satisfies the same Riccati differential equation (8.10b), repeated here as (13.24), as in Section 8.3:

$$\frac{\mathrm{d}\mathbf{V}}{\mathrm{d}T} = \mathbf{P} + \mathbf{A}'\mathbf{V} + \mathbf{V}\mathbf{A} - \mathbf{V}\mathbf{B}\mathbf{Q}^{-1}\mathbf{B}'\mathbf{V} \tag{13.24}$$

with initial value, as in equation (8.11b),

$$\mathbf{V}(0) = \mathbf{0}.$$

The additional term $w(T)$ satisfies a differential equation

$$\frac{\mathrm{d}w}{\mathrm{d}T} = \mathrm{tr}\,(\mathbf{V}\mathbf{B}\mathbf{Q}^{-1}\mathbf{B}'\mathbf{V}\boldsymbol{\Sigma}) + \mathrm{tr}\,(\mathbf{V}\boldsymbol{\Sigma}_1) \tag{13.25}$$

with initial value

$$w(0) = 0,$$

but does not affect the optimal control law. The optimal control law is the same as for the corresponding deterministic problem in Section 8.3 and is given by equations (8.9b) which are repeated here as equations (13.26):

$$\mathbf{u}(\mathbf{m}, \boldsymbol{\Sigma}, T) = -\mathbf{K}(T)\mathbf{m},$$

where

$$\mathbf{K}(T) = \mathbf{Q}^{-1}\mathbf{B}'\mathbf{V}. \tag{13.26}$$

The continuous-time problem is thus, like the discrete-time problem, both certainty equivalent and separable.

An example of the application of the foregoing results about optimal stochastic LQP-control is provided by considering the semi-free design problem illustrated in Fig. 13.8. This is the semi-free version of the fixed-configuration problem of Fig. 13.3 which was considered in Section 13.2 where

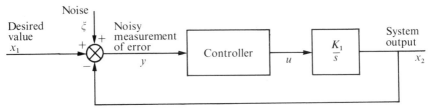

FIG. 13.8. Semi-free version of the problem in Fig. 13.3.

the controller was given the form:

$$u = K_c \times (\text{noisy measurement of error})$$

and the optimal value of K_c, minimizing the variance of the error, was found. The semi-free problem solved here finds the optimal form of controller as well as specifying optimal values for its parameters.

The desired value x_1 is regarded as the output of a linear dynamic system having transfer function $X/(1 + sT)$ and driven by white noise of unit magnitude. The stochastic variable x_1 is therefore described by a stochastic differential equation (12.34):

$$\mathrm{d}x_1 = -\frac{1}{T} x_1 \, \mathrm{d}t + \mathrm{d}w_1,$$

where w_1 is a Wiener process having covariance proportional to X^2/T^2. The white measurement noise ξ is similarly represented by a Wiener process having covariance proportional to N^2. The system output x_2 satisfies a first-order differential equation

$$\frac{\mathrm{d}x_2}{\mathrm{d}t} = K_1 u$$

and so the dynamics of the system can be written in the standard LQP-form with

$$\mathbf{A} = \begin{bmatrix} -\dfrac{1}{T} & 0 \\ 0 & 0 \end{bmatrix}, \qquad \mathbf{B} = \begin{bmatrix} 0 \\ K_1 \end{bmatrix}.$$

The performance criterion, introduced in Section 13.2 and based on variances of the error and the control, is

$$J = \sigma^2_{(x_1 - x_2)} + q\sigma^2_u,$$

and can be written in the standard quadratic form with

$$\mathbf{P} = \begin{bmatrix} 1 & -1 \\ -1 & 1 \end{bmatrix}, \qquad \mathbf{Q} = q.$$

This form of performance criterion implies that the system operates over infinite time and that the optimal control law is constant and given by the steady-state solution of the Riccati equations (13.24) for $\mathbf{V}$, which, with the above matrices, are

$$\frac{\mathrm{d}v_{11}}{\mathrm{d}T} = 1 - \frac{2}{T} v_{11} - \frac{K_1^2}{q} (v_{12})^2 = 0,$$

$$\frac{\mathrm{d}v_{12}}{\mathrm{d}T} = -1 - \frac{v_{12}}{T} - \frac{K_1^2}{q} v_{12}v_{22} = 0,$$

$$\frac{dv_{22}}{dT} = 1 - \frac{K_1^2}{q} (v_{22})^2 = 0 .$$

The certainty-equivalent optimal control law of equation (13.26) here,

$$u(\mathbf{m}, \mathbf{\Sigma}, T) = -q^{-1} \begin{bmatrix} 0 & K_1 \end{bmatrix} \begin{bmatrix} v_{11} & v_{12} \\ v_{12} & v_{22} \end{bmatrix} \begin{bmatrix} m_1 \\ m_2 \end{bmatrix}$$

$$= -\frac{K_1}{q} v_{12} m_1 - \frac{K_1}{q} v_{22} m_2,$$

depends only on v_{12} and v_{22} which have steady-state values

$$v_{22} = \sqrt{q}/K_1, \quad \hat{v}_{12} = -\left(\frac{1}{T} + \frac{K_1}{\sqrt{q}}\right)^{-1}$$

giving

$$u(\mathbf{m}, \mathbf{\Sigma}, \infty) = \frac{m_1}{(\sqrt{q} + q/K_1 T)} - \frac{m_2}{\sqrt{q}}.$$

The conditional mean $\mathbf{m}$ of the state $\mathbf{x}$ here is to be computed using the noisy observation y of error, which satisfies a stochastic differential equation

$$y \, dt = (x_1 - x_2) \, dt + dw_2$$

and is such that the matrix $\mathbf{C}$ is

$$\mathbf{C} = [1 \ -1].$$

The covariance matrix $\mathbf{\Sigma}$ in a system operating over infinite time, as in this example, goes to a steady-state value in the same way as does the matrix $\mathbf{V}$ of the solution function. So here, where the matrices $\mathbf{\Sigma}_1$ and $\mathbf{\Sigma}_2$ specifying the Wiener processes $\mathbf{w}$ are

$$\mathbf{\Sigma}_1 = \begin{bmatrix} X^2/T^2 & 0 \\ 0 & 0 \end{bmatrix}, \quad \mathbf{\Sigma}_2 = N^2,$$

equations (13.23b) give

$$\frac{d\sigma_{11}}{dt} = \frac{X^2}{T^2} - 2 \frac{\sigma_{11}}{T} - \frac{(\sigma_{11} - \sigma_{12})^2}{N^2} = 0,$$

$$\frac{d\sigma_{12}}{dt} = -\frac{\sigma_{12}}{T} + \frac{(\sigma_{11} - \sigma_{12})(\sigma_{22} - \sigma_{12})}{N^2} = 0,$$

$$\frac{d\sigma_{22}}{dt} = -\frac{(\sigma_{22} - \sigma_{12})^2}{N^2} = 0,$$

and the steady-state solutions are

$$\sigma_{12} = \sigma_{22} = 0,$$

$$\sigma_{11} = \frac{N^2}{T}\left\{\sqrt{\left(1 + \left(\frac{X}{N}\right)^2\right)} - 1\right\}.$$

The significance of these solutions is that uncertainty about the stochastic desired value x_1 persists but uncertainty about the output x_2, which depends only on the known control, u, can be eliminated. Using these solutions for the covariance matrix $\boldsymbol{\Sigma}$, equations (13.23a) for the conditional mean $\mathbf{m}$ become

$$\frac{dm_1}{dt} = -\frac{m_1}{T} + \frac{1}{T}\left\{\sqrt{\left(1 + \left(\frac{X}{N}\right)^2\right)} - 1\right\}(y - m_1 + m_2),$$

$$\frac{dm_2}{dt} = K_1 u.$$

These differential equations together with the certainty-equivalent optimal control law specify the optimal controller which is the solution to the semi-free design problem.

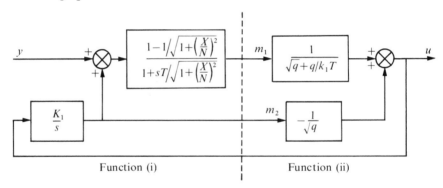

Function (i)   Function (ii)

FIG. 13.9. Block diagram of controller for example in Fig. 13.8.

Figure 13.9 is a block diagram of the optimal controller showing the two separate functions: (i) updating the conditional probability density and (ii) exercising control. The transfer function of this controller

$$\frac{U(s)}{Y(s)} = \frac{\left\{\sqrt{\left(1 + \left(\frac{X}{N}\right)^2\right)} - 1\right\}s}{K_1 + \frac{\sqrt{q}}{T}\sqrt{\left(1 + \left(\frac{X}{N}\right)^2\right)} + } $$

$$+ \left\{K_1 T + \sqrt{q} + \left(\sqrt{q} + \frac{q}{K_1 T}\right)\sqrt{\left(1 + \left(\frac{X}{N}\right)^2\right)}\right\}s + \left(T\sqrt{q} + \frac{q}{K_1}\right)s^2$$

is dynamic and of the same order, two, as the dynamics specified by the matrix
A. The minimal value of the performance criterion J which results from using
the optimal controller can be found from equation (13.25), as discussed in
connection with Fig. 13.4, by evaluating

$$J = \text{tr } (\mathbf{VBQ}^{-1}\mathbf{BV'\Sigma}) + \text{tr } (\mathbf{V\Sigma_1}). \tag{13.27}$$

For the example considered here equation (13.27) gives

$$J = \frac{K_1^2}{q} v_{12}^2 \sigma_{11} + \frac{X^2}{T^2} v_{11}$$

$$= \frac{N^2}{T} \left(\frac{1}{2} \left(\frac{X}{N} \right)^2 + \left(\frac{1}{1 + \sqrt{q/K_1 T}} \right)^2 \left\{ \sqrt{\left(1 + \left(\frac{X}{N} \right)^2 \right)} - 1 - \frac{1}{2} \left(\frac{X}{N} \right)^2 \right\} \right).$$

The solution to the semi-free problem can be compared with the solution
of Section 13.2 to the fixed-configuration problem when the cost of control is
neglected by letting q approach zero so that the optimal controller has transfer function

$$\frac{U(s)}{Y(s)} = \frac{1}{K_1} \left\{ \sqrt{\left(1 + \left(\frac{X}{N} \right)^2 \right)} - 1 \right\} \frac{s}{1 + sT}$$

as compared to the fixed-configuration controller with gain

$$K_c^* = \frac{1}{K_1 T} \left(\frac{X}{N} - 1 \right).$$

The resulting optimal value of the semi-free performance criterion,

$$J = \frac{N^2}{T} \left\{ \sqrt{\left(1 + \left(\frac{X}{N} \right)^2 \right)} - 1 \right\},$$

can be compared to the minimum fixed-configuration variance

$$(\sigma_e^2)^* = \frac{N^2}{T} \left(\frac{X}{N} - \frac{1}{2} \right)$$

as in Fig. 13.10. The dynamic semi-free controller can be used over the whole
range of values of signal-to-noise ratio X/N whereas the fixed-configuration
controller is only useful when X/N is greater than unity and then the semi-free
performance criterion is smaller, and therefore better. A further advantage of
the class of semi-free controllers is that they can keep the control signal u to
finite values so as to optimize a performance criterion including a cost of
control term, which the fixed-configuration controller with transfer function
K_c was unable to do.

The optimal semi-free controller in the above example is typical of con-

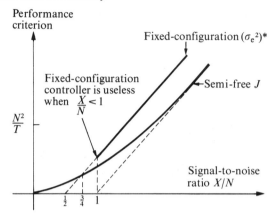

FIG. 13.10. Comparison of fixed-configuration and semi-free controllers.

trollers for the general class of stochastic LQP-problems in that its dynamic characteristics are associated with the function (i) of updating the conditional probability density for the state. The block diagram of the particular continuous-time controller in Fig. 13.9 and also the diagram in Fig. 13.7 of the general class of optimal discrete-time control systems both show that the function (ii) of exercising control requires only a gain matrix of the form

$$\mathbf{u} = -\mathbf{Km}$$

and that it is the function (i) of updating the conditional mean **m** which requires dynamic operations specified by the difference equations (13.14) or the differential equations (13.23). It follows that the integrators which are a well-known feature of real controllers as discussed in Chapters 2 to 5 are associated with this function (i) of estimating current values of uncertain states. Integral control can be regarded as a sub-optimal approximation to the optimal estimator of equations (13.23) which have the form

$$\dot{\mathbf{m}} = \text{function } (\mathbf{m}, \mathbf{\Sigma}, \mathbf{u}, \mathbf{y}),$$
$$\dot{\mathbf{\Sigma}} = \text{function } (\mathbf{\Sigma}) = \mathbf{0} \text{ in steady state}$$

of a set of integrators with outputs **m**.

The foregoing explanation of the role of integrators in feedback controllers is one of the fundamental results of stochastic LQP-control theory. Other results which have been established in this chapter are that feedback is essential in stochastic control problems; that LQP-problems with normal probability densities are separable in the technical sense that the argument of the optimal feedback control law is the conditional mean **m**; and that this optimal control law is the certainty-equivalent control law and is linear.

Problem 13.16

13.6. Bibliography

The method of Section 13.2 and Chapter 6, which uses transfer function analysis to solve fixed configuration problems, is closely related to the method developed for semi-free stationary stochastic LQP-problems by Kolmogorov (1941) and by Wiener (1949). These analytic methods are the subject of

NEWTON, GOULD, and KAISER (1957).

Applications of dynamic programming to stochastic problems are described in
BELLMAN (1961),
BELLMAN and KALABA (1965) Chapters 4 and 5,
FEL'DBAUM (1965),
AOKI (1967),
JACOBS (1967) Chapters 4 and 5,
BOUDAREL, DELMOS, and GUICHET (1971) Part 2,
KUSHNER (1971).

Optimal stochastic LQP-control is discussed in
TOU (1963),
LEE (1964) Chapter 5,
NOTON (1965) Chapter 6; (1972) Section 5.6,
AOKI (1967) Chapter II,
SAGE (1968) Section 11.4,
BRYSON and HO (1969) Chapter 14,
MEDITCH (1969) Chapters 9 and 10,
ÅSTRÖM (1970) Chapter 8,
TAKAHASHI, RABINS, and AUSLANDER (1970) Section 14.6,
KUSHNER (1971) Chapter 9 and Section 11.4,
KWAKERNAAK and SIVAN (1972) Chapter 5.

13.7. Problems

13.1. In the system of Fig. 13.2 which of the following quantities α are (a) stochastic variables? (b) random numbers? (c) deterministic numbers? (d) zero?

(i) $\alpha_1 = x - y + u$.

(ii) $\alpha_2 = u^2$.

(iii) $\alpha_3 = \int_{t_1}^{t_2} y^2 \, dt$, $t_2 > t_1 >$ present time $t = 0$.

(iv) $\alpha_4 = \dfrac{1}{t_2 - t_1} \alpha_3$.

(v) $\alpha_5 = E(\alpha_3)$.

(vi) $\alpha_6 = \lim_{t_2 \to \infty} (\alpha_4)$.

(vii) $\alpha_7 = E(x - y)$.

(viii) $\alpha_8 = \dfrac{1}{t_2 - t_1} \displaystyle\int_{t_1}^{t_2} (x - y)\, dt, \quad t_2 > t_1 > \text{present time } t = 0.$

(ix) $\alpha_9 = E(\alpha_8)$.

(x) $\alpha_{10} = E\big((x - y)^2\big)$.

13.2. Show that the following measures of the output y of a discrete-time linear system are identical.

(i) The variance σ_y^2 when the input x is a stochastic variable which can be regarded as the output of another linear system having z-transform function $G(z)$ and driven by a sequence of independent random variables of zero mean and unit variance.

(ii) The sum $\displaystyle\sum_{i=0}^{\infty} y^2(i)$ when the input x is a deterministic function having z-transform $G(z)$.

Does this correspondence also hold for continuous-time systems? What deterministic function corresponds, in this sense, to the Wiener process (integrated white noise). What random process corresponds in the same sense to the deterministic impulse function?

13.3. Prove that the value of gain K which minimizes the performance criterion

$$I = \sigma_e^2 + \sigma_u^2$$

for the continuous-time system shown in Fig. 13.11 when the input x is integrated white noise, satisfies the equation

$$2K^3 + K^2 = 1.$$

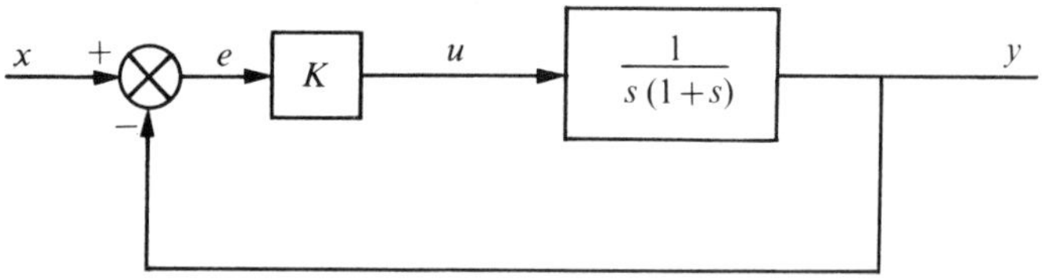

FIG. 13.11.

13.4. In an example of the system of Fig. 13.2 the controlled process has transfer function A (no dynamics, but gain A) and the controller is an integrator with transfer function K/s. The transfer functions of the linear dynamics generating stochastic desired values x, disturbances z and noises n are respectively

$$G_x(s) = \frac{X}{1 + s}, \qquad G_z(s) = \frac{Z}{s}, \qquad G_n(s) = N$$

and the independent white noises ξ_x, ξ_z, ξ_n are all of unit magnitude.

Show that the variance of the error $x - y$ is

$$\frac{AZ^2 + (X^2 + A^2Z^2)K + AN^2K^2 + A^2N^2K^3}{2K(1 + AK)}.$$

13.5. It is proposed to design the value of the controller gain K in the system of Problem 13.4 so as to minimize the variance of the error $x - y$.

(i) For the special case where there is no disturbance $(Z = 0)$ and where $A = 1$, show that the designed value is

$$K = \frac{X}{N} - 1$$

and comment on the signal-to-noise ratio. (This case corresponds to the problem of filtering the signal x from the noisy input $x + n$.)

(ii) For the special case where there is no noise $(N = 0)$, explain what is unsatisfactory about the designed value of K. Suggest two alternative ways in which the original problem might be modified so that the design would give an acceptable value when there is no noise.

13.6. Derive the functional recurrence equation specifying optimal control in the class of stationary discrete-time stochastic problems having dynamic equation

$$\mathbf{x}(i + 1) = \mathbf{G}\left(\mathbf{x}(i), \mathbf{u}(i), \boldsymbol{\xi}(i)\right),$$

and performance criterion

$$J = E\left(\sum_{i=1}^{N} H\left(\mathbf{x}(i), \mathbf{u}(i)\right)\right),$$

and where the probability density $p(\boldsymbol{\xi}(i))$ of the sets of random variables $\boldsymbol{\xi}(i)$ does not change with time.

Note how this equation is related to equation (8.6a) and to equation (13.7).

13.7. A stochastic controlled process has dynamic equation

$$x(i + 1) = ax(i) + \xi(i)u(i),$$

where the sequence of random variables $\xi(i)$ are drawn from a common probability distribution having non-zero mean β and variance σ^2. Use a functional recurrence equation to show that the value of the control $u(1)$ which minimizes

$$J = E\left(\sum_{i=1}^{2} x^2(i)\right)$$

is

$$u^*(i) = u_2(x) = -\frac{a\beta x}{\beta^2 + \sigma^2}.$$

13.8. If Problem 13.7 is modified to specify control over N discrete-time intervals by using performance criterion

$$J = E\left(\sum_{i=1}^{N} x^2(i)\right),$$

(i) show that the optimal value of J is

$$J = v_N x^2(1),$$

where

$$v_N = \frac{a^2\sigma^2}{\beta^2 + \sigma^2} v_{N-1} + 1;$$

(ii) state the condition which must be satisfied if there is to be a limiting value of v_N as N goes to infinity and find the limiting value. What does this imply about values of $x(i)$ as i goes to infinity?

(iii) Is the optimal control law here, the certainty-equivalent control?

13.9. The discrete-time position-control system of Problems 1.12 and 8.6 is subjected to a stochastic disturbance torque ξ such that the dynamic equations become

$$x_1(i + 1) = x_1(i) + x_2(i),$$
$$x_2(i + 1) = x_2(i) + u(i) + \xi(i),$$

where $\xi(i)$ is a sequence of independent random variables having zero mean and variance σ^2. Use numerical results from the iteration in Problem 8.6 to find the expected value

$$E\left(x_1^2(i) + x_2^2(i)\right)$$

when the system operates for a long time using the certainty-equivalent control law of Problem 8.6.

13.10. In a continuous-time tracking problem the controlled process has transfer function

$$\frac{Y(s)}{U(s)} = \frac{1}{s},$$

the desired value x is stochastic and generated (as in Problem 13.4) by white noise of unit amplitude driving a linear system having transfer function

$$G_x(s) = \frac{X}{1 + s},$$

and the performance criterion is the sum of the variance of the error $x - y$ and of the control u

$$J = \sigma_{x-y}^2 + \sigma_u^2.$$

(i) Define state variables

$$x_1 \equiv x, \qquad x_2 \equiv y$$

and use them to find the certainty-equivalent optimal control law.

(ii) The optimal value of J in this continuous-time problem is given, as in a discrete-time problem, by tr $\mathbf{V\Sigma}$. Find the optimal value of J for the above problem and compare it with the value $\min\limits_{K}\left\{\dfrac{X^2(1 + K^2)}{2(1 + K)}\right\}$ given by setting $N = 0$, $T = 1$, $K_1 = 1$, $q = 1$ in the fixed configuration worked example of Fig. 13.3 in Section 13.2.

13.11. A controlled process has dynamic equation

$$x(i + 1) = 0{\cdot}95x(i) + u(i) + \xi_1(i)$$

and output equation

$$y(i) = x(i) + \xi_2(i),$$

where $\xi_1(i)$ and $\xi_2(i)$ are sequences of independent normal random variables having zero mean values and variances

$$\sigma_1^2 = 1{\cdot}1, \qquad \sigma_2^2 = 2.$$

The initial value of x is

$$x(i_1) = 10$$

and the realizations of ξ_1 and ξ_2 over the stages

$$i = i_1, i_1 + 1, \ldots, i_1 + 9$$

are

$$\xi^{i_1} = \{0{\cdot}4, \ -0{\cdot}3, \ 1{\cdot}7, \ 0{\cdot}2, \ -0{\cdot}1, \ -0{\cdot}6, \ -0{\cdot}7, \ 0{\cdot}3, \ -1{\cdot}1,$$
$$- 1{\cdot}3, \ -0{\cdot}2, \ 0{\cdot}8, \ 1{\cdot}2, \ -0{\cdot}4, \ 0{\cdot}2, \ 0{\cdot}8, \ -0{\cdot}7, \ -2{\cdot}3\}.$$

(i) If there were no control action, $u(i) = 0$, what would be the set of numbers represented by $y^{i_1 + 9}$?

(ii) If the initial value $x(i_1)$ is known with certainty and subsequent values are estimated from y^i using equations (13.14) what will be the sequence of values of the variance of the conditional probability density $p(x(i) \mid y^i)$? How is the accuracy of this estimation affected by values of the control $u(i)$ and of the state $x(i)$?

(iii) What is the control law which will minimize the variance σ_x^2 of x over a long period of operation?

(iv) What will be the value of σ_x^2 when the control law of (iii) is used? How does this value compare with the value of σ_x^2 when there is no control action as in (i)? Why do these values differ from the steady-state value of the variance of $p(x(i) \mid y^i)$ in (ii)?

(v) What would be the first few values of $x(i)$ realized if the control law of (iii) were used?

(vi) What is the z-transfer function $U(z)/Y(z)$ of the controller using (iii) when the variance of $p(x(i) \mid y^i)$ has converged to its steady-state value? What is the root of the characteristic equation of the resulting feedback system?

13.12. Use equations (13.14a) and (13.20) to show that, when **B** is square and non-singular and **Q** is null, the optimal controller for discrete-time stochastic LQP-problems (specified by equations (13.2), (13.4), (13.6), (13.9)) with normal random variables can be described by the equation

$$\mathbf{u}^*(i) = -\mathbf{B}^{-1}\mathbf{A}\boldsymbol{\Sigma}(i)\mathbf{C}'\boldsymbol{\Sigma}_2^{-1}\mathbf{y}(i).$$

Use equations (13.2), (13.4), and (13.14b) to show that the resulting feedback system, when $\boldsymbol{\Sigma}(i)$ has converged to its steady-state value $\boldsymbol{\Sigma}$, has dynamics characterized by the matrix

$$\mathbf{A}\boldsymbol{\Sigma}(\mathbf{A}\boldsymbol{\Sigma}\mathbf{A}' + \boldsymbol{\Sigma}_1)^{-1}.$$

What is the significance of the assumptions about **B** and **Q**?

13.13. (i) Work through the derivation, from equation (13.16), of equation (13.17b).

(ii) Work through the derivation, from equation (13.18), of equations (13.19), (13.20), and (13.21).

13.14. Problem 13.7 is generalized to account for uncertainty about present states by introducing an output equation

$$y(i) = x(i) + \xi_2(i)$$

and rewriting the dynamic equation

$$x(i + 1) = ax(i) + \big(\beta + \xi_1(i)\big)u(i),$$

where $\xi_1(i)$ and $\xi_2(i)$ are sequences of independent normal random variables having zero mean values and variances σ_1^2, σ_2^2, respectively. The conditional probability density $p\big(x(i) \mid y^i\big)$ can be shown (Problem 14.4) to be normal with mean $m(i)$ and variance $\sigma^2(i)$ given by the difference equations

$$m(i) = am(i - 1) + \beta u(i - 1) + \frac{\sigma^2(i)}{\sigma_2^2}\big(y(i) - am(i - 1) - \beta u(i - 1)\big)$$

$$\sigma^2(i) = \frac{\sigma_2^2\big(a^2\sigma^2(i - 1) + \sigma_1^2 u^2(i - 1)\big)}{\sigma_2^2 + a^2\sigma^2(i - 1) + \sigma_1^2 u^2(i - 1)}.$$

(i) In what way are these equations for the conditional probability density significantly different from equations (13.14)?

(ii) Show that at time i, after $u(i)$ has been chosen but before $y(i + 1)$ can be observed, the conditional mean $m(i + 1)$ is a random variable having expected value

$$E\big(m(i + 1) \mid y^i, u(i)\big) = am(i) + \beta u(i)$$

and variance

$$\text{Variance}\big(m(i + 1) \mid y^i, u(i)\big) = \frac{\big(a^2\sigma^2(i) + \sigma_1^2 u^2(i)\big)^2}{\sigma_2^2 + a^2\sigma^2(i) + \sigma_1^2 u^2(i)}.$$

(iii) Use a functional recurrence equation to show that the value of the control $u(1)$ which minimizes

$$J = E\left(\sum_{i=1}^{2} x^2(i)\right)$$

is

$$u^*(1) = u_2(m, \sigma^2) = -\frac{\alpha\beta m}{\beta^2 + \sigma_1^2}.$$

(iv) Note that the optimal control law here has the same form as in Problem 13.7. Is it the certainty-equivalent control? Is this problem separable?

13.15. The discrete-time position-control system of Problems 1.12, 8.6, and 13.9 is generalized to account for measurement noise by introducing an output equation

$$y(i) = x_1(i) + \xi_2(i)$$

and rewriting the dynamic equations

$$x_1(i + 1) = x_1(i) + x_2(i),$$
$$x_2(i + 1) = x_2(i) + u(i) + \xi_1(i),$$

where $\xi_1(i)$ and $\xi_2(i)$ are sequences of independent normal random variables having zero mean values and unit variances. A terminal-control task is specified by the performance criterion

$$J = E\left(x_1^2(3) + x_2^2(3) + \sum_{i=1}^{3} u^2(i)\right).$$

(i) Use the fact that this LQP-problem is certainty equivalent to find the optimal control law

$$u^*(i) = u_{3-i+1}\big(\mathbf{m}(i)\big), \qquad i = 1, 2, 3,$$

where $\mathbf{m}(i)$ is the mean of the conditional probability density $p\big(\mathbf{x}(i) \mid y^i\big)$.

(ii) If the covariance matrix of the initial probability distribution for the state $\mathbf{x}(1)$ is

$$\Sigma(1) = \begin{bmatrix} 1 & 0 \\ 0 & 1 \end{bmatrix},$$

use equation (13.14b) to calculate $\Sigma(2)$ and $\Sigma(3)\big($N.B. the inverse of a 2×2-matrix is

$$\begin{bmatrix} a & b \\ c & d \end{bmatrix}^{-1} = \frac{1}{ad - bc}\begin{bmatrix} d & -b \\ -c & a \end{bmatrix}\bigg).$$

(iii) If the mean values of the initial probability distribution for the state $\mathbf{x}(1)$ are zero,

$$m_1(1) = 0, \qquad m_2(1) = 0,$$

express the optimal control $u^*(2)$, at time $i = 2$, as a feedback law in terms of the measurement $y(2)$.

(iv) Use equation (13.21) to calculate the optimal value of J when the initial mean values are zero as in (iii).

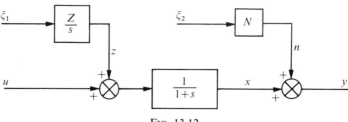

FIG. 13.12.

13.16. Figure 13.12 shows a continuous-time stochastic controlled process for which it is required to minimize the performance criterion

$$J = E\left(\int_0^T (x^2 + u^2)\, dt\right).$$

The white noises ξ_1, ξ_2 generating the stochastic disturbance z and measurement noise n are independent, normal, and of unit magnitude,

(i) find $\mathbf{A}$, $\mathbf{B}$, $\mathbf{C}$, $\mathbf{P}$, $\mathbf{Q}$, $\mathbf{\Sigma}_1$, $\mathbf{\Sigma}_2$ to represent the problem in terms of state variables

$$x_1 \equiv z, \qquad x_2 \equiv x$$

in the form used in Section 13.5;

(ii) show that the limiting solution $\mathbf{V}(T)$ to the Riccati equations (13.24) here is

$$\lim_{T \to \infty} \left(\mathbf{V}(T)\right) = \begin{bmatrix} \dfrac{T}{2} & 1 - \dfrac{1}{\sqrt{2}} \\ 1 - \dfrac{1}{\sqrt{2}} & \sqrt{2} - 1 \end{bmatrix}.$$

Find the certainty-equivalent control law $u(\mathbf{m})$. Explain why the coefficient v_{11} of $\mathbf{V}(T)$ goes to infinity with T;

(iii) find the steady-state value $\mathbf{\Sigma}$ of the covariance matrix of the conditional probability density for $\mathbf{x}$;

(iv) show that the optimal controller has transfer function of the form

$$\frac{U(s)}{Y(s)} = -K \frac{(1 + sT)}{1 + a_1 s + a_2 s^2}.$$

14. Estimating uncertain states

14.1. Introduction

In stochastic control systems where the present state is uncertain and described by a conditional probability distribution the controller has to perform two separate functions of (i) updating the conditional probabilities and (ii) generating a control based on the updated distribution. This separation was introduced in Section 13.4 where the function (ii) of generating controls for LQP-problems was discussed. The function (i) of updating the conditional probabilities is the subject of the present Chapter which introduces relevant results from the field of estimation theory.

The point of departure for deriving these results is Bayes's rule, which was introduced as equation (12.4) in Chapter 12 and is restated here:

$$p(\mathbf{x} \mid \mathbf{y}) = \frac{p(\mathbf{x})p(\mathbf{y} \mid \mathbf{x})}{\int_{\mathbf{x}} p(\mathbf{x})p(\mathbf{y} \mid \mathbf{x}) \, d\mathbf{x}} \qquad (14.1)$$

in terms of the continuous-valued states $\mathbf{x}$ and outputs $\mathbf{y}$ of stochastic control systems. Bayes's rule is the result in probability theory which specifies how noisy measurements $\mathbf{y}$ of a random variable $\mathbf{x}$ are to be processed so as to reduce uncertainty about the value of $\mathbf{x}$; it therefore provides the theoretical foundation for methods of analysing feedback signals in control systems. The significance of the various terms in equation (14.1) is discussed in Section 14.2.

In linear systems having equations such as (13.2) and (13.4), repeated here as (14.2) and (14.3):

$$\mathbf{x}(i + 1) = \mathbf{A}\mathbf{x}(i) + \mathbf{B}\mathbf{u}(i) + \boldsymbol{\xi}_1(i), \qquad (14.2)$$

$$\mathbf{y}(i) = \mathbf{C}\mathbf{x}(i) + \boldsymbol{\xi}_2(i) \qquad (14.3)$$

with normal probability distributions for $\boldsymbol{\xi}_1, \boldsymbol{\xi}_2$ and the initial value of $\mathbf{x}$, the conditional probability density $p(\mathbf{x} \mid \mathbf{y})$ given by Bayes's rule, equation (14.1), is also normal and so is completely specified by its mean and covariance which serve as sufficient statistics. The equations for updating these sufficient statistics were quoted as (13.14) and (13.23) in Chapter 13 and are derived here in Section 14.3. This solution to linear estimation problems is known as the Kalman filter.

In most other systems there are no sufficient statistics to summarize the conditional probability density function $p(\mathbf{x} \mid \mathbf{y})$ as it evolves according to Bayes's rule. In control systems it is usually impracticable for the controller to compute $p(\mathbf{x} \mid \mathbf{y})$ according to equation (14.1), and so alternative sub-optimal ways of estimating states have been developed. The alternatives are sub-optimal in so far as they differ from Bayes's rule which specifies the ideal way to generate information from measured data. They include linearization of non-linear system equations leading to the extended Kalman filter introduced in Section 14.4 and they include methods of statistical estimation, introduced in Section 14.5, which are not derived from probability theory and which need less prior information than does Bayes's rule. The filters based on these sub-optimal alternative methods are identical with the optimal Kalman filter for the special class of linear systems with normal random variables. In systems outside the special class having linear equations and normal random variables where all estimators are Kalman filters, questions arise about the quality of the various estimation procedures. The principal questions are:

(i) do the estimates converge?

(ii) does the expected value of estimates coincide with the true value (unbiased estimate) or is the estimate biased?

(iii) is the variance of the estimate as small as possible (efficient estimate) or is the estimate inefficient?

These are questions in estimation theory and beyond the scope of this book. General answers are seldom available and so the choice between statistical methods of estimation and the extended Kalman filter, or other alternative methods of state estimation when Bayes's rule is unrealizable, must usually be made on some empirical, rather than theoretical basis.

A particularly important class of non-linear systems is the class, already mentioned in Section 13.1, having equations such as (14.2) and (14.3) which look linear but become non-linear when elements of the matrices $\mathbf{A}$, $\mathbf{B}$, $\mathbf{C}$ are unknown constants which must be represented by additional state variables. The class is important because it is often convenient to represent practical control problems by apparently linear equations, either in the state-space form of equations (14.2) and (14.3) or as transfer functions, in which some parameters have unknown constant values. One sub-optimal approach to such problems is to divorce the two controller functions from each other; first estimating the unknown parameters without attempting to exercise control, as discussed in Section 14.6, and subsequently using a linear feedback control law based on the estimated values of the parameters without attempting efficient reduction of any remaining uncertainty about the parameters. This approach can be regarded as a mechanization of the design procedures used in Section I for deterministic linear systems; first identifying linear dynamic equations, then designing a suitable linear control law. Alternative sub-optimal controllers, which take account of the inherent non-linearity in this class of stochastic problems, are discussed in Chapter 15.

14.2. Bayes's rule

Bayes's rule is an equation

$$p(\mathbf{x} \mid \mathbf{y}) = \frac{p(\mathbf{x})p(\mathbf{y} \mid \mathbf{x})}{\int\limits_{\mathbf{x}} p(\mathbf{x})p(\mathbf{y} \mid \mathbf{x}) \, d\mathbf{x}} \tag{14.1}$$

relating three probability functions:

$p(\mathbf{x} \mid \mathbf{y})$ the conditional probability density for random variable $\mathbf{x}$ after a noisy measurement $\mathbf{y}$;

$p(\mathbf{x})$ the prior probability density for random variable $\mathbf{x}$ before observing the value of $\mathbf{y}$;

$p(\mathbf{y} \mid \mathbf{x})$ which is known as the *likelihood function* and specifies the relative probabilities of $\mathbf{y}$ taking its observed value as a function of the range of possible values of $\mathbf{x}$.

The significance of these terms is most conveniently discussed with reference to an example.

Consider the problem of deriving information about a single random variable x from a measurement y contaminated by additive noise ξ:

$$y = x + \xi.$$

Assume that before the measurement the uncertain value of x is described by a normal probability distribution having mean m_0 and variance σ_0^2; this is the prior probability distribution and is written

$$p(\mathbf{x}) = \frac{1}{\sigma_0 \sqrt{(2\pi)}} \exp\left(- \frac{(x - m_0)^2}{2\sigma_0^2}\right) \equiv N(m_0, \sigma_0^2),$$

where the notation $N(\cdot)$ is introduced for normal probability density functions. Assume also that the noise is independent of x and has normal probability density $p(\xi)$ with zero mean and variance σ_ξ^2,

$$p(\xi) = N(0, \sigma_\xi^2).$$

Now if the measurement y takes the particular value Y, the noise ξ must have taken the value $Y - x$, so the likelihood function is

$$p(y = Y \mid x) = p(\xi = Y - x) = \frac{1}{\sigma_\xi \sqrt{(2\pi)}} \exp\left(- \frac{(Y - x)^2}{2\sigma_\xi^2}\right).$$

Then equation (14.1) gives the conditional probability distribution

$$p(x \mid Y) = \frac{\sigma_0 \sigma_\xi 2\pi \exp\left(-\dfrac{(x - m_0)^2}{2\sigma_0^2} - \dfrac{(Y - x)^2}{2\sigma_\xi^2}\right)}{\sigma_0 \sigma_\xi 2\pi \displaystyle\int_{-\infty}^{\infty} \exp\left(-\dfrac{(x - m_0)^2}{2\sigma_0^2} - \dfrac{(Y - x)^2}{2\sigma_\xi^2}\right) dx}.$$

The measured value Y enters Bayes's rule only through the likelihood function $p(Y \mid x)$ which, together with the prior probability function $p(x)$ and the conditional probability function $p(x \mid Y)$, appears as a function of the uncertain random variable x. The name 'likelihood function' is used to distinguish $p(Y \mid x)$ from the related conditional probability function $p(y \mid X)$, which would be a function of y for some given value X of x. The integral in the denominator gives the prior probability density of the observed value Y and has a value which will be independent of the argument x of the other three terms in the equation; it can be regarded as a constant which guarantees that the conditional probability function $p(x \mid Y)$ satisfies the normalization condition of equation (12.10),

$$\int_{-\infty}^{\infty} p(x \mid Y) \, dx = 1.$$

Bayes's rule therefore has the form

  conditional probability $=$ constant $\times$ prior probability $\times$ likelihood.

$$(14.4)$$

When both the prior probability and the likelihood function are normal, as in the example here, only the exponential terms are functions of x and so Bayes's rule becomes

$$p(x \mid Y) = \text{constant} \times \exp\left(-\frac{(x - m_0)^2}{2\sigma_0^2} - \frac{(Y - x)^2}{2\sigma_\xi^2}\right).$$

This conditional probability can (by completing squares) be put in the form

$$p(x \mid Y) = \text{constant} \times \exp\left\{-\frac{\left(x - \dfrac{\sigma_\xi^2 m_0 + \sigma_0^2 Y}{\sigma_0^2 + \sigma_\xi^2}\right)^2}{\dfrac{2\sigma_0^2 \, 2\sigma_\xi^2}{\sigma_0^2 + \sigma_\xi^2}}\right\}$$

of a normal distribution:

$$p(x \mid Y) = \frac{1}{\sigma\sqrt{(2\pi)}} \exp\left(-\frac{(x - m)^2}{2\sigma^2}\right) \equiv N(m, \sigma^2)$$

having mean m and variance σ^2 given by

$$m = \frac{\sigma_\xi^2 m_0 + \sigma_0^2 Y}{\sigma_0^2 + \sigma_\xi^2}, \tag{14.5a}$$

$$\sigma^2 = \frac{\sigma_0^2 \sigma_\xi^2}{\sigma_0^2 + \sigma_\xi^2}. \tag{14.5b}$$

The example thus demonstrates an important property of normal distributions: that whenever the prior probability and the likelihood function are both normal the conditional distribution given by Bayes's rule is also normal.

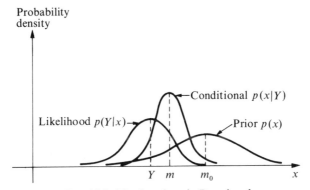

FIG. 14.1. The functions in Bayes's rule.

Figure 14.1 shows how the conditional probability density $p(x \mid Y)$ would be related to the prior probability density $p(x)$ and to the likelihood function $p(Y \mid x)$.

Equations (14.5) can be combined to give an alternative expression for the conditional mean m in the example:

$$m = m_0 + \frac{\sigma^2}{\sigma_\xi^2} (Y - m_0). \tag{14.6}$$

This equation (14.6) shows that the conditional mean is the sum of two terms. The first term m_0 is the mean of the prior probability distribution, that is the expected value of x before the measurement. The second term is proportional to the difference $(Y - m_0)$ between the observed value Y and the previously expected value m_0; it can be regarded as a correction to m_0 based on the new information provided by the measurement and is sometimes known as the *innovation* term. Equation (14.5b) for the conditional variance σ^2 can also be rewritten

$$\sigma^2 = k\sigma_0^2,$$

where

$$0 < k \equiv \frac{\sigma_{\xi}^2}{\sigma_0^2 + \sigma_{\xi}^2} < 1,$$

to show that it is always less than the variance σ_0^2 of the prior distribution, thus confirming that the measurement does reduce uncertainty. The constant k specifying the amount by which uncertainty is reduced depends only on the variances of the prior probability function and of the noise; it is independent of the observation Y and of the expected value m_0.

An important general feature of Bayes's rule is that it draws attention to the need for prior information in interpreting noisy measurements. The prior probability function and the likelihood function summarize prior information about values of the random variable x and about the noise causing y to differ from x; without such information the conditional distribution cannot be found. Figure 14.2 illustrates another example showing how the interpretation of noisy measurements depends on prior information; it shows

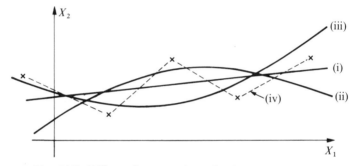

FIG. 14.2. Different interpretations of noisy measurements.

different conclusions about the relationship between two variables X_1 and X_2 based on a single set of measurements (the crosses in the figure) but using various prior informations as follows:

(i) the relationship is linear;
(ii) the relationship is non-linear with a local maximum;
(iii) the relationship is non-linear with a local minimum;
(iv) the measurement noise is negligible.

The shape of whatever graph summarizes the conclusions depends primarily on the prior information.

Problems 14.1–14.3

14.3. Estimation in linear dynamic systems (Kalman filter)

The example introduced in Section 14.2 is readily generalized to the problem

of estimating the state x of a first-order discrete-time controlled process having difference equation

$$x(i + 1) = ax(i) + bu(i) + \xi_1(i) \tag{14.7a}$$

and output equation

$$y(i) = cx(i) + \xi_2(i), \tag{14.7b}$$

where $\xi_1(i)$ and $\xi_2(i)$ are independent random variables drawn from normal
probability distributions having zero mean and variances σ_1^2 and σ_2^2 respectively, so that

$$p(\xi_1) = N(0, \sigma_1^2), \qquad p(\xi_2) = N(0, \sigma_1^2).$$

Provided that there is a normal prior probability distribution of x, Bayes's
rule will always give normal conditional probability distributions for x, and so
the conditional distribution for $x(i)$ after observation $y(i)$ has been processed
can be written, combining the notation of Section 13.4 with that introduced
here for normal distributions,

$$p\big(x(i) \mid y^i\big) = N\big(m(i), \sigma^2(i)\big).$$

In a stochastic control system of the general type discussed in Section 13.4
the information available for updating the conditional probability density for
$x(i)$ would be the previous conditional probability density $p(x(i - 1) \mid y^{i-1})$
together with the value U given to the previous control $u(i - 1)$ and the value
Y of the present measurement $y(i)$. The appropriate prior probability for
Bayes's rule is then

$$p\big(x(i)\big) = p\big(x(i) \mid y^{i-1}, u(i - 1) = U\big),$$

a normal distribution with mean $m_0(i)$ and variance $\sigma_0^2(i)$ which are determined by equation (14.7a) together with the results of equations (12.19) and
(12.20) about the mean and variance of linear combinations of independent
random variables,

$$\left. \begin{aligned} m_0 &= am(i - 1) + bU \\ \sigma_0^2 &= a^2\sigma^2(i - 1) + \sigma_1^2 \end{aligned} \right\}^{\dagger} . \tag{14.8}$$

---

† In equations (14.8), and throughout most of Section 14.3, the time argument i of the
statistics $m_0(i)$, $\sigma_0^2(i)$ of the prior probability distribution for $x(i)$ is suppressed in order
to simplify the appearance of the equations. The more conventional notation of estimation theory, which would distinguish between prior statistics and current conditional
statistics by writing

the prior statistics $m_0(i)$, $\sigma_0^2(i)$ as $m(i \mid i - 1)$, $\sigma^2(i \mid i - 1)$,
the current statistics $m(i)$, $\sigma^2(i)$ as $m(i \mid i)$, $\sigma^2(i \mid i)$,

is introduced in Section 14.5.

The likelihood function, determined by equation (14.7b) together with the normal probability density of the noise ξ_2, is

$$p\big(Y \mid x(i)\big) = \frac{1}{\sigma_2 \sqrt{(2\pi)}} \exp\left(- \frac{(Y - cx(i))^2}{2\sigma_2^2}\right).$$

The conditional probability given by Bayes's rule, using equation (14.4) and concentrating on the exponential terms of the normal distributions, is then

$$p\big(x(i) \mid y^i\big) = \text{constant} \times \exp\left(- \frac{(x(i) - m_0)^2}{2\sigma_0^2} - \frac{(Y - cx(i))^2}{2\sigma_2^2}\right)$$

and this can (by completing squares) be put in the form

$$p\big(x(i) \mid y^i\big) = \text{constant} \times \exp\left\{- \frac{\left(x(i) - \dfrac{\sigma_2^2 m_0 + \sigma_0^2 c\, Y}{c^2\sigma_0^2 + \sigma_2^2}\right)^2}{\dfrac{2\sigma_0^2\sigma_2^2}{c^2\sigma_0^2 + \sigma_2^2}}\right\}$$

of a normal distribution having mean $m(i)$ and variance $\sigma^2(i)$ given by

$$m(i) = \frac{\sigma_2^2 m_0 + \sigma_0^2 c\, Y}{c^2\sigma_0^2 + \sigma_2^2}, \tag{14.9a}$$

$$\sigma^2(i) = \frac{\sigma_0^2 \sigma_2^2}{c^2\sigma_0^2 + \sigma_2^2}. \tag{14.9b}$$

The arguments used to derive these equations (14.9) which specify how the conditional probability for the state is to be updated in a first-order process are in no way restricted to first-order systems. In the general high-order discrete-time linear system of equations (14.2) and (14.3) with independent normal random variables ξ_1 and ξ_2 having zero means and covariance matrices Σ_1 and Σ_2 respectively, so that

$$p(\xi_1) = N(\mathbf{0}, \Sigma_1), \qquad p(\xi_2) = N(\mathbf{0}, \Sigma_2)$$

and with a normal prior probability distribution for $\mathbf{x}$, the conditional probability distribution for $\mathbf{x}(i)$ is also normal:

$$p\big(\mathbf{x}(i) \mid y^i\big) = N\big(\mathbf{m}(i), \Sigma(i)\big)$$

with mean $\mathbf{m}(i)$ and covariance matrix $\Sigma(i)$. The appropriate prior probability for Bayes's rule is then†

$$p\big(\mathbf{x}(i)\big) = p\big(\mathbf{x}(i) \mid y^{i-1}, \mathbf{u}\,(i-1)\big),$$

† The convention introduced in previous examples of representing realized values of $u(i-1)$ and $y(i)$ by capitals U, Y is not usable here where capitals represent matrices. The notation therefore ceases to distinguish between the name of a variable and its realized value.

a normal distribution with mean $\mathbf{m}_0$ and covariance matrix $\mathbf{\Sigma}_0$, as in equations (13.15),

$$\mathbf{m}_0 = \mathbf{A}\mathbf{m}(i-1) + \mathbf{B}\mathbf{u}(i-1),$$
$$\mathbf{\Sigma}_0 = \mathbf{A}\mathbf{\Sigma}(i-1)\mathbf{A}' + \mathbf{\Sigma}_1. \tag{14.10}$$

The likelihood function, determined by equation (14.3) and using the expression of equation (12.16) for the multi-dimensional normal distribution, is

$$p(\mathbf{y}(i) \mid \mathbf{x}(i)) = \text{constant} \times \exp\left\{-\tfrac{1}{2}(\mathbf{y}(i) - \mathbf{C}\mathbf{x}(i))'\mathbf{\Sigma}_2^{-1}(\mathbf{y}(i) - \mathbf{C}\mathbf{x}(i))\right\}.$$

The conditional probability given by Bayes's rule, using equation (14.4) and concentrating on the exponential terms of the normal distributions, is then

$$p(\mathbf{x}(i) \mid y^i) = \text{constant} \times \exp\left\{-\tfrac{1}{2}(\mathbf{x}(i) - \mathbf{m}_0)'\mathbf{\Sigma}_0^{-1}(\mathbf{x}(i) - \mathbf{m}_0) - \right.$$
$$\left. - \tfrac{1}{2}(\mathbf{y}(i) - \mathbf{C}\mathbf{x}(i))'\mathbf{\Sigma}_2^{-1}(\mathbf{y}(i) - \mathbf{C}\mathbf{x}(i))\right\},$$

which can be rearranged in a form

$$p(\mathbf{x}(i) \mid y^i) = \text{constant} \times \exp\left\{-\tfrac{1}{2}(\mathbf{x}'(i)(\mathbf{\Sigma}_0^{-1} + \mathbf{C}'\mathbf{\Sigma}_2^{-1}\mathbf{C})\mathbf{x}(i) - \right.$$
$$\left. - (\mathbf{m}_0'\mathbf{\Sigma}_0^{-1} + \mathbf{y}'(i)\mathbf{\Sigma}_2^{-1}\mathbf{C})\mathbf{x}(i) - \mathbf{x}'(i)(\mathbf{\Sigma}_0^{-1}\mathbf{m}_0 + \mathbf{C}'\mathbf{\Sigma}_2^{-1}\mathbf{y}(i))\right\}$$

corresponding to a normal distribution having mean $\mathbf{m}(i)$ and covariance matrix $\mathbf{\Sigma}(i)$ given by

$$\mathbf{m}(i) = (\mathbf{\Sigma}_0^{-1} + \mathbf{C}'\mathbf{\Sigma}_2^{-1}\mathbf{C})^{-1}(\mathbf{\Sigma}_0^{-1}\mathbf{m}_0 + \mathbf{C}'\mathbf{\Sigma}_2^{-1}\mathbf{y}(i)), \tag{14.11a}$$

$$\mathbf{\Sigma}(i) = (\mathbf{\Sigma}_0^{-1} + \mathbf{C}'\mathbf{\Sigma}_2^{-1}\mathbf{C})^{-1}. \tag{14.11b}$$

These equations (14.11) can be combined to show that the conditional mean $\mathbf{m}$ is the sum of two terms as at equation (14.6), an expected value $\mathbf{m}_0$ and an innovation proportional to the difference $\mathbf{y}(i) - \mathbf{C}\mathbf{m}_0$ between the actual value $\mathbf{y}(i)$ and the expected value $\mathbf{C}\mathbf{m}_0$ of the measurement $\mathbf{y}$,

$$\mathbf{m}(i) = \mathbf{m}_0 + \mathbf{K}(i)(\mathbf{y}(i) - \mathbf{C}\mathbf{m}_0), \tag{14.12a}$$

where

$$\mathbf{K}(i) \equiv \mathbf{\Sigma}(i)\mathbf{C}'\mathbf{\Sigma}_2^{-1}. \tag{14.12b}$$

Equations (14.12) and (14.11b) can be combined with equations (14.10) to eliminate $\mathbf{m}_0$ and $\mathbf{\Sigma}_0$ and give

$$\mathbf{m}(i) = \mathbf{A}\mathbf{m}(i-1) + \mathbf{B}\mathbf{u}(i-1) + $$
$$+ \mathbf{\Sigma}(i)\mathbf{C}'\mathbf{\Sigma}_2^{-1}\{\mathbf{y}(i) - \mathbf{C}(\mathbf{A}\mathbf{m}(i-1) + \mathbf{B}\mathbf{u}(i-1))\}, \tag{14.13a}$$

$$\mathbf{\Sigma}(i) = \{(\mathbf{\Sigma}_1 + \mathbf{A}\mathbf{\Sigma}(i-1)\mathbf{A}')^{-1} + \mathbf{C}'\mathbf{\Sigma}_2^{-1}\mathbf{C}\}^{-1}. \tag{14.13b}$$

Equations (14.13) are the solution to the general problem of state estimation in the class of discrete-time linear systems specified by equations (14.2) and

(14.3) and having normal random variables. They have already been quoted
as equations (13.14) in Chapter 13. They are in a recursive form such that
updated values of the conditional mean $\mathbf{m}(i)$ and covariance matrix $\boldsymbol{\Sigma}(i)$ can
be calculated from previous values $\mathbf{m}(i-1)$, $\boldsymbol{\Sigma}(i-1)$. For this purpose the
amount of matrix inversion needed to update the covariance matrix can be
reduced by using a matrix identity known as the matrix inversion lemma
which is applicable to expressions such as that on the right of equation
(14.11b):

$$(\boldsymbol{\Sigma}_0^{-1} + \mathbf{C}'\boldsymbol{\Sigma}_2^{-1}\mathbf{C})^{-1} \equiv \boldsymbol{\Sigma}_0 - \boldsymbol{\Sigma}_0\mathbf{C}'(\boldsymbol{\Sigma}_2 + \mathbf{C}\boldsymbol{\Sigma}_0\mathbf{C}')^{-1}\mathbf{C}\boldsymbol{\Sigma}_0.$$

The equation for updating the covariance matrix can therefore be written

$$\text{where} \qquad \left.\begin{aligned} \boldsymbol{\Sigma}(i) &= \boldsymbol{\Sigma}_0 - \boldsymbol{\Sigma}_0\mathbf{C}'(\boldsymbol{\Sigma}_2 + \mathbf{C}\boldsymbol{\Sigma}_0\mathbf{C}')^{-1}\mathbf{C}\boldsymbol{\Sigma}_0, \\ \boldsymbol{\Sigma}_0 &= \boldsymbol{\Sigma}_1 + \mathbf{A}\boldsymbol{\Sigma}(i-1)\mathbf{A}'. \end{aligned}\right\} \qquad (14.13\text{c})$$

The advantage of equations (14.13c) is that only one inversion is required of a
$p \times p$-matrix where p, the number of outputs $\mathbf{y}$, is often less than n, the number of states $\mathbf{x}$, in place of the two inversions of $n \times n$-matrices in equation
(14.13b).

Equations (14.13) can be regarded as difference equations specifying the
dynamic system shown in Fig. 14.3 which has as its inputs the variables
$\mathbf{u}(i-1)$ and $\mathbf{y}(i)$ and has as its states the sufficient statistics $\mathbf{m}(i)$, $\boldsymbol{\Sigma}(i)$

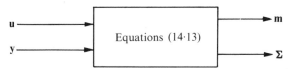

FIG. 14.3. The Kalman filter.

specifying the normal conditional probability density $p(\mathbf{x}(i) \mid y^i)$ which describes values of the uncertain state $\mathbf{x}(i)$. There are many problems, including
the stochastic LQP-control problem discussed in Section 13.4, where the conditional mean $\mathbf{m}(i)$ is used as an estimate of the state $\mathbf{x}(i)$. For such problems
equations (14.13) can be regarded as specifying a filter which has as its output
the estimate $\mathbf{m}(i)$. The covariance matrix $\boldsymbol{\Sigma}(i)$ which is also generated by the
filter specifies how values of $\mathbf{x}(i)$ are distributed about the conditional mean
$\mathbf{m}(i)$ and so can be regarded as providing information about the accuracy of
$\mathbf{m}(i)$ as an estimate of $\mathbf{x}(i)$. The equations were originally derived by R. E.
Kalman to give least-squares estimates of $\mathbf{x}(i)$ (as discussed in Section 14.5)
and are known as the Kalman filter; the weighting coefficient $\mathbf{K}(i)$ of the
innovation term in equations (14.12a) and (14.13a) is similarly known as the
Kalman gain. It can be shown† that the stochastic component $\mathbf{y}(i) - \mathbf{Cm}_0$ of
the innovations generated by a Kalman filter is always statistically indepen-

† See for example Åström (1970) Chapter 7.

dent in time, in the same way that the variables $\xi(i)$ in equations (14.2) and (14.3) are independent.

The purpose of the Kalman filter is to reduce uncertainty about the state $\mathbf{x}(i)$. This uncertainty is measured by the covariance matrix $\mathbf{\Sigma}(i)$ and so conclusions about the performance of the filter can be derived from discussion of the equation (14.13b) which specifies how $\mathbf{\Sigma}(i)$ evolves in time.

(i) It can be shown† that the covariance equation (14.13b) converges to a unique solution provided the matrices $\mathbf{A}$ and $\mathbf{C}$ are such that the system specified by equations (14.2) and (14.3) is observable in the sense of Section 7.4. This steady-state value $\mathbf{\Sigma}$ of the covariance matrix represents the minimum uncertainty that can be achieved, even after an infinite number of observations; it is specified by a steady-state version of equation (14.13b):

$$\mathbf{\Sigma}^{-1} = (\mathbf{\Sigma}_1 + \mathbf{A}\mathbf{\Sigma}\mathbf{A}')^{-1} + \mathbf{C}'\mathbf{\Sigma}_2^{-1}\mathbf{C} \tag{14.14}$$

but would often be more conveniently evaluated by simulating equations (14.13c) on a computer and letting them converge to the steady-state, as with equations (8.12) and (8.10).

In the first-order example of equations (14.7), equation (14.14) is sufficiently simple that it can be solved in closed form. For the particular case where the coefficients a and c are both unity,

$$a = 1, \qquad c = 1,$$

the steady-state value of the variance of the conditional probability distribution for x is

$$\sigma^2 = \frac{\sigma_1^2}{2}\left(\sqrt{\left(1 + 4\frac{\sigma_2^2}{\sigma_1^2}\right)} - 1\right),$$

which can be approximated for small values of the ratio σ_2/σ_1 by

$$\sigma^2 \simeq \frac{\sigma_1^2}{2}\left(1 + 2\frac{\sigma_2^2}{\sigma_1^2} - 1\right) = \sigma_2^2 \quad \text{when } \sigma_2 \ll \sigma_1,$$

and for large values of σ_2/σ_1 by

$$\sigma^2 \simeq \frac{\sigma_1^2}{2}\left(2\frac{\sigma_2}{\sigma_1}\right) = \sigma_1\sigma_2 \quad \text{when } \sigma_2 \gg \sigma_1.$$

The first of these two approximations shows that when the uncertainty due to measurement noise ξ_2 is small the performance of the Kalman filter tends to be limited by the measurement noise, and that when there is no measurement noise ($\sigma_2 = 0$) the variance σ^2 goes to zero and uncertainty about x can be completely eliminated. The second approximation shows that when the uncertainty due to disturbance ξ_1 is small the performance is limited by both the disturbance and the measurement noise, but when there is no disturbance ($\sigma_1 = 0$) the variance σ^2 again goes to zero.

† See Kalman (1963b).

These conclusions based on analysis of the first-order example are also true for the general high-order case of equation (14.13b).

(ii) The existence of a steady-state value Σ of the covariance matrix implies that there is also a steady-state value $\mathbf{K}$ of the Kalman gain defined by equation (14.12b). These steady-state values depend on the matrices $\mathbf{A}$, $\mathbf{C}$, Σ_1, Σ_2 only, they are independent of the initial value, say $\Sigma(0)$, of the co-variance matrix. The corresponding Kalman filter is not time-varying and is given by equations (14.12a) or (14.13a) which are linear. This stationary linear filter is much more convenient for practical purposes than is the general time-varying filter. It is optimal, in the sense that Bayes's rule represents optimal data processing, whenever the transient effects of prior information about $\mathbf{x}$ have become overlayed by the new information provided by the measurements $\mathbf{y}(i)$ as is the case after the filter has been operating for a long time.

(iii) Neither $\mathbf{u}(i-1)$ nor $\mathbf{y}(i)$ appears in the covariance equation (14.13b) and therefore once an initial value, say $\Sigma(0)$, has been specified the matrix $\Sigma(i)$ for all future times is completely determined and could be pre-computed. This implies that the rate of reduction of uncertainty is independent of the realizations of the random variables $\xi_1(i)$, $\xi_2(i)$ and of the values taken by the controls $\mathbf{u}$ and the outputs $\mathbf{y}$. It is one of the reasons why the stochastic LQP-control system of Section 13.4 is certainty equivalent: this is discussed further in Section 15.2.

The arguments used to derive the Kalman filter equations (14.12) and (14.13) do not require that the system of equations (14.2) and (14.3) be stationary, and so the Kalman filter can be generalized for non-stationary systems where the matrices $\mathbf{A}$, $\mathbf{B}$, $\mathbf{C}$, Σ_1, Σ_2 which specify the system may be time-varying. It can also be generalized to give the conditional probability density for the state $\mathbf{x}$ at times previous and subsequent to the time i of the current measurement $\mathbf{y}(i)$; these alternative problems, which are known respectively as smoothing and prediction, are beyond the scope of this book.[†]

All the results about Kalman filters for discrete-time systems have continuous-time equivalents, and the conclusions about performance carry over. For example the Kalman filter giving the conditional mean $\mathbf{m}$ and covariance matrix Σ for the state $\mathbf{x}$ of a continuous-time system having stochastic differential equations

$$\mathbf{dx} = \mathbf{A}\mathbf{x}\,dt + \mathbf{B}\mathbf{u}\,dt + d\mathbf{w}_1,$$
$$\mathbf{y}\,dt = \mathbf{C}\mathbf{x}\,dt + d\mathbf{w}_2$$

---

† See Sage and Melsa (1971) Section 8.3. A bibliography on smoothing is given by Meditch (1973).

has already been quoted as equations (13.23) and is repeated here as equations (14.15):

$$\frac{d\mathbf{m}}{dt} = \mathbf{Am} + \mathbf{Bu} + \mathbf{K}(\mathbf{y} - \mathbf{Cm}), \tag{14.15a}$$

where $\mathbf{K}$ is the Kalman gain of equation (14.12b)

$$\mathbf{K} \equiv \boldsymbol{\Sigma}\mathbf{C}'\boldsymbol{\Sigma}_2^{-1},$$

and

$$\frac{d\boldsymbol{\Sigma}}{dt} = \boldsymbol{\Sigma}_1 + \mathbf{A}\boldsymbol{\Sigma} + \boldsymbol{\Sigma}\mathbf{A}' - \boldsymbol{\Sigma}\mathbf{C}'\boldsymbol{\Sigma}_2^{-1}\mathbf{C}\boldsymbol{\Sigma}. \tag{14.15b}$$

The continuous-time equations (14.15) are stated here without proof because mathematical techniques beyond the scope of this book are required to derive them directly from Bayes's rule.† An alternative derivation is indicated in Section 14.5 where equation (14.15a) is shown to solve the classical problem of weighted least-squares estimation of the state $\mathbf{x}$ of the linear continuous-time system.

   Least-squares estimation is one of several classical estimation problems which for linear systems, both continuous time and discrete time, have Kalman filter equations as their solution.‡ The discussion of Section 14.5 shows that least-squares estimation of the state of a linear system is an LQP-problem and this is reflected in the fact that the covariance equation (14.15b) of the Kalman filter for continuous-time estimation is a Riccati equation identical in form to the equation (8.10b) of optimal deterministic LQP-control. The correspondence, known as duality, between covariance equations of the Kalman filter and equations of optimal deterministic LQP-control also exists in discrete time. It is demonstrated by eliminating the covariance matrix $\boldsymbol{\Sigma}$ of the conditional probability distributions in equations (14.13c) to show that the covariance matrix $\boldsymbol{\Sigma}_0$ of the prior probability distributions satisfies a difference equation

$$\boldsymbol{\Sigma}_0(i + 1) = \boldsymbol{\Sigma}_1 + \mathbf{A}\boldsymbol{\Sigma}_0(i)\mathbf{A}' - \mathbf{A}\boldsymbol{\Sigma}_0(i)\mathbf{C}'(\boldsymbol{\Sigma}_2 + \mathbf{C}\boldsymbol{\Sigma}_0(i)\mathbf{C}')^{-1}\mathbf{C}\boldsymbol{\Sigma}_0(i)\mathbf{A}'$$

identical in form to the equation (8.10a) of LQP-control. Duality implies that results about LQP-control can be translated into results about estimation and vice versa. Details of the translation are given by comparing the co-variance equations of estimation problems specified by matrices $\mathbf{A}$, $\mathbf{C}$, $\boldsymbol{\Sigma}_1$, $\boldsymbol{\Sigma}_2$ with the dual equations (8.10) of LQP-control problems specified by matrices $\mathbf{A}_c$, $\mathbf{B}_c$, $\mathbf{P}$, $\mathbf{Q}$ to show that

$$\mathbf{A}_c' \text{ is the dual of } \mathbf{A},$$
$$\mathbf{B}_c' \text{ is the dual of } \mathbf{C},$$

† See Bucy and Joseph (1968) Chapter 4 or Jazwinsky (1970) Section 7.4.

‡ Discussion of various ways of deriving Kalman filter equations as solutions to problems in estimation theory is given by Jazwinski (1970) Chapter 7.

$$\mathbf{P} \text{ is the dual of } \boldsymbol{\Sigma}_1,$$
$$\mathbf{Q} \text{ is the dual of } \boldsymbol{\Sigma}_2,$$

$$\left.\begin{array}{c}\text{the solution}\\ \text{function } \mathbf{V}\end{array}\right\} \text{ is the dual of } \left\{\begin{array}{l}\boldsymbol{\Sigma} \text{ in continuous time,}\\ \boldsymbol{\Sigma}_0 \text{ in discrete time.}\end{array}\right.$$

One example of the use of duality is that it translates the condition of equation (7.9) for observability of $\mathbf{A}$ and $\mathbf{C}$ into the condition of equation (7.8) for controllability of $\mathbf{A}_c$ and $\mathbf{B}_c$; in either case this is the condition for convergence of the dynamic equations for $\boldsymbol{\Sigma}$ or $\mathbf{V}$. Another aspect of duality is that computer programs and results for LQP-control problems can be used for estimation problems, and vice versa.

In practice Kalman filters are found to be sensitive to the matrices $\mathbf{A}$, $\mathbf{B}$, $\mathbf{C}$, $\boldsymbol{\Sigma}_1$, $\boldsymbol{\Sigma}_2$. To realize a filter, in the form of equations (14.13) or (14.15), requires knowledge of the matrices for the system whose state is to be estimated and such knowledge is not always available. It can then happen that a filter is realized on the assumption that the matrices take certain values which are in fact incorrect; under these circumstances the filter can suffer from a phenomenon known as divergence which leads to incorrect values for the conditional mean $\mathbf{m}$ and covariance $\boldsymbol{\Sigma}$. Discussion of divergence is beyond the scope of this book.†

Problems 14.4–14.10

14.4. Estimation by linearization in non-linear systems (extended Kalman filter)

State estimation in the general class of non-linear systems specified by equations (13.1) and (13.3), repeated here as equations (14.16),

$$\mathbf{x}(i+1) = \mathbf{G}_1\big(\mathbf{x}(i), \mathbf{u}(i), \boldsymbol{\xi}_1(i), i\big), \tag{14.16a}$$

$$\mathbf{y}(i) = \mathbf{G}_2\big(\mathbf{x}(i), \boldsymbol{\xi}_2(i), i\big), \tag{14.16b}$$

is a problem having no general explicit solution. Bayes's rule in the form of equation (14.4) specifies an expression for updating the conditional probability density $p\big(\mathbf{x}(i) \mid \mathbf{y}(i)\big)$ of the current state $\mathbf{x}(i)$:

$$p\big(\mathbf{x}(i) \mid y^i\big) = \text{constant} \times p\big(\mathbf{x}(i) \mid y^{i-1}, \mathbf{u}(i-1)\big) \times p\big(\mathbf{y}(i) \mid \mathbf{x}(i)\big),$$

but neither the prior probability $p\big(\mathbf{x}(i) \mid y^{i-1}, \mathbf{u}(i-1)\big)$ nor the likelihood function $p\big(\mathbf{y}(i) \mid \mathbf{x}(i)\big)$ in this expression can generally be evaluated. The prior probability can be written, using a generalization of equations (12.3), as an integral

$$p\big(\mathbf{x}(i) \mid y^{i-1}, \mathbf{u}(i-1)\big) = \int_{x(i-1)} p\big(\mathbf{x}(i) \mid \mathbf{x}(i-1), \mathbf{u}(i-1)\big) \times$$
$$\times \ p\big(\mathbf{x}(i-1) \mid y^{i-1}\big) \, \mathrm{d}\mathbf{x}(i-1)$$

† See Jazwinski (1970) Chapters 7 and 8 or Fitzgerald (1971).

in which the first term $p\big(\mathbf{x}(i) \mid \mathbf{x}(i-1), \mathbf{u}(i-1)\big)$ depends on the probability density of the disturbance $\boldsymbol{\xi}_1$ and can only be evaluated if the function $\mathbf{G}_1$ has an inverse $\mathbf{G}_1^{-1}$ such that a value of $\boldsymbol{\xi}_1(i-1)$ can be found corresponding to specific values of the arguments $\mathbf{x}(i)$, $\mathbf{x}(i-1)$, $\mathbf{u}(i-1)$. Even where this first term in the integral for prior probability can be evaluated the integral itself can seldom be evaluated analytically unless the second term $p(\mathbf{x}(i-1) \mid y^{i-1})$, which is the previous conditional probability density of the state $\mathbf{x}(i-1)$, is normal. The likelihood function depends on the probability density of the noise $\boldsymbol{\xi}_2$ and can only be evaluated if the function $\mathbf{G}_2$ has an inverse $\mathbf{G}_2^{-1}$ such that a value of $\boldsymbol{\xi}_2(i)$ can be found corresponding to specific values of the arguments $\mathbf{x}(i)$ and $\mathbf{y}(i)$. In non-linear systems the inverses $\mathbf{G}_1^{-1}$ and $\mathbf{G}_2^{-1}$ sometimes do exist, but the conditional probability density for the state $\mathbf{x}$ scarcely ever remains normal so Bayes's rule cannot be realized in practice and it becomes necessary to use approximate methods of state estimation.

One type of approximation is a filter which generates a sequence of estimates $\hat{\mathbf{x}}(i)$ of the state $\mathbf{x}(i)$ according to an equation

$$\hat{\mathbf{x}}(i) = \text{prediction of } \mathbf{x}(i) + \text{constant} \times \text{innovation},$$

having the same form as the Kalman filter equation (14.12a) for the conditional mean. This approximation is derived by linearizing the non-linear functions $\mathbf{G}_1$, $\mathbf{G}_2$ about values of the state based on the estimate $\hat{\mathbf{x}}(i-1)$ of the previous state $\mathbf{x}(i-1)$ and about the mean values of the independent random variables $\boldsymbol{\xi}_1$, $\boldsymbol{\xi}_2$. The linearization assumes that the difference $\mathbf{x}(i-1) - \hat{\mathbf{x}}(i-1)$ between true and estimated values of the state and the differences between the random variables ξ_1, ξ_2 and their mean values are small enough to justify a first-order Taylor series expansion of the non-linear functions $\mathbf{G}_1$, $\mathbf{G}_2$ in equations (14.16). Equation (14.16a) then gives

$$\mathbf{x}(i) \simeq \mathbf{G}_1\big(\hat{\mathbf{x}}(i-1), \mathbf{u}(i-1), 0, i-1\big) + \mathbf{A}\big(\mathbf{x}(i-1) - \hat{\mathbf{x}}(i-1)\big) +$$
$$+ \; \mathbf{B}_1\boldsymbol{\xi}_1(i-1) = \hat{\mathbf{x}}_0 + \mathbf{A}\big(\mathbf{x}(i-1) - \hat{\mathbf{x}}(i-1)\big) + \boldsymbol{\beta}_1, \quad (14.17a)$$

where

(i) $\hat{\mathbf{x}}_0 \equiv \mathbf{G}_1\big(\hat{\mathbf{x}}(i-1), \mathbf{u}(i-1), 0, i-1\big)$ is the prediction of $\mathbf{x}(i)$ using the previous estimate $\hat{\mathbf{x}}(i-1)$ and the known value $\mathbf{u}(i-1)$ of the control at the previous stage, and assuming that the unknown value of $\boldsymbol{\xi}_1(i-1)$ was its mean, zero;

(ii) $\mathbf{A}$ is a Jacobian matrix of partial derivatives of $\mathbf{G}_1$ with respect to the states $\mathbf{x}$ evaluated at the point of linearization. The typical element of $\mathbf{A}$ is

$$a_{ij} \equiv \frac{\partial g_{1,i}}{\partial x_j}\big(\hat{\mathbf{x}}(i-1), \mathbf{u}(i-1), 0, i-1\big);$$

(iii) $\mathbf{B}_1$ is a Jacobian matrix of partial derivatives of $\mathbf{G}_1$ with respect to the disturbance variables ξ_1 evaluated at the point of linearization. The typical element of $\mathbf{B}_1$ is

$$b_{1,ij} \equiv \frac{\partial g_{1,i}}{\partial \xi_{1,j}} \big(\hat{\mathbf{x}}(i-1), \mathbf{u}(i-1), 0, i-1\big);$$

(iv) β_1 is a set of random variables with mean value zero and covariance matrix $\mathbf{B}_1\Sigma_1\mathbf{B}_1'$, where Σ_1 is the covariance matrix of the random variables $\xi_1(i-1)$.

Similarly equation (14.16b) becomes

$$\mathbf{y}(i) = \mathbf{G}_2(\hat{\mathbf{x}}_0, 0, i) + \mathbf{C}(\mathbf{x}(i) - \hat{\mathbf{x}}_0) + \beta_2, \qquad (14.17b)$$

where

(i) $\mathbf{G}_2(\hat{\mathbf{x}}_0, 0, i)$ is a prediction of $\mathbf{y}(i)$ using $\hat{\mathbf{x}}_0$, the prediction of $\mathbf{x}(i)$ from (i) above, and assuming that $\xi_2(i)$ will take its mean value of zero;

(ii) $\mathbf{C}$ is the matrix of partial derivatives of $\mathbf{G}_2$ with respect to the states $\mathbf{x}$ evaluated at the point of linearization. The typical element of $\mathbf{C}$ is

$$c_{ij} \equiv \frac{\partial g_{2,i}}{\partial x_j} (\hat{\mathbf{x}}_0, 0, i);$$

(iii) β_2 is a set of random variables with mean value zero and covariance matrix $\mathbf{B}_2\Sigma_2\mathbf{B}_2'$, where Σ_2 is the covariance matrix of the random variables $\xi_2(i)$, and $\mathbf{B}_2$, the matrix of partial derivatives of $\mathbf{G}_2$ with respect to the noise variables ξ_2, has typical element

$$b_{2,ij} \equiv \frac{\partial g_{2,i}}{\partial \xi_{2,j}} (\hat{\mathbf{x}}_0, 0, i).$$

The linearized equations (14.17) can be rewritten

$$\mathbf{x}(i) - \hat{\mathbf{x}}_0 = \mathbf{A}\big(\mathbf{x}(i-1) - \hat{\mathbf{x}}(i-1)\big) + \beta_1 \qquad (14.17a)$$

$$\mathbf{y}(i) - \mathbf{G}_2(\hat{\mathbf{x}}_0, 0, i) = \mathbf{C}\big(\mathbf{x}(i) - \hat{\mathbf{x}}_0\big) + \beta_2 \qquad (14.17b)$$

in a form which shows that the difference $\mathbf{x}(i) - \hat{\mathbf{x}}_0$ between the true and the predicted value of the state is the sum of two sets of random variables, $\mathbf{A}\big(\mathbf{x}(i-1) - \hat{\mathbf{x}}(i-1)\big)$ and β_1 in equation (14.17a), and that the innovation $\mathbf{y}(i) - \mathbf{G}_2(\hat{\mathbf{x}}_0, 0, i)$ can be regarded as a noisy measurement of a term, $\mathbf{C}\big(\mathbf{x}(i) - \hat{\mathbf{x}}_0\big)$ in equation (14.17b), proportional to the difference $\mathbf{x}(i) - \hat{\mathbf{x}}_0$. Equations (14.17) thus belong to the general class of linear equations (14.2) and (14.3) for which the Kalman filter is appropriate when all the random variables have normal distributions. The random variables in equations (14.17) are generally not normal, but the approximate filter is derived on the assumption that they are all normal and that the estimate $\hat{\mathbf{x}}(i-1)$ is the conditional mean of the state $\mathbf{x}(i-1)$ so that

$$p\big(\mathbf{x}(i-1) - \hat{\mathbf{x}}(i-1)\big) \simeq N\big(0, \boldsymbol{\Sigma}(i-1)\big),$$

the covariance matrix $\boldsymbol{\Sigma}(i-1)$ being generated as part of the filter output. The normal distributions assumed for the random variables $\boldsymbol{\beta}_1, \boldsymbol{\beta}_2$ are

$$p(\boldsymbol{\beta}_1) \simeq N(0, \mathbf{B}_1\boldsymbol{\Sigma}_1\mathbf{B}_1'),$$

$$p(\boldsymbol{\beta}_2) \simeq N(0, \mathbf{B}_2\boldsymbol{\Sigma}_2\mathbf{B}_1').$$

With these assumptions the Kalman filter, equations (14.12) and (14.13), gives an estimate of the difference $\mathbf{x}(i) - \hat{\mathbf{x}}_0$:

$$\text{estimate of } (\mathbf{x}(i) - \hat{\mathbf{x}}_0) = \mathbf{K}(i)\big(\mathbf{y}(i) - \mathbf{G}_2(\hat{\mathbf{x}}_0, \mathbf{0}, i)\big)$$

and hence the estimate $\hat{\mathbf{x}}(i)$ of the state $\mathbf{x}(i)$ is

$$\hat{\mathbf{x}}(i) = \hat{\mathbf{x}}_0 + \mathbf{K}(i)\big(\mathbf{y}(i) - \mathbf{G}_2(\hat{\mathbf{x}}_0, \mathbf{0}, i)\big), \tag{14.18a}$$

where the gain $\mathbf{K}(i)$ is given, as in equation (14.12b), by

$$\mathbf{K}(i) = \boldsymbol{\Sigma}(i)\mathbf{C}'(\mathbf{B}_2\boldsymbol{\Sigma}_2\mathbf{B}_2')^{-1} \tag{14.18b}$$

and the covariance matrix $\boldsymbol{\Sigma}(i)$ is generated, as in equations (14.13c), by

$$\boldsymbol{\Sigma}(i) = \boldsymbol{\Sigma}_0 - \boldsymbol{\Sigma}_0\mathbf{C}'(\mathbf{B}_2\boldsymbol{\Sigma}_2\mathbf{B}_2' + \mathbf{C}\boldsymbol{\Sigma}_0\mathbf{C}')^{-1}\mathbf{C}\boldsymbol{\Sigma}_0,$$

where

$$\boldsymbol{\Sigma}_0 = \mathbf{B}_1\boldsymbol{\Sigma}_1\mathbf{B}_1' + \mathbf{A}\boldsymbol{\Sigma}(i-1)\mathbf{A}'. \tag{14.18c}$$

The approximate filter of equations (14.18) is known as the extended Kalman filter because it extends the Kalman filter to linearizable non-linear problems. The question of what constitutes a linearizable non-linear system has no satisfactory theoretical answer; it is a question about convergence of the covariance equation (14.18c) in which the matrices $\mathbf{A}, \mathbf{B}_1, \mathbf{B}_2, \mathbf{C}$ of partial derivatives depend on the estimates $\hat{\mathbf{x}}$ and must be recomputed at every stage; it is also a question about the validity of assuming that the random variables are normally distributed and that the estimate $\hat{\mathbf{x}}$ corresponds to the conditional mean of the state $\mathbf{x}$. It is known that performance can be sensitive to the initial values of the estimates $\hat{\mathbf{x}}$ and the matrix $\boldsymbol{\Sigma}$. In practice the use of an extended Kalman filter would be considered justifiable if it was found to perform well; one indication of good performance would be that the inno-vation terms $\mathbf{y}(i) - \mathbf{G}_2(\hat{\mathbf{x}}_0, \mathbf{0}, i)$ were independent, as in the Kalman filters of Section 14.3. Methods for improving the approximation by including second-order terms in the Taylor series expansion of the non-linear functions $\mathbf{G}_1, \mathbf{G}_2$, or by iterating the linearization at each stage using the estimate of equations (14.18) as a new point of linearization to modify the matrices $\mathbf{A}, \mathbf{B}_1, \mathbf{B}_2, \mathbf{C}$ so that equations (14.18) give an improved estimate, are beyond the scope of this book.†

† See Jazwinski (1970) Chapters 8 and 9.

There exists a continuous-time equivalent to the discrete-time extended Kalman filter derived above. The general class of non-linear continuous-time systems has stochastic differential equations

$$dx = g_1(x, u, t)\, dt + B_1(x, u, t)\, dw_1,\qquad (14.19a)$$

$$y\, dt = g_2(x, t)\, dt + B_2(x, t)\, dw_2,\qquad (14.19b)$$

in which the random forcing terms $B_1\, dw_1$, $B_2\, dw_2$ must be separate from the general non-linear terms g_1, g_2. The extended Kalman filter for this class of systems generates an estimate $\hat{x}$ according to the differential equation

$$\frac{d\hat{x}}{dt} = g_1(\hat{x}, u, t) + K\big(y - g_2(\hat{x}, t)\big),\qquad (14.20a)$$

where the gain K is given, as before, by

$$K = \Sigma C'(B_2\Sigma_2 B_2')^{-1}\qquad (14.20b)$$

and the covariance-like matrix Σ is generated by the differential equation

$$\frac{d\Sigma}{dt} = B_1\Sigma_1 B_1' + A\Sigma + \Sigma A' - \Sigma C'(B_2\Sigma_2 B_2')^{-1}C\Sigma.\qquad (14.20c)$$

In equations (14.20) A and C are Jacobian matrices of partial derivatives of g_1 and g_2 with respect to x; B_1 and B_2 are the matrices appearing separately in equations (14.19), and A, B_1, B_2, C are all evaluated at the point of linearization. The matrices Σ_1 and Σ_2 specify the covariance matrices of the independent Wiener processes w_1 and w_2.

Problems 14.11, 14.12

14.5. Statistical methods of estimation

Bayes's rule is the the fundamental result in probability theory which specifies how to process noisy measurements of a random process; it is derived on the assumption that prior information in the form of probability density functions of all random variables is available and in the absence of such information it cannot be used. When Bayes's rule is unrealizable it is as frequently because of the absence of this prior probabilistic information as because of the computational difficulties mentioned in Section 14.4. In such situations uncertain states can be estimated by alternative methods which are described as statistical, rather than probabilistic, because they are not derived from probability theory. They make no attempt to generate the conditional probability density function and they are not generally optimal from the point of view of probability theory; their advantage is that they can often be used for estimation in systems where Bayes's rule is unrealizable.

Statistical methods of estimation are based on the idea that the quality of an estimate $\hat{\mathbf{x}}$ of a state $\mathbf{x}$ can be measured by some scalar criterion, which can be evaluated for all possible values of $\hat{\mathbf{x}}$, and that the value of $\hat{\mathbf{x}}$ giving the best estimate of $\mathbf{x}$ is that which gives the criterion its optimum (minimum or maximum) value. This procedure is analogous to the fundamental procedure of optimal control theory, which is to define a performance criterion and then choose controls to optimize the criterion. Estimation criteria are introduced here for the simple system of the example in Section 14.2 where a state x is measured by

$$y = x + \xi.$$

The estimate $\hat{x}$ of x depends, as discussed in Section 14.2, on what prior information is available.

(i) If the prior information consists of a previous estimate $\hat{x}_0$ of x together with knowledge that the mean value of the random variable ξ was zero an estimation criterion would be the sum of squares

$$I \equiv (\hat{x} - \hat{x}_0)^2 + \hat{\xi}^2,$$

where $\hat{\xi}$ is the estimate of ξ implied by the measured value Y of y and the estimate $\hat{x}$ of x according to

$$\hat{\xi} = Y - \hat{x}.$$

Minimizing this criterion ensures that the estimate of each random variable is, in a least-squares sense, as close as possible to its predicted value. The criterion can be written

$$I_{\text{LS}} \equiv (\hat{x} - \hat{x}_0)^2 + (Y - \hat{x})^2,$$

and the minimizing, least-squares estimate $\hat{x}$ is given by ordinary calculus:

$$\hat{x}_{\text{LS}} = \tfrac{1}{2}(\hat{x}_0 + Y).$$

(ii) If further prior information was available giving values of the variance σ_0^2, σ_ξ^2 of the random variables x, ξ, it could be used to specify a weighted least-squares criterion

$$I_{\text{WLS}} \equiv \frac{(\hat{x} - \hat{x}_0)^2}{\sigma_0^2} + \frac{(Y - \hat{x})^2}{\sigma_\xi^2},$$

in which sensitivity to the difference between predicted and estimated values of the random variables increases as variance decreases. The minimizing, weighted least-squares estimate $\hat{x}$ is

$$\hat{x}_{\text{WLS}} = \frac{\sigma_\xi^2 \hat{x}_0 + \sigma_0^2 Y}{\sigma_0^2 + \sigma_\xi^2}$$

and has the same form as the Bayes's rule equation (14.5a) for updating the conditional mean when the prior information consists of normal prior probability densities for the random variables.

(iii) If there was no prior information about the predicted value $\hat{x}_0$ of x, the criterion would be

$$I \equiv \xi^2,$$

and the minimizing estimate would be

$$\hat{x} = Y.$$

(iv) If there was no prior information about the mean value of the random variable ξ, the criterion would be

$$I \equiv (\hat{x} - \hat{x}_0)^2,$$

and the minimizing estimate would be

$$\hat{x} = \hat{x}_0.$$

(v) If there was no prior information at all, it would not be possible to define a criterion or to make an estimate.

The development of statistical estimation procedures for discrete-time dynamic systems, such as control systems, is simplified by using a conventional notation of estimation theory:

$\hat{x}(i \mid j) \equiv$ estimate at time j of the value $x(i)$ of a stochastic variable x at time i.

With this notation the foregoing simple example can be generalized to provide a recursive statistical procedure for estimating the state $\mathbf{x}(i)$ in the general class of discrete-time control systems specified by equations (14.16):

$$\mathbf{x}(i + 1) = \mathbf{G}_1\big(\mathbf{x}(i), \mathbf{u}(i), \boldsymbol{\xi}_1(i), i\big), \tag{14.16a}$$

$$\mathbf{y}(i) = \mathbf{G}_2\big(\mathbf{x}(i), \boldsymbol{\xi}_2(i), i\big). \tag{14.16b}$$

The generalized estimation criterion is chosen to be quadratic in the estimated variables for the same reasons, of analytic convenience, that control performance criteria are chosen to be quadratic whenever possible. The estimation criterion is

$$I \equiv \hat{\boldsymbol{\xi}}_1'(i - 1 \mid i)\mathbf{P}\hat{\boldsymbol{\xi}}_1(i - 1 \mid i) + \hat{\boldsymbol{\xi}}_2'(i \mid i)\mathbf{Q}\hat{\boldsymbol{\xi}}_2(i \mid i) + \big(\hat{\mathbf{x}}(i - 1 \mid i) -$$
$$- \hat{\mathbf{x}}(i - 1 \mid i - 1)\big)'\mathbf{R}\big(\hat{\mathbf{x}}'(i - 1 \mid i) - \hat{\mathbf{x}}(i - 1 \mid i - 1)\big), \tag{14.21}$$

where $\mathbf{P}$, $\mathbf{Q}$, $\mathbf{R}$ are symmetric non-singular weighting matrices. Here $\hat{\boldsymbol{\xi}}_1(i - 1 \mid i)$, $\hat{\boldsymbol{\xi}}_2(i \mid i)$, $\hat{\mathbf{x}}(i - 1 \mid i)$ are estimates of the random variables $\boldsymbol{\xi}_1(i - 1)$, $\boldsymbol{\xi}_2(i)$, $\mathbf{x}(i - 1)$ whose values are implied by equations (14.16) when an estimate $\hat{\mathbf{x}}(i \mid i)$ of $\mathbf{x}(i)$ is determined. This requires that the functions $\mathbf{G}_1$, $\mathbf{G}_2$ of equations (14.16) have inverses $\mathbf{G}_1^{-1}$, $\mathbf{G}_2^{-1}$ such that estimates of the random variables $\boldsymbol{\xi}$ corresponding to estimates of $\mathbf{x}$ and given values of $\mathbf{u}$ and $\mathbf{y}$ can be found from equations of the form

$$\hat{\boldsymbol{\xi}}_1(i - 1 \mid i) = \mathbf{G}_1^{-1}\big(\hat{\mathbf{x}}(i \mid i), \hat{\mathbf{x}}(i - 1 \mid i), \mathbf{u}(i - 1), i - 1\big), \tag{14.22a}$$

$$\hat{\boldsymbol{\xi}}_2(i \mid i) = \mathbf{G}_2^{-1}\big(\hat{\mathbf{x}}(i \mid i), \mathbf{y}(i), i\big). \tag{14.22b}$$

If the inverses of equations (14.22) do not exist, statistical methods based on the criterion of equation (14.21) cannot be used. Generation of an estimate $\hat{x}(i - 1 \mid i)$ of the previous state $x(i - 1)$ is an essential part of the estimation of the current state $x(i)$ because they both appear in the dynamic equation (14.16a) and so they must both be estimated simultaneously. The prior estimate $\hat{x}(i - 1 \mid i - 1)$ is not essential for statistical estimation but is often available in control systems as the result of estimation at the previous stage; if it were not available the weighting matrix $\mathbf{R}$ would be given the value zero. The weighting matrices $\mathbf{P}$, $\mathbf{Q}$, $\mathbf{R}$ reflect prior information.

(i) If the prior information consisted only of expected values of the random variables, the weighting matrices would be identity matrices,

$$\mathbf{P} = \mathbf{I}, \qquad \mathbf{Q} = \mathbf{I}, \qquad \mathbf{R} = \mathbf{I},$$

specifying a least-squares criterion. .

(ii) If further prior information was available giving values of the co-variance matrices $\boldsymbol{\Sigma}_1$, $\boldsymbol{\Sigma}_2$, $\boldsymbol{\Sigma}(i - 1)$ of the random variables $\boldsymbol{\xi}_1(i - 1)$, $\boldsymbol{\xi}_2(i)$, $x(i - 1) - \hat{x}(i - 1 \mid i - 1)$, the weighting matrices would be the inverses of the covariance matrices,

$$\mathbf{P} = \boldsymbol{\Sigma}_1^{-1}, \qquad \mathbf{Q} = \boldsymbol{\Sigma}_2^{-1}, \qquad \mathbf{R} = \boldsymbol{\Sigma}^{-1}(i - 1),$$

specifying a weighted least-squares criterion.

(iii) If there was no prior information about one set of random variables, for example no prior estimate $\hat{x}(i - 1 \mid i - 1)$ the corresponding matrix would be null, for example $\mathbf{R}$ might be null as mentioned above.

(iv) Sometimes there is prior information about the probability distributions of the independent random variables $\boldsymbol{\xi}$, but not about the state x. In this situation the estimates $\hat{x}$ can be chosen to maximize the joint likelihood of the estimated random variables $\hat{\boldsymbol{\xi}}$, which is given by

$$p(\hat{\boldsymbol{\xi}}_1, \hat{\boldsymbol{\xi}}_2) = p(\hat{\boldsymbol{\xi}}_1)p(\hat{\boldsymbol{\xi}}_2)$$

provided that $\boldsymbol{\xi}_1$ and $\boldsymbol{\xi}_2$ are independent of each other. If the known prior probability distributions are normal,

$$p(\boldsymbol{\xi}_1) = N(0, \boldsymbol{\Sigma}_1), \qquad p(\boldsymbol{\xi}_2) = N(0, \boldsymbol{\Sigma}_2),$$

the joint likelihood

$$p\big(\hat{\boldsymbol{\xi}}_1(i - 1 \mid i), \hat{\boldsymbol{\xi}}_2(i \mid i)\big) =$$
$$= \text{constant} \times \exp\big(-\tfrac{1}{2}\big(\hat{\boldsymbol{\xi}}_1'(i - 1 \mid i)\boldsymbol{\Sigma}_1^{-1}\hat{\boldsymbol{\xi}}_1(i - 1 \mid i) + \hat{\boldsymbol{\xi}}_2'(i \mid i)\boldsymbol{\Sigma}_2^{-1}\hat{\boldsymbol{\xi}}_2(i \mid i)\big)\big)$$

is maximized by minimizing the quadratic sum ( ) inside the exponential; this provides a maximum-likelihood criterion in the form of equation (14.21) with

$$\mathbf{P} = \boldsymbol{\Sigma}_1^{-1'}, \qquad \mathbf{Q} = \boldsymbol{\Sigma}_2^{-1}, \qquad \mathbf{R} = 0.$$

If the known probability distributions were not normal a maximum-likelihood criterion could still be used but it would not have the quadratic form of equation (14.21).

Equations (14.21) and (14.22) combine to give an expression

$$I = \big\{\mathbf{G}_1^{-1}\big(\hat{x}(i\mid i),\, \hat{x}(i-1\mid i),\, \mathbf{u}(i-1),\, i-1\big)\big\}'\mathbf{PG}_1^{-1}\{\} \;+$$
$$+\big\{\mathbf{G}_2^{-1}\big(\hat{x}(i\mid i),\, \mathbf{y}(i),\, i\big)\big\}'\mathbf{QG}_2^{-1}\{\} \;+$$
$$+\big(\hat{x}(i-1\mid i) - \hat{x}(i-1\mid i-1)\big)'\mathbf{R}\big(\hat{x}(i-1\mid i) - \hat{x}(i-1\mid i-1)\big) \quad (14.23)$$

which, for known values of the previous control $\mathbf{u}(i-1)$, the current output $\mathbf{y}(i)$, and the prior estimate $\hat{x}(i-1\mid i-1)$, can be minimized with respect to $\hat{x}(i\mid i)$ and $\hat{x}(i-1\mid i)$. The minimizing values provide current estimates of the current state and the previous state. Improved estimates could be obtained by iterating the minimization using the new estimate $\hat{x}(i-1\mid i)$ of the previous state in place of the prior estimate $\hat{x}(i-1\mid i-1)$. For most non-linear systems the minimizations here have to be done numerically using a computer, so the number of states which can be estimated in real time in a feedback control system is usually limited to low values. The question of whether least-squares statistical estimation is a suitable procedure for a given non-linear problem does not generally have a simple theoretical answer. In practice use of the procedure would be considered justifiable if it performed well; one indication of good performance would be that the differences, known as *residuals*, between current estimates such as $\hat{\xi}_1(i-1\mid i)$, $\hat{\xi}_2(i\mid i)$, $\hat{x}(i-1\mid i)$, and their previously expected values $\big($zero, zero, $\hat{x}(i-1\mid i-1)\big)$ were independent, like the innovations in a Kalman filter.

For the class of linear systems of equations (14.2) and (14.3),

$$\mathbf{x}(i+1) = \mathbf{A}\mathbf{x}(i) + \mathbf{B}\mathbf{u}(i) + \xi_1(i), \qquad (14.2)$$
$$\mathbf{y}(i) = \mathbf{C}\mathbf{x}(i) + \xi_2(i), \qquad (14.3)$$

least-squares estimation becomes an LQP-problem and the minimizing values of $\hat{x}(i\mid i)$ and $\hat{x}(i-1\mid i)$ can be found by calculus. They are

$$\hat{x}(i\mid i) = \mathbf{A}\hat{x}(i-1\mid i-1) + \mathbf{B}\mathbf{u}(i-1) +$$
$$+ \mathbf{K}\big\{\mathbf{y}(i) - \mathbf{C}\big(\mathbf{A}\hat{x}(i-1\mid i-1) + \mathbf{B}\mathbf{u}(i-1)\big)\big\}, \quad (14.24)$$

where

$$\mathbf{K} = \big(\mathbf{I} + (\mathbf{P}^{-1} + \mathbf{A}\mathbf{R}^{-1}\mathbf{A}')\mathbf{C}'\mathbf{QC}\big)^{-1}(\mathbf{P}^{-1} + \mathbf{A}\mathbf{R}^{-1}\mathbf{A}')\mathbf{C}'\mathbf{Q}$$

and

$$\hat{x}(i-1\mid i) = \hat{x}(i-1\mid i-1) + \mathbf{R}^{-1}\mathbf{A}'\mathbf{C}'\mathbf{Q}\big(\mathbf{y}(i) - \mathbf{C}\hat{x}(i\mid i)\big).$$

Equation (14.24) is a recursive filtering equation having the same convenient form as the Kalman filter equation (14.13a) which gives the conditional mean when the probability distributions are all known and normal. For the particular case of weighted least-squares estimation, (ii) above, the gain $\mathbf{K}$ in equation (14.24) is identical with the Kalman gain of equations (14.12b) and (14.13b), and the statistical estimation is identical with the Kalman filter equation for the conditional mean. This correspondence between the solutions to statistical and to probabilistic problems only exists for linear systems with normal random variables; it is another aspect of the duality between estimation problems and LQP-control problems which is discussed in Section 14.3.

For the general class of continuous-time control systems having stochastic differential equations (14.19):

$$d\mathbf{x} = \mathbf{g}_1(\mathbf{x}, \mathbf{u}, t)\, dt + \mathbf{B}_1(\mathbf{x}, \mathbf{u}, t)\, d\mathbf{w}_1, \tag{14.19a}$$

$$\mathbf{y}\, dt = \mathbf{g}_2(\mathbf{x}, t)\, dt + \mathbf{B}_2(\mathbf{x}, t)\, d\mathbf{w}_2, \tag{14.19b}$$

the general estimation criterion of equation (14.21) can be written

$$I = \widehat{d\mathbf{w}_1'\mathbf{P}\, d\mathbf{w}_1} + \widehat{d\mathbf{w}_2'\mathbf{Q}\, d\mathbf{w}_2} +$$
$$+ \big(\hat{\mathbf{x}}(t - dt) - \hat{\mathbf{x}}_0(t - dt)\big)'\mathbf{R}\big(\hat{\mathbf{x}}(t - dt) - \hat{\mathbf{x}}_0(t - dt)\big), \tag{14.25}$$

where $\widehat{d\mathbf{w}}_1$ and $\widehat{d\mathbf{w}}_2$ are estimates of the Wiener process increments of equations (14.19), $\hat{\mathbf{x}}(t - dt)$ is an estimate of the state $\mathbf{x}(t - dt)$ at the beginning of the time increment dt, and $\hat{\mathbf{x}}_0(t - dt)$ is the prior estimate of $\mathbf{x}(t - dt)$ which is usually available in control systems. The weighting matrices $\mathbf{P}$, $\mathbf{Q}$, $\mathbf{R}$ reflect prior information as before and it is assumed that the matrices $\mathbf{B}_1$, $\mathbf{B}_2$ in equations (14.19) are non-singular so that values of the increments $d\mathbf{w}$ corresponding to given values of $\mathbf{x}$, $d\mathbf{x}$, $\mathbf{u}$, $\mathbf{y}$ can be found from equations of the form

$$d\mathbf{w}_1 = \mathbf{B}_1^{-1}\big(d\mathbf{x} - \mathbf{g}_1(\mathbf{x}, \mathbf{u}, t)\, dt\big), \tag{14.26a}$$

$$d\mathbf{w}_2 = \mathbf{B}_2^{-1}\big(\mathbf{y} - \mathbf{g}_2(\mathbf{x}, t)\big)\, dt, \tag{14.26b}$$

where the increment $d\mathbf{x}$ in equation (14.26a) is the difference

$$d\mathbf{x} = \mathbf{x}(t) - \mathbf{x}(t - dt). \tag{14.26c}$$

For known values of the control $\mathbf{u}$, the output $\mathbf{y}$, and the prior estimate $\hat{\mathbf{x}}_0(t - dt)$, equations (14.25) and (14.26) combine to given an expression

$$I = \{\hat{\mathbf{x}}(t) - \hat{\mathbf{x}}(t - dt) - \mathbf{g}_1(\hat{\mathbf{x}}(t), \mathbf{u}, t)\, dt\}'(\mathbf{B}_1^{-1})'\mathbf{P}\mathbf{B}_1^{-1}\{\hat{\mathbf{x}}(t) - \hat{\mathbf{x}}(t - dt) -$$
$$- \mathbf{g}_1(\ )\, dt\} + \{\mathbf{y} - \mathbf{g}_2(\hat{\mathbf{x}}(t), t)\}'\, dt(\mathbf{B}_2^{-1})'\mathbf{Q}\mathbf{B}_2^{-1}\{\mathbf{y} - \mathbf{g}_2(\ )\} +$$
$$+ \{\hat{\mathbf{x}}(t - dt) - \hat{\mathbf{x}}_0(t - dt)\}'\mathbf{R}\{\hat{\mathbf{x}}(t - dt) - \hat{\mathbf{x}}_0(t - dt)\} \tag{14.27}$$

which can, in principle, be minimized with respect to $\hat{\mathbf{x}}(t - dt)$ and $\hat{\mathbf{x}}(t)$. The minimizing values usually converge to the prior estimate $\hat{\mathbf{x}}_0(t - dt)$ as the time interval dt goes to zero so the minimization does not lead directly to a new estimate $\hat{\mathbf{x}}(t)$. However the increment

$$\hat{\mathbf{x}}(t) - \hat{\mathbf{x}}_0(t - dt)$$

can be of order dt so that the limit

$$\lim_{dt \to 0} \frac{\hat{x}(t) - \hat{x}_0(t - dt)}{dt} = \frac{d\hat{x}}{dt}$$

exists and defines a filter in the form of an integrator having output $\hat{x}(t)$.

For the class of linear systems having stochastic differential equations

$$dx = Ax \, dt + Bu \, dt + dw_1,$$
$$y \, dt = Cx \, dt + dw_2,$$

least-squares estimation becomes an LQP-problem and the minimizing values of $\hat{x}(t - dt)$ and $\hat{x}(t)$ in equation (14.27) can be found by calculus. The minimizing value of $\hat{x}(t)$ can be put in a form

$$\hat{x}(t) = \hat{x}_0(t - dt) + A\hat{x}(t) \, dt + Bu \, dt$$
$$+ (R^{-1} + P^{-1})((I - A \, dt)')^{-1}C'Q(y - C\hat{x}(t)) \, dt^2$$

which confirms that it converges to $\hat{x}_0(t - dt)$ as dt goes to zero, and also gives the ratio

$$\frac{\hat{x}(t) - x_0(t - dt)}{dt} = A\hat{x}(t) + Bu +$$
$$+ (R^{-1} + P^{-1})((I - A \, dt)')^{-1}C'Q(y - C\hat{x}(t)) \, dt.$$

For the particular case of weighted least-squares estimation, the correspondence between the weighting matrices P, Q, and the covariance matrices of the Wiener process increments would be

$$P = \frac{1}{dt}\Sigma_1^{-1}, \qquad Q = \frac{1}{dt}\Sigma_2^{-1},$$

and the matrix R would correspond to the inverse of the covariance matrix Σ for the state x; then the above ratio is

$$\frac{\hat{x}(t) - \hat{x}_0(t - dt)}{dt} = A\hat{x}(t) + Bu +$$
$$+ (\Sigma^{-1} + dt\Sigma_1)((I - A \, dt)')^{-1}C'\Sigma_2^{-1}(y - C\hat{x}(t)),$$

and in the limit as dt goes to zero this becomes

$$\frac{d\hat{x}}{dt} = A\hat{x}(t) + Bu + \Sigma C'\Sigma_2^{-1}(y - C\hat{x}(t)). \qquad (14.28)$$

Equation (14.28) defines a filter in the form of an integrator having output $\hat{x}(t)$ and is identical with the Kalman filter equation (14.15a) for the conditional mean. A Riccati equation corresponding to the Kalman filter covariance equation (14.15b) can also be derived from analysis of this LQP problem.†

† See Jazwinski (1970), Section 7.4.

This derivation of the Kalman filter equations is the principal point of interest here in continuous-time statistical estimation; little is known about the practicability of other estimation procedures based on minimizing the general criterion of equation (14.27). Discrete-time statistical estimation based on the general criterion of (14.23) is clearly practicable and is used, for example, in parameter estimation as discussed in Section 14.6.

Problems 14.13–14.16

14.6. Linear mathematical models for parameter estimation

In some practical situations the mathematical model used to represent a controlled process is uncertain and can be improved by estimation procedures. Such situations arise for example in the control of chemical or of economic processes where relationships between outputs such as product quality or profit and the available controls **u** are often so complex that they can only be determined experimentally on the process. One example of a procedure for determining mathematical models is the measurement of frequency responses of continuous-time systems mentioned in Section 1.2: this particular procedure is not discussed here because its relationship to the stochastic theory of control in the presence of uncertain mathematical models is not as clear as in the case of procedures for discrete-time systems.

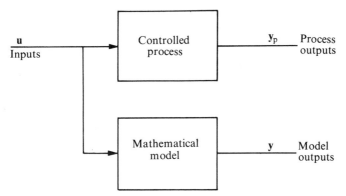

FIG. 14.4. Mathematical model represents the process if $y \simeq y_p$.

Figure 14.4 illustrates the relationship between a controlled process and the mathematical model, such as equations (14.16), which represents it. Both the process and the model have the same number of inputs **u** and outputs **y** and the model is said to represent the process insofar as its outputs **y** in response to inputs **u** are similar to the process outputs $\mathbf{y}_P$ in response to the same inputs. The input and output variables **u** and **y** are defined by the controlled process, but the nature of the functions **G** in the mathematical model and the numbers of variables **x** and ξ representing dynamic effects and uncertainty are in a sense arbitrary. They must initially be decided on the basis of prior information about the controlled process but could subsequently be modified to give improved representation by taking account of additional information provided by observations of process behaviour.

Estimation procedures for improving mathematical models are like other procedures of estimation theory in that the quantities estimated are values of a finite set of variables, such as parameters in the functions **G** of equations (14.16). The forms of functional relationships cannot usually be determined by estimation procedures and so the form of the functions **G** cannot be found from observations of process behaviour but must be determined, if necessary by assumption, on the basis of prior information. The commonest assumption is, as mentioned in Section 14.1, that the functions **G** are linear so that equations (14.16) take the form of equations (14.2) and (14.3):

$$\mathbf{x}(i + 1) = \mathbf{A}\mathbf{x}(i) + \mathbf{B}\mathbf{u}(i) + \boldsymbol{\xi}_1(i), \tag{14.2}$$

$$\mathbf{y}(i) = \mathbf{C}\mathbf{x}(i) + \boldsymbol{\xi}_2(i), \tag{14.3}$$

where some or all the elements of the matrices **A**, **B**, **C** and the covariance matrices $\boldsymbol{\Sigma}_1$, $\boldsymbol{\Sigma}_2$ are unknown constants to be estimated.

From the point of view of stochastic control theory, unknown constant parameters of the functions **G** must be regarded as additional state variables. The uncertain matrices **A**, **B**, **C** can be named $\mathbf{X}_A$, $\mathbf{X}_B$, $\mathbf{X}_C$, to show that their elements are state variables, and the uncertain covariance matrices $\boldsymbol{\Sigma}_1$, $\boldsymbol{\Sigma}_2$ can be accounted for by introducing matrices $\mathbf{X}_1$, $\mathbf{X}_2$ of unknown constant states and writing the stochastic variables $\boldsymbol{\xi}_1(i)$, $\boldsymbol{\xi}_2(i)$ in the form

$$\boldsymbol{\xi}_1(i) = \mathbf{X}_1\boldsymbol{\eta}_1(i), \qquad \boldsymbol{\xi}_2(i) = \mathbf{X}_2\boldsymbol{\eta}_2(i),$$

where $\boldsymbol{\eta}_1(i)$, $\boldsymbol{\eta}_2(i)$ are sets of uncorrelated, independent random sequences having zero means and identity covariance matrices so that the covariance matrices of $\boldsymbol{\xi}_1$, $\boldsymbol{\xi}_2$ become

$$\boldsymbol{\Sigma}_1 = \mathbf{X}_1\mathbf{X}_1', \qquad \boldsymbol{\Sigma}_2 = \mathbf{X}_2\mathbf{X}_2'.$$

With this change of notation equations (14.2) and (14.3) become

$$\left.\begin{aligned}
\mathbf{x}(i + 1) &= \mathbf{X}_A(i)\mathbf{x}(i) + \mathbf{X}_B(i)\mathbf{u}(i) + \mathbf{X}_1(i)\boldsymbol{\eta}_1(i) \\
\mathbf{X}_A(i + 1) &= \mathbf{X}_A(i) \\
\mathbf{X}_B(i + 1) &= \mathbf{X}_B(i) \\
\mathbf{X}_C(i + 1) &= \mathbf{X}_C(i) \\
\mathbf{X}_1(i + 1) &= \mathbf{X}_1(i) \\
\mathbf{X}_2(i + 1) &= \mathbf{X}_2(i) \\
\mathbf{y}(i) &= \mathbf{X}_C(i)\mathbf{x}(i) + \mathbf{X}_2(i)\boldsymbol{\eta}_2(i)
\end{aligned}\right\} \tag{14.29}$$

and can be seen to represent a non-linear system as has already been mentioned in Section 13.1. Little is known about optimal control of such non-linear stochastic systems. An obvious sub-optimal approach is to divorce the problem of estimating the unknown constant parameters from the problem of exercising efficient control: the parameters are estimated over some initial period of time, while little attempt is made to exercise control, and the resulting estimates are subsequently taken as true values in the linear model of equations (14.2) and (14.3) so that a control law can be found using the methods of Chapter 13. Provided that the estimates of the parameters converge to their true values, and provided that the cost of poor control while the estimates converge is small compared to the subsequent cost of control, this sub-optimal approach can give good performance.

The parameter estimation problem here is non-linear, because of the non-linearities in equation (14.29), so Bayes's rule cannot generally be realized and sub-optimal estimation procedures must be used. Two exceptional cases where the parameter estimation problem becomes linear are as follows.

(i) Where the only uncertain parameters are elements of the matrix $\mathbf{B}$. The term $\mathbf{X}_B\mathbf{u}$ in equations (14.29) is not non-linear in uncertain variables, as the product terms such as $\mathbf{X}_A\mathbf{x}$ or $\mathbf{X}\boldsymbol{\eta}$ are, so in this case there is no non-linearity. The uncertain states $\mathbf{x}$ and $\mathbf{X}_B$ here can be estimated using a Kalman filter which is similar to those of Section 14.3 except that its covariance equations depend on the controls $\mathbf{u}(i)$.

(ii) Where dynamics of the controlled process are represented in terms of impulse responses. For example, it has been shown in Chapter 1 that a single-variable discrete-time dynamic system can be represented either by a difference equation

$$y(i) + a_{n-1}y(i-1) + \cdots + a_0 y(i-n) = b_n u(i) + \cdots + b_0 u(i-n),$$

having $2n + 1$ coefficients $a_0, \ldots, a_{n-1}, b_0, \ldots, b_n$, or by the corresponding impulse response function

$$y(i) = g_0 u(i) + g_1 u(i-1) + g_2 u(i-2) + \cdots,$$

having an infinite number of coefficients $g_0, g_1, g_2, \ldots$ which in stable systems can be truncated after some finite settling time. The difference equation is non-linear in the coefficients $a_0, \ldots, a_{n-1}$ and the variable y, whereas the impulse response is linear in the coefficients $g_0, g_1, \ldots$ and y, and so might be preferable as a basis for estimation even though the number of parameters to be estimated could be larger than $2n + 1$. In practice impulse-response representations are not much used, partly because the difference equation or state space representation is needed as a basis for designing control laws and partly because they do not lead to linear estimation problems when random disturbances ξ_1 are included in the dynamic equations.

Sub-optimal estimation procedures for the non-linear majority of parameter estimation problems mostly include some numerical computation, such as minimization of the estimation criterion of equation (14.23), and can therefore become unrealizable if the number n_1 of parameters to be estimated is too.

large. The maximum value of n_1 in a linear mathematical model having n state variables $\mathbf{x}$ and representing a process having m inputs $\mathbf{u}$ and p outputs $\mathbf{y}$ is

$$
\begin{aligned}
n_1 = \quad & n^2, & \text{elements of } \mathbf{A}, \\
+ \; & nm, & \text{elements of } \mathbf{B}, \\
+ \; & np, & \text{elements of } \mathbf{C}, \\
+ \; & \frac{n(n + 1)}{2}, & \text{elements of } \mathbf{\Sigma}_1, \\
+ \; & \frac{p(p + 1)}{2}, & \text{elements of } \mathbf{\Sigma}_2.
\end{aligned}
$$

It depends primarily on the number n of state variables $\mathbf{x}$, which is usually chosen to be as small as is consistent with adequate representation. This choice of n must usually be made empirically. One approach is to set n initially to the lowest possible value, the larger of m or p, and then to decide whether or not this value is large enough on the basis of whether or not the innovations or residuals computed in the resulting estimation seem to be independent. If the residuals seem not to be independent the value of n can be increased and the estimation repeated.

Once a value for n is determined the total number of parameters to be estimated can be minimized by using suitable canonical forms in the linear mathematical model. The companion matrix which was introduced in Section 7.5 as part of the state space representation of single-variable ($m = 1$, $p = 1$) systems provides a canonical form for the matrix $\mathbf{A}$ in single-variable systems and also in controllable multi-variable systems.† The desirable feature of the companion matrix here is that it has only n uncertain elements whereas the completely general form of $\mathbf{A}$ has n^2 uncertain elements. There are no general canonical forms for the matrices $\mathbf{B}$ and $\mathbf{C}$ except that in observable systems it can be convenient to assume that each output $\mathbf{y}$ corresponds to only one component of the state vector $\mathbf{x}$ so that $\mathbf{C}$ is partitioned into a null matrix and an identity matrix:

$$
\mathbf{C} = \begin{bmatrix} \mathbf{0} & \begin{matrix} 1 & & \\ & \cdot & \\ & & 1 \end{matrix} \end{bmatrix}. \tag{14.30}
$$

In mathematical models of stochastic systems the number of variables ξ representing uncertainty is arbitrary in the same way that the number of state variables $\mathbf{x}$ is arbitrary. It is sometimes convenient to use a mathematical model where the same random variables ξ appear in both the dynamic

† The companion matrix does not provide a canonical form for a class of multi-variable systems known as *non-cylic*; see Mayne (1972).

equations and the output equations. Such a model arises if the Kalman filter equation (14.12a) is rewritten

$$\hat{x}(i \mid i) = \hat{x}(i \mid i - 1) + K\xi(i),$$

where

$$\xi(i) \equiv y(i) - C\hat{x}(i \mid i - 1)$$

is the innovation; introducing notation

$$x_1(i) \equiv \hat{x}(i \mid i - 1) = A\hat{x}(i - 1 \mid i - 1) + Bu(i - 1),$$

$$K_1 \equiv AK,$$

it follows that

$$x_1(i + 1) = Ax_1(i) + Bu(i) + K_1\xi(i), \qquad (14.31a)$$
$$y(i) = Cx_1(i) + \xi(i). \qquad (14.31b)$$

Equations (14.31) specify a mathematical model having the same inputs u, outputs y, matrices A, B, and C as the model of equations (14.2) and (14.3); its state x_1 corresponds to a one-step-ahead prediction of the state x in the earlier model. Its significance as a canonical form is that $\xi(i)$ can be eliminated to give a dynamic equation

$$x_1(i + 1) = (A - KC)x_1(i) + Bu(i) + K_1y(i)$$

in which there is no uncertainty about the future state $x_1(i + 1)$: the covariance matrix in a Kalman filter estimating x_1 would therefore converge to zero steady-state value and the equation (14.12a) estimating x_1 would become

$$\hat{x}_1(i \mid i) = \hat{x}_1(i \mid i - 1)$$
$$= (A - KC)\hat{x}_1(i - 1 \mid i - 1) + Bu(i - 1) + K_1y(i - 1). \qquad (14.32)$$

Thus x_1 can be estimated, without the need to solve a covariance equation, once A, B, C, and K_1 are known.

For single-variable systems the canonical form of equations (14.31), using a companion matrix for A and the form of equation (14.30) for C, gives

$$x(i + 1) = \begin{bmatrix} 0 & \cdots & 0 & -\alpha_1 \\ 1 & & & -\alpha_2 \\ & \ddots & & \vdots \\ & & 1 & -\alpha_n \end{bmatrix} x(i) + \begin{bmatrix} \beta_1 \\ \beta_2 \\ \vdots \\ \beta_n \end{bmatrix} u(i) + \begin{bmatrix} \gamma_1 \\ \gamma_2 \\ \vdots \\ \gamma_n \end{bmatrix} \xi(i), \qquad (14.33a)$$

$$y(i) = x_n(i) + \xi(i). \qquad (14.33b)$$

Eliminating the state variables x from equations (14.33) translates this canonical form into a difference equation of order n:

$$y(i) + a_{n-1}y(i - 1) + \cdots + a_0y(i - n) = b_{n-1}u(i - 1) + \cdots$$
$$+ b_0u(i - n) + \xi(i) + c_{n-1}\xi(i - 1) + \cdots + c_0\xi(i - n), \qquad (14.34)$$

where the coefficients a, b, c are

$$a_k = \alpha_{k+1},$$
$$b_k = \beta_{k+1},$$
$$c_k = \alpha_{k+1} + \gamma_{k+1}.$$

Difference equations like equation (14.34) have been used as a basis for parameter estimation in practice.†

In situations where the problem of estimating parameters of a mathematical model, which is sometimes referred to as system identification, is divorced from the problem of exercising control there may be no great urgency to generate parameter estimates as rapidly as possible and so no need to use recursive estimation procedures giving updated estimates of the parameters at every interval. The alternative to recursive estimation is non-recursive estimation where data is gathered over a finite operating time, say N discrete-time intervals, before any estimation is done. This has the advantage that estimates are based on more data than is used in recursive procedures so that better estimates can be produced in non-linear estimation problems having no sufficient statistics. The disadvantages of non-recursive estimation are that time must elapse while data is gathered before any estimation can be done and that the computation of estimates is usually more complex than with recursive procedures. The statistical estimation procedures introduced in Section 14.5 are suitable for non-recursive parameter estimation although the computations become more complex than in recursive estimation because the quadratic criteria of equation (14.21) must be summed over some extended period of observations ($i = 1, \ldots, N$) and then minimized with respect to estimates of the unknown parameters, and of the state variables $\mathbf{x}(i)$ and independent random variables $\xi(i)$ at all values of i, simultaneously. Discussion of non-recursive estimation procedures and their associated mathematical programming problems is beyond the scope of this book.‡

Problems 14.17–14.19

14.7. Bibliography

Kalman filters are discussed in
  Lee (1964) Chapter 3,
  Sorenson (1966),
  Aoki (1967) Sections II.3, V.3D, and VI.5,
  Bucy and Joseph (1968),
  Sage (1968) Chapters 9 and 10,
  Bryson and Ho (1969) Chapters 12 and 13,
  Meditch (1969) Chapters 5 to 8,
  Åström (1970) Chapter 7,
  Jazwinski (1970) Chapter 7,
  Takahashi, Rabins, and Auslander (1970) Sections 13.4 and 13.5,
  Kushner (1971) Sections 9.5 and 11.4.2,
  Sage and Melsa (1971a) Chapter 7,

† See for example Åström, Bohlin, and Wensmark (1965).

‡ At the time of writing the application of these procedures to control problems is an active field of research, see Åström and Eykhoff (1971).

KWAKERNAAK and SIVAN (1972) Sections 4.3, 4.4, and 6.5,
and extended Kalman filters in
AOKI (1967) Section V.3F,
JAZWINSKI (1970) Chapter 8,
SAGE and MELSA (1971a) Section 9.4.
Introductory books on estimation theory which describe its application to estimating states of linear dynamic systems and to Kalman filtering are
DEUTSCH (1969),
LIEBELT (1967),
NAHI (1969).
More advanced books which include applications of estimation theory to estimating states of non-linear dynamic systems† are
JAZWINSKI (1970),
SAGE and MELSA (1971a).
Estimation of the parameters of linear mathematical models of dynamic systems
is discussed in
LEE (1964) Chapter 4,
ÅSTRÖM, BOHLIN and WENSMARK (1965),
BOX and JENKINS (1970),
ÅSTRÖM and EYKHOFF (1971),
SAGE and MELSA (1971b),
GRAUPE (1972).

14.8. Problems

14.1. Derive equation (14.1) from equation (12.2). (Consider two events A and B defined by
$$A \equiv \text{the value } X \text{ of } x \text{ is in the range } x < X < x + \delta x,$$
$$B \equiv \text{the value } Y \text{ of } y \text{ is in the range } y < Y < y + \delta y.)$$

14.2. A measurement y is related to an uncertain constant x by
$$y = cx + \xi,$$
where ξ is an independent noise with
$$p(\xi) = N(0, \sigma_\xi^2),$$
and the prior probability distribution for x is
$$p(x) = N(0, \sigma_0^2).$$
Start from equation (14.1) and prove (including the necessary completion of squares) that when the measured value of y is Y the resulting conditional probability for x is
$$p(x \mid Y) = N \left(\frac{\sigma_0^2 c\, Y}{c^2 \sigma_0^2 + \sigma_\xi^2}, \ \frac{\sigma_0^2 \sigma_\xi^2}{c^2 \sigma_0^2 + \sigma_\xi^2} \right).$$

---

† At the time of writing this is an active field of research, see *Proceedings of symposium on non-linear estimation theory and its applications*, University of California, San Diego, 1970, 1971, 1972, 1973, published by Western Periodicals Co., Hollywood.

14.3. A set of N measurements $y(i)$ is related to an uncertain constant x by
$$y(i) = x + \xi(i), \quad i = 1, \ldots, N,$$
where $\xi(i)$ is a sequence of independent random variables with
$$p(\xi(i)) = N(0, \sigma_\xi^2).$$
Two schemes for deriving an estimate x from the measurements $y(i)$ are

(i) average the measurements to give

$$\hat{x}_1 = \frac{1}{N} \sum_{i=1}^{N} y(i);$$

(ii) assume a normal prior probability distribution
$$p(x) = N(m_0, \sigma_0^2)$$
and regard equations (14.5) as Kalman filter equations to be iterated N times leading to
$$\hat{x}_2 = m(N).$$

Calculate and compare the variances to be expected of the two estimates.

14.4. Derive the difference equations quoted in Problem 13.14 for the mean $m(i)$ and variance $\sigma^2(i)$ of the conditional probability density $p(x(i) \mid y^i)$ there.

14.5. A stochastic system has dynamic equation
$$x(i + 1) = x(i) + \xi(i),$$
where $\xi(i)$ is a sequence of independent, zero-mean, normal random variables having variance σ^2, and non-linear output equation
$$y(i) = x^2(i).$$
The conditional probability distribution $p(x(i) \mid y^i)$ has only two possible values, as shown in Fig. 14.5, and can be summarized by two sufficient statistics, w and q.

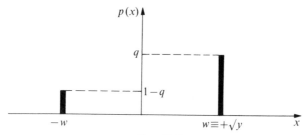

FIG. 14.5.

Use Bayes's rule to show that $q(i)$, the probability that $x(i)$ has the positive value $+\sqrt{y}\,(i)$, is given by the difference equation

$$q(i) = \frac{P_1 q(i - 1) + P_2 (1 - q(i - 1))}{P_1 + P_2},$$

where

$$P_1 = \exp\left(-\frac{\left(w(i) - w(i-1)\right)^2}{2\sigma^2}\right),$$

$$P_2 = \exp\left(-\frac{\left(w(i) + w(i-1)\right)^2}{2\sigma^2}\right).$$

14.6. Draw a block diagram specifying the dynamic system (Kalman filter) whose output is the conditional mean m of the state x and whose inputs are the control u and the output y, of the first-order controlled process of equations (14.7). Assume that the observation is over a long time, so that the covariance takes its steady-state value, and that the coefficients a, b, and c are all unity.

Comment on the relative importance of prediction and measurement when

$$\sigma_2 = \sqrt{2}\,\sigma_1.$$

14.7. Draw a block diagram specifying the dynamic system (Kalman filter) whose output is the conditional mean m of the state x and whose inputs are the input u and the output y of the continuous-time system shown in Fig. 14.6. Assume that

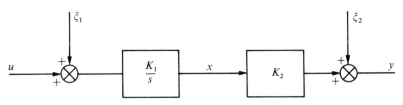

FIG. 14.6.

the observation is over a long time, so that the covariance takes its steady-state value, and that ξ_1, ξ_2 are independent, normal, white noises of magnitude σ_1^2, σ_2^2 respectively.

14.8. A continuous-time measurement $y(t)$ is related to an uncertain constant x by

$$y(t) = x + \xi(t),$$

where $\xi(t)$ is independent normal white noise of magnitude σ_ξ^2 and the prior probability distribution for x (at time $t = 0$) is

$$p(x) = N\big(m(0), \sigma^2(0)\big).$$

Show that the Kalman filter which estimates x includes an integrator having time-varying gain $\left(t + \sigma_\xi^2/\sigma^2(0)\right)^{-1}$. What initial condition should be set on this integrator? What is the limiting value of the variance of the estimate as time t goes to infinity?

14.9. A general class of non-stationary, linear, discrete-time systems has dynamic equation

$$x(i+1) = A(i)x(i) + B(i)u(i) + \xi_1(i),$$

and output equation

$$y(i) = C(i)x(i) + \xi_2(i),$$

where $\xi_1(i)$, $\xi_2(i)$ are sequences of independent random variables having probability distributions

$$p(\xi_1(i)) = N(0, \Sigma_1(i)), \qquad p(\xi_2(i)) = N(0, \Sigma_2(i)).$$

Repeat the argument of Section 14.3, starting with a first-order example, to find Kalman filter equations, corresponding to equations (14.13), for this class of systems.

14.10. A second-order discrete-time system has difference equations

$$x_1(i + 1) = x_2(i) + \xi_1(i),$$
$$x_2(i + 1) = x_1(i) + x_2(i) + \xi_2(i),$$

and a single output

$$y(i) = x_2(i),$$

where $\xi_1(i)$, $\xi_2(i)$ are sequences of independent normal random variables having zero means and unit variances. Use duality and the results of Problems 8.6 and 8.7 to find the steady-state covariance matrix of the Kalman filter estimating the states x_1, x_2.

Explain why uncertainty persists here in the absence of measurement noise.

14.11. A non-linear system has the dynamic equations (discussed in Section 13.1),

$$x_1(i + 1) = x_2(i)x_1(i) + x_3(i)u(i) + \xi_1(i),$$
$$x_2(i + 1) = x_2(i),$$
$$x_3(i + 1) = x_3(i),$$

and output equation

$$y(i) = x_1(i) + \xi_2(i),$$

where $\xi_1(i)$ $\xi_2(i)$ are sequences of independent random variables having zero means and variances σ_1^2, σ_2^2 respectively. Write down equations specifying an extended Kalman filter to estimate the states $x_1(i)$, $x_2(i)$, $x_3(i)$.

14.12. A continuous-time non-linear system has dynamic equation

$$dx = (e^x - ux)\, dt - u\, dw_1$$

and output equation

$$y\, dt = + \sqrt{x}\, dt + dw_2,$$

where w_1, w_2 are independent Wiener processes having variances over a time interval Δ of $\sigma_1^2 \Delta$ and $\sigma_2^2 \Delta$ respectively. Find the pair of simultaneous first-order differential equations specifying the extended Kalman filter whose inputs are the control u and the output y and whose output is an estimate $\hat{x}$ of the state x.

14.13. A controlled process has dynamic equation

$$x(i + 1) = ax(i) + bu(i) + \xi_1(i)$$

and output equation

$$y(i) = cx(i) + \xi_2(i),$$

where $\xi_1(i)$, $\xi_2(i)$ are sequences of independent, zero-mean, random variables.

There is no information about variances of any of the random variables but at each time i the values Y, U of the present output $y(i)$ and the previous control $u(i - 1)$ are known. Specify the criterion for least-squares estimation of the states $x(i)$, $x(i - 1)$ and derive the resulting equation giving $\hat{x}(i \mid i)$ in each of the following cases:

   (i) no prior information about $x(i - 1)$;
   (ii) given a prior estimate $\hat{x}(i - 1 \mid i - 1)$ of $x(i - 1)$.
Show that prior information about the variances of ξ_1 and ξ_2 would not affect the estimates in case (i).

14.14. Confirm that the equation giving $\hat{x}(i \mid i)$ in Problem 14.13(ii) is identical with the Kalman filter equation giving the conditional mean when the random variables ξ_1, ξ_2 and the prior estimate $\hat{x}(i - 1 \mid i - 1)$ all have unit variance. If there were prior information that the variances of ξ_1, ξ_2, $\hat{x}(i - 1 \mid i - 1)$ were σ_1^2, σ_2^2, $\sigma^2(i - 1)$ respectively, what would be the equation giving the weighted least-squares estimate $\hat{x}(i \mid i)$?

   What information about the estimates of x is provided by the Kalman filter but not by the least-squares estimation procedure of Section 14.5?

14.15. Specify the criterion for recursive least-squares estimation, in the absence of prior information about variances, of the states of the non-linear system of Problem 14.11. Show that the estimation procedure using this criterion generates no new information about $x_1(i - 1)$ or x_2 and find the equations giving $\hat{x}_1(i \mid i)$, $\hat{x}_3(i - 1 \mid i)$. (The algebra here can be simplified by writing

$$\hat{x}_1(i - 1 \mid i - 1) \equiv X_1, \qquad \hat{x}_2(i - 1 \mid i - 1) \equiv X_2, \qquad \hat{x}_3(i - 1 \mid i - 1) \equiv X_3,$$
$$\hat{x}_2(i - 1 \mid i) \equiv \hat{x}_2, \qquad \hat{x}_3(i - 1 \mid i) \equiv \hat{x}_3, \qquad u(i - 1) \equiv U, \qquad y(i) \equiv Y.)$$

   Comment on how the relative importances of prediction and measurement depend on U in the two estimation equations. Compare the statistical estimation procedure with the extended Kalman filter of Problem 14.11.

14.16. A controlled process has stochastic differential equation

$$dx = (ax + bu)\, dt + dw_1$$

and output equation

$$y\, dt = cx\, dt + dw_2,$$

where w_1, w_2 are independent Wiener processes having variances over a time interval Δ of $\sigma_1^2 \Delta$, $\sigma_2^2 \Delta$ respectively. Start from equation (14.27) and derive the differential equation of a filter which will generate weighted least-square estimates of x when the variance of these estimates has the constant value σ^2. Check that this filter is identical with a Kalman filter and find the value of σ^2 on the assumption that all random variables are normal.

14.17. A stochastic system has dynamic equations

$$x_1(i + 1) = ax_1(i) + x_2(i)u(i) + \xi_1(i),$$
$$x_2(i + 1) = x_2(i)$$

and output equation

$$y(i) = x_1(i) + \xi_2(i),$$

where x_2 is an unknown constant and $\xi_1(i)$, $\xi_2(i)$ are sequences of independent normal random variables having zero means and variances σ_1^2, σ_2^2 respectively. Find matrices $\mathbf{A}(i)$, $\mathbf{B}(i)$, etc. which will represent this system as a second-order non-stationary linear system in the class of Problem 14.9. Check that the covariance equations of the Kalman filter which estimates x_1 and x_2 depend on the variable u.

14.18. Derive equation (14.34) from equations (14.33).

14.19. What is the number of parameters needed to specify a linear mathematical model of a system having m inputs $\mathbf{u}$ and p outputs $\mathbf{y}$ using the canonical form of equations (14.31) with a companion matrix $\mathbf{A}$, a partitioned null and diagonal matrix $\mathbf{C}$, and n state variables $\mathbf{x}$? What is the difference between this number and the maximum number there could be in a linear model having the same numbers of variables?

15. Structure of stochastic controllers

15.1. Introduction

In Chapter 13 analytic solutions to stochastic LQP-problems with normal random variables were obtained by considering dynamic programming equations of optimal control for a general class of discrete-time stochastic problems and solving the equations for the special class of LQP-problems. The general class of problems is specified by equations (13.1) and (13.3), repeated here as (15.1),

$$x(i + 1) = G_1(x(i), u(i), \xi_1(i), i), \qquad (15.1a)$$
$$y(i) = G_2(x(i), \xi_2(i), i), \qquad (15.1b)$$

and by the performance criterion of equations (13.6), repeated here as (15.2):

$$J = E \sum_{i=i_1}^{i_2} H(x(i), u(i), i). \qquad (15.2)$$

Equation (15.1a) describes the dynamics of the controlled process as a set of first-order Markov dynamic equations in which the independent random variables $\xi_1(i)$ account for uncertainty about the future. The output equation (15.1b) introduces independent random variables $\xi_2(i)$ which account for uncertainty about the present state $x(i)$. When the present state is uncertain its values are described by a conditional probability density function $p(x(i) \mid y^i)$ and the general dynamic programming analysis is in terms of functionals

$f_N(p(x \mid y), i_2) \equiv$ Value of J using optimal control over N stages finishing at time i_2 and starting from uncertain state $x(i_2 - N + 1)$ described by
$p(x(i_2 - N + 1) \mid y^{i_2 - N + 1}) = p(x \mid y),$

$u_N(p(x \mid y), i_2) \equiv$ Optimal control $u^*(i_2 - N + 1)$ for a process finishing at time i_2 and starting from uncertain state $x(i_2 - N + 1)$ described by
$p(x(i_2 - N + 1) \mid y^{i_2 - N + 1}) = p(x \mid y),$

having the conditional probability density function as an argument. The analysis leads to functional recurrence equations which are intractable unless the conditional probability density function can be specified by sufficient statistics so that the solution is characterized by functions rather than functionals; this seldom happens outside the class of linear systems with normal random variables.

The absence of sufficient statistics together with the insolubility of equations specifying optimal control in non-LQP-problems frustrates most attempts to derive general theoretical results about optimal control of stochastic systems. It is known however that the control depends on the current conditional probability density for the state and that updating the conditional probability density is an essential function to be performed by controllers in stochastic systems. Such controllers therefore perform dual functions:

(i) estimate the current state by updating the conditional probability density function;

(ii) realize a control law which exercises control using the resulting information about the current state,

as introduced in Section 13.4, and have the structure shown in Fig. 15.1.

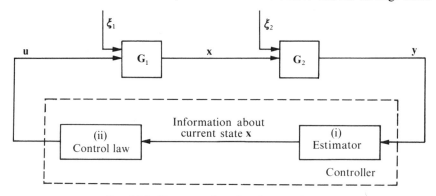

FIG. 15.1. Structure of stochastic control system.

This representation of controller structure provides a basis for discussing sub-optimal controllers to perform satisfactorily even though the estimator and the control law are not necessarily optimal and the information transmitted between the two is less than the complete conditional probability density function.

One of the principal features of controllers having the structure of Fig. 15.1 is the possibility of interactions between the dual functions of estimation and control. Two forms of interaction are possible: one is that uncertainty about the current state could result in cautious control which takes account of current uncertainty by exercising control less strongly than if the state were known; the other is that the control law might introduce some probing action in order to reduce future uncertainty more rapidly than would otherwise happen. Both interactions are absent in the class of LQP-problems having normal random variables, which is one reason why analytic results about optimal controllers for such problems were obtainable in Chapter 13.

The properties of separability and certainty equivalence which were introduced in Chapter 13, together with a third property of neutrality introduced in Section 15.2, are used in Section 15.3 to characterize the controller interactions in terms of intentional error, probing, and caution. The potential application of this characterization to the design of sub-optimal control laws for the general class of non-linear stochastic control problems is outlined in Section 15.4. The discussion here in Chapter 15 is in terms of discrete-time systems.

15.2. Properties of stochastic control problems

Interaction between the dual functions of controllers for stochastic systems is described by three properties, separability, neutrality, and certainty equivalence, defined as follows.

(i) *Separability*
All stochastic control problems are separable in the sense that the controller performs the two separate functions shown in Fig. 15.1. The property of separability, which was introduced in Section 13.4, refers to the information transmitted from the estimator to the control law. A control problem has the property of separability if, in the optimal controller, the only information that need be transmitted is an estimate $\hat{x}(i)$ of the current state $x(i)$. The general problem of equations (15.1) and (15.2) is separable in this sense if its optimal control law can be expressed in the form

$$\mathbf{u}^*(i) = \text{function of } \big(\hat{x}(i)\big). \tag{15.3}$$

The estimate $\hat{x}$ is often the conditional mean of the current state. The property implies that the optimal control is independent of any measure of the accuracy of the estimate, such as the covariance of the conditional probability density, and thus indicates absence of the type of interaction associated with cautious control.

(ii) *Neutrality*
A stochastic control problem is said to be neutral if the rate at which uncertainty about current states is reduced is independent of the controls. For this purpose uncertainty about a random variable can be measured by the negative entropy† of its probability distribution; the appropriate definition for the

---

† Negative entropy is the fundamental concept of information theory, see Shannon and Weaver (1949).

negative entropy $s(i)$ of the conditional probability distribution for the current state $\mathbf{x}(i)$ in a stochastic control system is

$$s(i) \equiv \int_{\mathbf{x}(i)} p\big(\mathbf{x}(i) \mid y^i\big) \, \log \, \big(p(\mathbf{x}(i) \mid y^i)\big) \, \mathrm{d}\mathbf{x}(i).$$

The general problem of equations (15.1) and (15.2) is said to be neutral when the negative entropy $s(i)$ is independent of all past controls $\mathbf{u}(i_1), \ldots, \mathbf{u}(i-1)$. The negative entropy of a normal probability distribution having covariance matrix $\boldsymbol{\Sigma}$ is

$$s = \tfrac{1}{2} \log |\boldsymbol{\Sigma}| + \text{constant},$$

so that a control system with normal probability distributions would be neutral if the covariance matrix $\boldsymbol{\Sigma}(i)$ of the conditional probability distribution was independent of all past controls. The LQP-problems of Section 13.4 with covariance matrices given by the Kalman filter equations (13.14b) and (13.23b) which are independent of past controls, as mentioned in Section 14.3, satisfy this definition of neutrality. The property implies that the rate of reduction of uncertainty is independent of the control and thus indicates absence of the type of interaction associated with probing.

(iii) *Certainty equivalence*

The property of certainty equivalence refers to the form of the function in equation (15.3) specifying the optimal control law for a separable problem. A separable stochastic control problem is said to be certainty equivalent when the function in equation (15.3) is identical with the function specifying the optimal feedback control law for an equivalent deterministic problem having no uncertainty about the state. For this purpose uncertainty is eliminated from the general stochastic problem of equations (15.1) and (15.2) by replacing the random variables $\boldsymbol{\xi}$ by their mean values of zero so that the stochastic problem reduces to the general deterministic problem of equations (9.1a) and (9.2a):

$$\mathbf{x}(i+1) = \mathbf{G}_1\big((\mathbf{x}(i), \mathbf{u}(i), \mathbf{0}, i)\big), \qquad (9.1a)$$

$$I = \sum_{i=i_1}^{i_2} H\big(\mathbf{x}(i), \mathbf{u}(i), i\big), \qquad (9.2a)$$

for which the optimal control can be expressed as the feedback law illustrated in Fig. 15.2 and written in the form

$$\mathbf{u}^c(i) = \text{function of } \big(\mathbf{x}(i)\big). \qquad (15.4)$$

The stochastic problem is certainty equivalent when the control law in Fig. 15.1 has the same functional form as the control law in Fig. 15.2, that is when the functions in equations (15.3) and (15.4) are identical. The LQP-problems

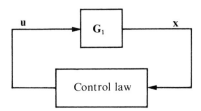

FIG. 15.2. Deterministic equivalent of system in Fig. 15.1.

of Chapter 13 all satisfy this definition of certainty equivalence. The property is stronger than either separability or neutrality and usually indicates that there is no interaction between the dual functions of the controller.

The above definitions cannot generally be used to investigate separability or neutrality or certainty equivalence of specific control problems. To investigate neutrality would require analytic expressions for the negative entropy $s(i)$, and hence for the conditional probability density function, which are generally non-existent for the reasons given in Section 14.4. To investigate separability or certainty equivalence it would be necessary also to find the optimal control law, which is generally impossible for the reasons given in Section 15.1. The special class of LQP-problems with normal random variables has already been shown to have all three properties; also it is known that the properties of separability and neutrality are preserved when the restriction to quadratic performance criteria is relaxed, and that the property of separability is preserved when the restriction on linearity is relaxed to the extent that equation (15.1a) may be non-linear in the control **u**. The significance of the properties for the general class of all other stochastic control problems is that they provide a basis for discussing interactions between the dual functions of the controller and can lead to practical sub-optimal control laws.

15.3. Intentional errors, probing, and caution

The definition of certainty equivalence leads to the idea of a certainty-equivalent control law defined, for the general system of Fig. 15.1, by

$$\mathbf{u}^c(i) = \text{function of } \big(\hat{\mathbf{x}}(i)\big), \tag{15.5}$$

where the function in equation (15.5) is identical in form to that in equation (15.4) which is the optimal control law for the equivalent deterministic problem of Fig. 15.2. In a certainty-equivalent problem, where there is usually no interaction between the dual functions of the controller, this certainty-equivalent control $\mathbf{u}^c$ is the optimal control $\mathbf{u}^*$. In other problems the certainty-equivalent control $\mathbf{u}^c$ gives some indication of what the control might be if interactions were neglected and control were exercised without any regard for the possibility of allowing for uncertainty.

The difference between the optimal control **u*** and the certainty-equivalent control **u**ᶜ provides some indication of the optimal allowance to be made for uncertainty. It is described as intentional error,

$$\text{Intentional error} \equiv \mathbf{u}^* - \mathbf{u}^c. \tag{15.6}$$

An example of a problem where intentional error arises has dynamic equation

$$x(i + 1) = ax(i) + \big(b + \xi(i)\big)u(i),$$

where a and b are known constants and $\xi(i)$ is a sequence of independent, zero-mean, random variables having variance σ^2. If there is no uncertainty about the current state $x(i)$ and if it is required to minimize the performance criterion

$$J = E \sum_{i=1}^{\infty} x^2(i),$$

the optimal control†

$$u^*(i) = -\frac{ab}{b^2 + \sigma^2} x(i)$$

differs from the certainty-equivalent control

$$u_c(i) = -\frac{a}{b} x(i)$$

in that it uses less feedback gain,

$$\left| \frac{ab}{b^2 + \sigma^2} \right| < \left| \frac{a}{b} \right|.$$

The optimal control thus appears to take account of the form of interaction associated with cautious control. The other form of interaction, which might introduce probing to affect the rate of reduction of uncertainty, is absent because there is no uncertainty about states and the problem is neutral.

A second example where intentional errors arise has dynamic equations

$$x_1(i + 1) = x_1(i) + x_2(i)u(i)$$
$$x_2(i + 1) = x_2(i),$$

in which x_2 represents an unknown constant. Uncertainty about current states is included in the form of an output equation

$$y(i) = x_1(i) + \xi(i),$$

where $\xi(i)$ is a sequence of independent, zero-mean, normal random variables having variance σ_ξ^2; the probability distribution for the states is also normal. The function of state estimation in the controller for this problem can be

† See Problems 13.7 and 13.8.

done by a Kalman filter which generates five sufficient statistics (m_1, m_2, σ_{11}, σ_{12}, σ_{22}) specifying the normal conditional probability distribution†:

$$p\big(\mathbf{x}(i) \mid y^i\big) = N\left(\begin{bmatrix} m_1 \\ m_2 \end{bmatrix}, \quad \begin{bmatrix} \sigma_{11} & \sigma_{12} \\ \sigma_{12} & \sigma_{22} \end{bmatrix}\right).$$

The elements σ_{11}, σ_{12}, σ_{22} of the covariance matrix here depend on past values of the controls, so the problem is not neutral. Dynamic programming equations in terms of the five sufficient statistics are in principle soluble, and numerical solutions giving the optimal control law to minimize the performance criterion

$$J = E \sum_{i=1}^{N} x^2(i)$$

have been computed‡ for values of N up to seven. Figure 15.3 shows how the

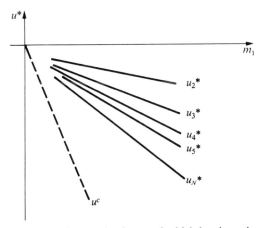

FIG. 15.3. Example of an optimal control which has intentional errors.

optimal control u^* was found to depend on the conditional mean m_1 of the state x_1 and on the number N of stages-to-go for some fixed combination of values of the other sufficient statistics (m_2, σ_{11}, σ_{12}, σ_{22}). The certainty-equivalent control

$$u^c = -\frac{m_1}{m_2}$$

is also shown and it can be seen that there is intentional error which decreases in magnitude as the number of stages-to-go increases.

† See Problem 14.17.
‡ Patchell (1971).

The nature of the intentional errors in Fig. 15.3 can be clarified by considering the case where there are only two stages-to-go ($N = 2$) and the control sequence has only one non-zero member

$$u(1) = u_2^*, \; u(2) = 0.$$

In this case the control problem is neutral because control is implemented only once and there is no possibility of benefiting from any reduction of uncertainty and so no reason to introduce any probing action. The difference between u_2^* and u^c can therefore be attributed to caution, as in the previous example. When the number N of stages-to-go is greater than two, the control problem ceases to be neutral and the possibility exists of benefiting from rapid reduction of uncertainty about the unknown constant x_2. Figure 15.3 shows that the magnitude of the optimal control increases with N in a way which could thus be attributed to probing action. As N goes to infinity the optimal control in this example goes to the certainty-equivalent control and the intentional errors go to zero. The transient nature of the intentional errors here is a feature of this particular example in which uncertainty about the unknown constant goes to zero as operating time goes to infinity; in other examples where uncertainty persists intentional errors may also persist.

In both of the above examples the intentional errors are related to the two possible forms of controller interaction by identifying caution as the intentional error in a neutral problem. This way of analysing interactions and intentional errors is applicable to any problem where an appropriate neutral control law can be defined. The idea of a neutral control law here is similar to that of the certainty-equivalent control law: just as the certainty-equivalent control law, defined with reference to an equivalent deterministic problem, is the control law which would be optimal if all uncertainty were neglected, so the neutral control law, defined with reference to an equivalent neutral problem, is the control law which would be optimal if all possibility of introducing probing action were neglected. Equivalent neutral problems are not, however, as easily specified as the equivalent deterministic problems of Fig. 15.2.

For discrete-time problems the equivalent neutral problem is sometimes provided, as in the second example above, by the two-stage version of the original problem. This is given by setting

$$i_2 = i_1 + 1$$

in the performance criterion of equation (15.2). The resulting two-stage problem is neutral in the sense that, because control is exercised only once, there is no possibility of benefiting from further reduction of uncertainty and so no reason to introduce probing action. At the time of writing the conditions under which equivalent neutral problems can be specified in this way have not been fully clarified. They probably include stationarity of the functions $\mathbf{G}_1$, $\mathbf{G}_2$, $\mathbf{H}$ in equations (15.1) and (15.2), and controllability in the sense that it should be possible to transfer the state $\mathbf{x}$ to any point in state space in a single stage.

If an equivalent neutral problem can be defined, its optimal control law can be taken as the neutral control $\mathbf{u}^n$ for the original problem and used as a means of identifying the two different types of interaction. This is done by defining

$$\text{probing} \equiv \mathbf{u}^* - \mathbf{u}^n, \tag{15.7}$$

$$\text{caution} \equiv \mathbf{u}^n - \mathbf{u}^c. \tag{15.8}$$

Here probing is defined as the difference between the optimal control $\mathbf{u}^*$ and the neutral control $\mathbf{u}^n$ which is optimal in every respect except that it neglects probing action: similarly caution is defined as the difference between the neutral control $\mathbf{u}^n$ and the certainty-equivalent control $\mathbf{u}^c$ which neglects both forms of interaction, caution and probing.

In the second example above, the two-stage-to-go optimal control $u^*_{2.}$ is the neutral control. It can be found analytically to be

$$u^n = u^*_2 = -\frac{m_1 m_2 + \sigma_{12}}{m_2^2 + \sigma_{22}^2}.$$

Figure 15.4 illustrates how the definitions of equations (15.7) and (15.8) are then used to identify caution and probing in the optimal control.

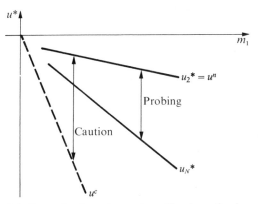

FIG. 15.4. Example of caution and probing in optimal control.

15.4. Structure of control laws

Intentional error, probing, and caution are theoretical concepts and cannot generally be evaluated in non-LQP or non-normal problems where they may be non-zero. Their significance is that they lead to a general theoretical result about the structure of control laws to perform function (ii) in the controller for the general class of stochastic systems of Fig. 15.1. The result is obtained by combining the definitions of equations (15.6), (15.7), and (15.8) to give

$$\mathbf{u}^* = \mathbf{u}^c + \text{intentional error} \tag{15.9a}$$

$$= \mathbf{u}^c + \text{caution} + \text{probing} \tag{15.9b}$$

$$= \mathbf{u}^n + \text{probing}. \tag{15.9c}$$

It shows that a control law can be regarded as the sum of three components, the certainty-equivalent control $\mathbf{u}^c$ plus caution plus probing as in equation (15.9b), or can be regarded as the sum of two components, the neutral control $\mathbf{u}^n$ plus probing as in equation (15.9c).

Equations (15.9) offer a basis for constructing practical sub-optimal control laws for stochastic problems where the optimal control cannot be found. If an equivalent neutral problem can be specified and solved the sub-optimal control law can have the form of equation (15.9c):

$$\mathbf{u}^{\text{sub–opt}} = \mathbf{u}^n + \text{probing}^{\text{sub–opt}}. \tag{15.10a}$$

When it is impossible to find the neutral control $\mathbf{u}^n$ the sub-optimal control law can have the form of equation (15.9b):

$$\mathbf{u}^{\text{sub–opt}} = \mathbf{u}^c + \text{caution}^{\text{sub–opt}} + \text{probing}^{\text{sub–opt}}. \tag{15.10b}$$

The certainty-equivalent component $\mathbf{u}^c$ in equation (15.10b) may also be sub-optimal because a system where the optimal control law cannot be found is often one where Bayes's rule cannot be realized so that the estimator performing function (i) in the controller is often sub-optimal and the argument $\hat{\mathbf{x}}$ in equation (15.5) has to be a sub-optimal estimate.

The development of controllers which use the sub-optimal control laws of equations (15.10) together with sub-optimal estimation procedures such as those described in Sections 14.4 and 14.5 is, at the time of writing, an active area of research.

15.5. Bibliography

The possibility of interactions between the dual functions of estimation and control in a stochastic controller is introduced in
    BELLMAN (1961) Chapter 16,
    FEL'DBAUM (1965) Chapter 6.
Discussion of the properties of separability, neutrality, and certainty equivalence has been mainly in periodicals rather than in books; the literature is surveyed and the properties are described in
    PATCHELL and JACOBS (1971).
Intentional errors, probing, and caution are the subject of research reported in
    JACOBS and PATCHELL (1972),
    HUGHES and JACOBS (1974).

Answers to problems

CHAPTER 1

1.1. (i), (iii), (iv) linear in y, (vi) linear in y.

1.2. m/k_2, equation becomes non-linear.

1.3. Ratio is $1/(1 + K)$.

1.4. (i) 0·625, 2·5 rad/unit time.
(ii) 0·5, 5 rad/unit time.
(iii) 2, 1·25 rad/unit time.

1.5. $1·4 \times 10^5$ mA V^{-1}, $4·33 \times 10^{-5}$.

1.8. $11·5\,\pi$ rad s^{-1}, 260.

1.9. (i) $10(-0·2)^i + \frac{10}{3}\left(1 - (-0·2)^i\right)$.
(ii) $10(0·2)^i + 5\left(1 - (0·2^i)\right)$.

1.11. $1/4a$.

1.12. (i) $K = 1/4ab$, $k = 1/b$.
(ii) X.
(iii) no.

1.15. $(i + 1)a^i$.

1.16. (i) $x(0) = 0$, $x(1) = 0$, $x(2) = \dfrac{X}{4}$, $x(3) = \dfrac{X}{2}$, $x(4) = \dfrac{11}{16} X$,

$$x(5) = \frac{13}{16} X, \ldots$$

(ii) $x(i) = X\left(1 - (\tfrac{1}{2})^{i-1} - \tfrac{1}{4}(i-1)(\tfrac{1}{2})^{i-2}\right)$.

1.20 $at - a(T_2 - T_1)(1 - e^{-t/T_2})$.

1.22. (ii) 0·067.

1.24. (i) $y(0) = 0$, $y(1) = 1$, $y(2) = -a$, $y(3) = a^2$, $y(4) = -a^3, \ldots$
(ii) $y(0) = 0$, $y(1) = 0$, $y(2) = \tfrac{1}{4}$, $y(3) = \tfrac{1}{4}$, $y(4) = \tfrac{3}{16}$, $y(5) = \tfrac{1}{8}, \ldots$

1.25. $\dfrac{1 - e^{-Ts}}{s}$.

1.27. (i) $\dfrac{KTXz(z - \mathrm{e}^{-T/T_1})}{(z - 1)\{(z - 1)(z - \mathrm{e}^{-T/T_1}) + KT(z - \mathrm{e}^{-T/T_1}) - KT_1(1 - \mathrm{e}^{-T/T_1})(z - 1)\}}$.

   (ii) $\dfrac{\frac{1}{2}KT^2X(z + 1)(z - \mathrm{e}^{-T/T_1})}{(z - 1)\{(z - 1)(z - \mathrm{e}^{-T/T_1}) + KT(z - \mathrm{e}^{-T/T_1}) - KT_1(1 - \mathrm{e}^{-T/T_1})(z - 1)\}}$.

   (iii) $\dfrac{KTXz(z - \mathrm{e}^{-T/T_1})}{(z - 1)\{(z - 1)(z - \mathrm{e}^{-T/T_1}) + K(T/T_1)z\}}$.

1.28. $GX(z)/X(z)\{1 + GH(z)\}$.

CHAPTER 2

2.1. $y = \dfrac{K_1K_3x - K_2L}{1 + K_1K_3}$, $K_3 = 571$ m^3 rad^{-1}, $x = 22.3$.

2.2. (i) $y = \dfrac{K_1x + K_2z}{K_1 + K_2}$.

   (ii) $K_1 = K_2^2/2Q$, $\omega_0 = K_2/\sqrt{(2)}Q$.

2.3. (i) $e = x/(1 + K)$, $\omega_0' = \omega_0\sqrt{(1 + K)}$, $\xi' = \xi/\sqrt{(1 + K)}$.

   (ii) $e = 0$.

2.4. $x = 78\frac{1}{2}$ V, $K = 1.215$ Nm V^{-1}, $T = 9.5$ s.

2.5. $\dfrac{\partial y}{\partial K_1} = \dfrac{(1 + K_1K_3)K_3x - (K_1K_3x - K_2L)K_3}{(1 + K_1K_3)^2}$.

   (i) 50%.

   (ii) 25%.

2.6. No relationship for constant x and ramp z.

2.7. Differentiating the error.

2.8. fa/K.

CHAPTER 3

3.1. (ii) and (iii) ((ii) by inspection).

3.3. No.

3.6. $0 < K < 2$.

3.7. $\frac{1}{2}$ s, $\frac{1}{3}$ s,

3.8. $\dfrac{(a_1 - a_2 - a_3)(a_2 + a_3)}{a_1^2 a_2 a_3}$.

3.9. $0.000\ 105 < K < 0.003\ 86$.

3.10. 20 units of time.

3.11. (i) 0.49.

   (ii) 1.5.

3.13. 1.25.

3.14. 0.0841 units of time.

CHAPTER 4

4.1. (iii) with $a > 1$, (iv).

4.2. $F^2/2J$.

4.3. 3.

4.4. -0.47, yes.

4.6. 1·51.

4.7. 2·06, 2·6 units of time.

4.8. (i) (a) 1·25,   (b) 0·0636.
      (ii) (a) ∞,   (b) 0·212.
      (iii) (a) 1·955.   (b) 0·137.

4.9. (iv) 0·11.
      (v) 3 units of time, 0·55.
      (vi) 0·62, 0.088, 3·6 units of time.

4.10. $1/4K$.

4.11. $2\dfrac{T_1}{T}(1 + e^{-T/T_1})$.

CHAPTER 5

5.1. (i) 0·08, 157 units of time.
      (ii) 4.

5.2. (i) 2·5.
      (ii) Steady-state errors halved.

5.3. 3.

5.5. Choose zeros of both phase advance and phase lag to cancel poles of $G(s)$.
      Choose gain to give damping ratio $1/\sqrt{2}$.

5.6. Integration in form of phase lag combined with phase advance (assuming
      $T_1 > T_2$) leads to

$$H(s) \simeq \frac{1}{36K_1K_2K_3T_2^2 s}(1 + 6sT_2)(1 + sT_1).$$

CHAPTER 6

6.2. $1/4\zeta$.

6.3. 0·5.

6.4. $\sqrt{(aK)}$.

6.5. 5/14, 10·9 dB.

6.6. $\dfrac{4 + 10\alpha + 9\alpha^2 + 4\alpha^3 + \alpha^4}{2\alpha + \alpha^2 - 2\alpha^3 - \alpha^4}$ where $\alpha = aK$, roots $s_{1,2} = \frac{1}{2} \pm \mathrm{j}0.387$.

6.7. $(q)^{-1/2}$, $(2J\sqrt{q})^{1/2}$, $1/\sqrt{2}$.

CHAPTER 7

7.1.   (i) $A = -\dfrac{1}{T}$, $B = \dfrac{1}{T}$.

(ii) $\mathbf{A} = \begin{bmatrix} 0 & 1 \\ -\dfrac{1}{T_1 T_2} & -\dfrac{T_1 + T_2}{T_1 T_2} \end{bmatrix}$, $\mathbf{B} = \begin{bmatrix} 0 \\ \dfrac{K}{T_1 T_2} \end{bmatrix}$.

(iii) $\mathbf{A} = \begin{bmatrix} 0 & 1 & 0 & 0 \\ 0 & 0 & 1 & 0 \\ 0 & 0 & 0 & 1 \\ 0 & 0 & 0 & 0 \end{bmatrix}$, $\mathbf{B} = \begin{bmatrix} 0 \\ 0 \\ 0 \\ 10 \end{bmatrix}$.

(iv) $\mathbf{A} = \begin{bmatrix} 0 & 1 & 0 \\ 0 & 0 & 1 \\ -8 & -14 & -7 \end{bmatrix}$, $\mathbf{B} = \begin{bmatrix} 0 \\ 0 \\ 4 \end{bmatrix}$.

(v) $\mathbf{A} = \begin{bmatrix} 0 & 1 & 0 & 0 & 0 & 0 & 0 & 0 & 0 & 0 \\ & & & \cdot & & & & & & \\ & & & & \cdot & & & & & \\ & & & & & \cdot & & & & \\ & & & & & & \cdot & & & \\ & & & & & & & \cdot & & \\ & & & & & & & & 1 & \\ 0 & 0 & 0 & 0 & 0 & 0 & 0 & 0 & 0 & 0 \end{bmatrix}$, $\mathbf{B} = \begin{bmatrix} 0 \\ 0 \\ \cdot \\ \cdot \\ \cdot \\ \cdot \\ \cdot \\ 1 \end{bmatrix}$.

7.2.   (i) $-1/T$.
(ii) $-1/T_1$, $-1/T_2$.
(iii) 0, 0, 0, 0.
(iv) -1, -2, -4.
(v) $\exp(j2r\pi/9)$, $r = 0, 1, \ldots, 8$.

7.3.   (i) $\mathbf{S} = -\dfrac{1}{T}$, $B^* = \dfrac{1}{T}$.

(ii) $\mathbf{S} = \begin{bmatrix} -\dfrac{1}{T_1} & \\ & -\dfrac{1}{T_2} \end{bmatrix}$, $\mathbf{B}^* = \dfrac{K}{T_1 - T_2} \begin{bmatrix} 1 \\ -1 \end{bmatrix}$.

(iii) System has repeated eigenvalues, answer to 7.1(iii) is already in Jordan canonical form.

7.4. $u(1) = -\left(\dfrac{1}{ab} x_1(1) + \dfrac{2}{b} x_2(1)\right)$, $u(2) = \dfrac{1}{ab} x_1(1) + \dfrac{1}{b} x_2(1)$.

7.5. $\mathbf{B}^* = \dfrac{k}{a_1 - a_2} \begin{bmatrix} 1 \\ -1 \end{bmatrix}$ has no zero rows, so controllable.

7.6. $\mathbf{M}_c = \begin{bmatrix} 0 & 0 & 0 & 10 \\ 0 & 0 & 10 & 0 \\ 0 & 10 & 0 & 0 \\ 10 & 0 & 0 & 0 \end{bmatrix}$ non-singular, so controllable.

7.7. Uncontrollable, unobservable.

7.9.
(i) $\mathbf{x}(i) = \begin{bmatrix} y(i-3) \\ y(i-2) \\ y(i-1) \end{bmatrix}$, $\mathbf{A} = \begin{bmatrix} 0 & 1 & 0 \\ 0 & 0 & 1 \\ -4 & -3 & -2 \end{bmatrix}$, $\mathbf{B} = \begin{bmatrix} 0 \\ 0 \\ 5 \end{bmatrix}$.

(ii) $\mathbf{x}(i) = \begin{bmatrix} -4y(i-1) + 5u(i-1) \\ -3y(i-1) - 4y(i-1) + 5u(i-2) \\ y(i) \end{bmatrix}$,

$\mathbf{A} = \begin{bmatrix} 0 & 0 & -4 \\ 1 & 0 & -3 \\ 0 & 1 & -2 \end{bmatrix}$, $\mathbf{B} = \begin{bmatrix} 5 \\ 0 \\ 0 \end{bmatrix}$.

7.10.
$\mathbf{A} = \begin{bmatrix} 0 & 0 & -1 \\ 1 & 0 & 4 \\ 0 & 1 & 1 \end{bmatrix}$, $\mathbf{B} = \begin{bmatrix} 0 \\ 13 \\ 3 \end{bmatrix}$.

7.11. $\mathbf{M}_c = \begin{bmatrix} a_0 a_1 b_2 & -a_0 b_2 \\ (a_1^2 - a_0)b_2 & -a_1 b_2 \end{bmatrix}$.

$|\mathbf{M}_c| = -a_0^2 b_2^2$ is zero and system uncontrollable if
(i) $b_2 = 0$ in which case control u cannot affect system;
(ii) $a_0 = 0$ in which case system is really first order and uncontrollability arises because order of state space equations is too high.

CHAPTER 8

8.1. (i) $u_N(x) = \dfrac{a + a^2 + \cdots + a^{N-1}}{1 + a + \cdots + a^{N-1}} x$, $N > 1$.

$u_1(x) = 0$.

(ii) $u_N(x) = x$, $N > 1$.

$u_1(x) = 0$.

8.2. (i) $u(1) = -0.924x$, $u(2) = -0.718x$, $u(3) = 0$.

(ii) $u(1) = -1.06x$, $u(2) = -1.34x$, $u(3) = -2.85x$.

8.3. $u(t) = -\dfrac{e^{2(T-t)} - 1}{e^{2(T-t)} + 1} x(t)$.

8.5. $\dfrac{U(s)}{Y(s)} = -0.934(1 + 0.395s)$, damping ratio $= 0.707$.

8.6. $\mathbf{K}_1 = [0\ 0]$, $\mathbf{K}_2 = [0\ 1]$, $\mathbf{K}_3 = [0.5\ 1.5]$, $\mathbf{K}_4 = [0.6\ 1.6]$,
$\mathbf{K}_5 = [0.615\ 1.615]$, $\mathbf{K}_6 = [0.618\ 1.618]$, $\mathbf{K}_7 = [0.618\ 1.618]$.
Roots are $s = 0$, $s = 0.382$.

8.7. (ii) Using control from 7.4, $I = \mathbf{x}' \begin{bmatrix} 3 & 2 \\ 2 & 3 \end{bmatrix} \mathbf{x}$ which is greater than greatest

value using optimal control and given by $\mathbf{V}_\infty = \begin{bmatrix} 2.618 & 1.618 \\ 1.618 & 2.618 \end{bmatrix}$.

8.8. $\mathbf{u}(i) = -(\mathbf{B}'\mathbf{PB} + \mathbf{Q})^{-1}\mathbf{B}'\mathbf{PAx}(i)$, $\mathbf{B}$ square and non-singular and $\mathbf{Q}$ null.

8.9. $\dfrac{d\mathbf{V}}{dT} = \mathbf{P} + \mathbf{K}'\mathbf{QK} + 2\mathbf{V}(\mathbf{A} - \mathbf{BK})$.

8.10. $u(t) = -\dfrac{x}{1 + T - t}$.

CHAPTER 9

9.1. $\dot{x} = -x - p/2$, $\dot{p} = +p - 2x$; $K(t_2) = 0$.

9.2. $I = \mathbf{x'Vx}$ so $\mathbf{p} \equiv \dfrac{\partial I}{\partial \mathbf{x}} = 2\mathbf{Vx}$

9.3. $x(t_1)$ given, $\dot{x}(t_2) + x^3(t_2) = 0$.

9.4. $K(1) = -\dfrac{2e}{e^2 - 1}\left(e - \dfrac{x(0)}{x(1)}\right)$.

9.6. (i) $u_1 = 0$, $u_2 = -\tau(x - \tau x^3)/(1 + \tau^2)$, ...
      (ii) $u_1 = x^3 - x/\tau$, ...
      In both cases algebra becomes impossible so solve numerically:
      $$f_N(x) = \min_u \{x^2 + u^2 + f_{N-1}(x - \tau x^3 + \tau u)\}.$$

9.7. $p_2 = -Ct + D$, $\dfrac{dx_2}{dx_1} = -1 \pm 1/x_2$.

CHAPTER 10

10.3. $2 \cdot 67 \times 10^{-3}$ rad.

10.4. $9 \cdot 95$ rad.

10.5. A/k_2, 0.

10.7. Amplitude approximately $1\frac{1}{4}$, period approximately 5 units of time.

CHAPTER 11

11.2. $a_{11} + a_{22} < 0$, $a_{11}a_{22} - a_{12}a_{21} > 0$ (*cf.* Fig. 10.22).

11.3. (ii) $-1 < x_1 < 1$ and $-\sqrt{2} < x_2 < \sqrt{2}$.

11.4. Stability depends on sign of a. Stronger than linearization.

11.5. $\omega \approx \sqrt{(1/T_1T_2)}$, $a \approx 4AKT_1T_2/\pi(T_1 + T_2)$.

11.9. $a \approx 3 \cdot 2$, $\omega \approx 7$.

11.10. 2×10^{-2} rad, $28 \cdot 3$ rad s^{-1}.

11.11. $\dfrac{4A}{\pi a}(1 + 2\sqrt{(1 - B^2/a^2)} + 2\sqrt{(1 - 4B^2/a^2)})$.

CHAPTER 12

12.2. (i) $0 \cdot 495$.
       (ii) $0 \cdot 355$.

12.3. (i) $0 \cdot 18$.
       (ii) $0 \cdot 66$, $0 \cdot 293$, $0 \cdot 05$.
       (iii) Nothing, all classes equally probable.

12.7. (i) $\dfrac{1}{y}$, $1 \leqslant y \leqslant e$.

(ii) $\dfrac{1}{2\sqrt{y}}, 0 \leqslant y \leqslant 1$.

(iii) $\dfrac{a}{\pi(a^2 + y^2)}$.

12.9.   (i) $\dfrac{a}{2}, \dfrac{a^2}{12}$.

(ii) $4\frac{1}{2}, 8\frac{1}{4}$.

(iii) $0, b^2/6$.

(iv) $pn, p(1 - p)n$.

(v) m, σ^2.

(vi) Non-existent

12.13. 14·2.

12.15.   (i) 11.

(iii) 9·5.

12.16. $11(1 - \alpha + \alpha^2)$.

12.19.   (ii) $\dfrac{x(0)}{2}\dfrac{e^t - e^{-t}}{t}$, ∞ or 0, $\dfrac{x^2(0)}{2}\dfrac{e^{2t+\tau} - e^{-(2t+\tau)}}{2t + \tau}$, ∞ or 0.

$\sigma_x^2 = \dfrac{x^2(0)}{4}\dfrac{(t - 1)\,e^{2t} - (t + 1)\,e^{-2t} + 2}{t^2}$.

(iv) $x(0)$, $x(0)$, $x^2(0) + \sin t \sin (t + \tau)$, $x^2(0) + \frac{1}{2}\Omega^2 \cos \tau$,
$\sigma_x^2 = (\sin t)^2$.

(vi) $0, 0, \frac{1}{2}\cos \tau, \frac{1}{2}\cos \tau, \sigma_x^2 = \frac{1}{2}$.

(vii) $0, 0, \frac{1}{2}\cos \tau, \frac{1}{2}\Omega_1^2 \cos \tau, \sigma_x^2 = \frac{1}{2}$.

12.20.   (i) (vi), (vii).

(ii) (v), (vi).

(iii) (vi).

12.21. $\dfrac{N^2K^2}{2T}\,e^{-|\tau|/T}, \dfrac{N^2K^2}{2T}$.

12.22. $\dfrac{N^2K^2(1 + 2\zeta T)}{4\zeta(1 + 2\zeta T + T^2)}$.

CHAPTER 13

13.1.   (i) (a).

(ii) (a).

(iii) (b).

(iv) (b).

(v) (c).

(vi) (c).

(vii) (c) and (d).

(viii) (b).

(ix) (c) and (d).

(x) (c).

13.2.   Yes. Step function. White noise.

13.5.   (i) $\dfrac{X}{N} > 1$.

(ii) $K = \infty$. Include either more dynamics (up to third order) or cost of control.

13.6. $f_N(\mathbf{x}) = \min_{\mathbf{u}} \left\{ H(\mathbf{x}, \mathbf{u}) + \int_{\xi} f_{N-1}\big(\mathbf{G}(\mathbf{x}, \mathbf{u}, \boldsymbol{\xi})\big) p(\boldsymbol{\xi})\, d\boldsymbol{\xi} \right\}$;

$f_1(\mathbf{x}) = \min_{\mathbf{u}} \{ H(\mathbf{x}, \mathbf{u}) \}$.

13.8.   (ii) $\dfrac{a^2\sigma^2}{\beta^2 + \sigma^2} < 1; \; v = \dfrac{\beta^2 + \sigma^2}{\beta^2 + (1 - a^2)\sigma^2}$; $x(i) \to 0$ as $i \to \infty$.

(iii) No.

13.9. $2\cdot618\sigma^2$.

13.10.   (i) $u^* = \tfrac{1}{2}x_1 - x_2$.

(ii) $0\cdot375X^2 < 0\cdot414X^2$.

13.11.   (i) $\{0, 0, 0, 0, 0, 0, 0, 0, 0, 0, 8\cdot6, 8\cdot9, 11\cdot1, 11\cdot2, 9\cdot0, 8\cdot5, 8\cdot0, 6\cdot4, 3\cdot3\}$.

(ii) $0, 0\cdot71, 0\cdot93, 0\cdot99, 1, 1, \ldots$ Not affected by u or x.

(iii) $u^* = -0\cdot95 \, m$.

(iv) $2 < 11$.

(v) $10, 0\cdot4, 0\cdot38, 1\cdot98, 0\cdot77$.

(vi) $-0\cdot475$.

13.14.   (i) $\sigma^2(i)$ depends on $u(i - 1)$.

(iv) No, Yes.

13.15.   (i) $u^*(1) = -\big(0\cdot4m_1(1) + m_2(1)\big), \; u^*(2) = -0\cdot5m_2(2), \; u^*(3) = 0$.

(ii) $\boldsymbol{\Sigma}(2) = \dfrac{1}{3}\begin{bmatrix} 2 & 1 \\ 1 & 5 \end{bmatrix}, \; \boldsymbol{\Sigma}(3) = \dfrac{1}{12}\begin{bmatrix} 9 & 6 \\ 6 & 20 \end{bmatrix}$.

(iii) $u^*(2) = -\tfrac{1}{6}y(2)$.

(iv) $6\tfrac{5}{12}$.

13.16.   (i) $\begin{bmatrix} 0 & 0 \\ 1 & -1 \end{bmatrix}, \begin{bmatrix} 0 \\ 1 \end{bmatrix}, [0 \; 1], \begin{bmatrix} 0 & 0 \\ 0 & 1 \end{bmatrix}, 1, \begin{bmatrix} Z^2 & 0 \\ 0 & 0 \end{bmatrix}, N^2$.

(ii) $u^* = -(\sqrt{2} - 1)\left(\dfrac{1}{\sqrt{2}} m_1 + m_2 \right)$.

(iii) $\sigma_{11} = NZ\sqrt{(1 + 2Z/N)}, \; \sigma_{12} = NZ, \; \sigma_{22} = N^2(\sqrt{(1 + 2Z/N)} - 1)$.

CHAPTER 14

14.3. $\dfrac{\sigma^2}{N} > \left(\dfrac{1}{\sigma_0^2} + \dfrac{N}{\sigma^2} \right)^{-1}$.

14.6. $M(z) = \dfrac{(1 - K)}{z - (1 - K)} U(z) + \dfrac{Kz}{z - (1 - K)} Y(z)$,

where $K = \dfrac{1}{2}\dfrac{\sigma_1^2}{\sigma_2^2} \left(\sqrt{\left(1 + 4\dfrac{\sigma_2^2}{\sigma_1^2}\right)} - 1 \right)$.

14.7. $M(s) = \dfrac{K_1/KK_2}{1 + s/KK_2} U(s) + \dfrac{1/K_2}{1 + s/KK_2} Y(s)$,

where $K = K_1\sigma_1/\sigma_2$.

14.8. $m(0)$, 0.

14.9. $\mathbf{m}(i) = \mathbf{A}(i - 1)\mathbf{m}(i - 1) + \mathbf{B}(i - 1)\mathbf{u}(i - 1)$

$\qquad + \mathbf{\Sigma}(i)\mathbf{C}'(i)\mathbf{\Sigma}_2^{-1}(i)\{\mathbf{y}(i) - \mathbf{C}(i)\big(\mathbf{A}(i - 1)\mathbf{m}(i - 1) + \mathbf{B}(i - 1)\mathbf{u}(i - 1)\big)\}.$

$\mathbf{\Sigma}(i) = \mathbf{\Sigma}_0 - \mathbf{\Sigma}_0\mathbf{C}'(i)\big(\mathbf{\Sigma}_2(i) + \mathbf{C}(i)\mathbf{\Sigma}_0\mathbf{C}'(i)\big)^{-1}\mathbf{C}(i)\mathbf{\Sigma}_0,$

where $\mathbf{\Sigma}_0 = \mathbf{\Sigma}_1(i - 1) + \mathbf{A}(i - 1)\mathbf{\Sigma}(i - 1)\mathbf{A}'(i - 1).$

14.10. $\mathbf{\Sigma} = \begin{bmatrix} 1{\cdot}618 & 0 \\ 0 & 0 \end{bmatrix}.$

14.11. $\hat{\mathbf{x}}(i) = \begin{bmatrix} \hat{x}_2(i - 1)\hat{x}_1(i - 1) + \hat{x}_3(i - 1)u(i - 1) \\ \hat{x}_2(i - 1) \\ \hat{x}_3(i - 1) \end{bmatrix} + \dfrac{1}{\sigma_2^2}\begin{bmatrix} \sigma_{11}(i) \\ \sigma_{12}(i) \\ \sigma_{13}(i) \end{bmatrix} \times$

$\qquad \times \big(y(i) - \hat{x}_2(i - 1)\hat{x}_1(i - 1) - \hat{x}_3(i - 1)u(i - 1)\big).$

$\mathbf{\Sigma}(i)$ given by equations (14.18c) with

$\mathbf{A} = \begin{bmatrix} \hat{x}_2(i - 1) & \hat{x}_1(i - 1) & u(i - 1) \\ 0 & 1 & 0 \\ 0 & 0 & 1 \end{bmatrix}; \quad \mathbf{B}_1 = \begin{bmatrix} 1 & 0 & 0 \\ 0 & 0 & 0 \\ 0 & 0 & 0 \end{bmatrix}, \quad \mathbf{B}_2 = 1$

$\mathbf{C} = [1\ 0\ 0], \quad \mathbf{\Sigma}_1 = \begin{bmatrix} \sigma_1^2 & 0 & 0 \\ 0 & 0 & 0 \\ 0 & 0 & 0 \end{bmatrix}, \quad \mathbf{\Sigma}_2 = \sigma_2^2.$

14.12. $\dfrac{\mathrm{d}\hat{x}}{\mathrm{d}t} = \mathrm{e}^{\hat{x}} - \hat{x}u - \dfrac{\sigma^2}{2\sigma_2^2}\left(1 - \dfrac{y}{\sqrt{\hat{x}}}\right).$

$\dfrac{\mathrm{d}\sigma^2}{\mathrm{d}t} = u^2\sigma_1^2 + 2\sigma^2(\mathrm{e}^{\hat{x}} - u) - \dfrac{\sigma^4}{4\hat{x}\sigma_2^2}.$

14.13. (i) $\hat{x}(i \mid i) = \dfrac{Y}{c}.$

$\qquad$ (ii) $\hat{x}(i \mid i) = \hat{x}(i \mid i - 1) + \dfrac{(1 + a^2)c}{1 + (1 + a^2)c^2}\big(Y - c\hat{x}(i \mid i - 1)\big),$

$\qquad$ where $\hat{x}(i \mid i - 1) = a\hat{x}(i - 1 \mid i - 1) + bU.$

14.14. Equation (14.13), nothing about variance of $\hat{x}(i \mid i)$.

14.15. $\hat{x}_1(i \mid i) = \hat{x}_1(i \mid i - 1) + \dfrac{1 + U^2}{2 + U^2}\big(Y - \hat{x}_1(i \mid i - 1)\big),$

$\qquad \hat{x}_3 = X_3 + \dfrac{U}{2 + U^2}\big(Y - \hat{x}_1(i \mid i - 1)\big),$

$\qquad$ where $\hat{x}_1(i \mid i - 1) = X_1X_2 + X_3U.$

14.16. $\sigma^2 = \dfrac{a\sigma_2^2}{c^2}\left\{1 + \sqrt{\left(1 + \dfrac{\sigma_1^2}{a^2\sigma_2^2}\right)}\right\}.$

14.17. $\mathbf{A}(i) = \begin{bmatrix} a & u(i) \\ 0 & 1 \end{bmatrix}, \quad \mathbf{B}(i) = \mathbf{0}.$

14.19. $n(1 + m + p) + \dfrac{p}{2}(p + 3), \dfrac{n}{2}(3n - 1) - p.$

Bibliography

BEFORE 1940

MAXWELL, J. C. (1868). On governors. *Proc. Royal Society* **16**, 270–283.

BOOLE, G. (1872). *A treatise on the calculus of finite differences* (2nd edn). MacMillan.

LIAPUNOV, A. A. (1892). *Problème Général de la stabilité du mouvement*. Reprinted by Princeton University Press, 1949.

POINCARÉ, H. (1892). *Les methodes nouvelles de la Mécanique céleste*, vol. 1. Gauthier-Villars.

PIAGGIO, H. T. H. (1920). *Differential equations*. Bell.

NYQUIST, H. (1932). Regeneration theory. *Bell System Technical Journal*, **11**. Reprinted in *Mathematical trends in control theory* (ed. R. Bellman and R. Kalaba). Dover, 1964.

MILNE-THOMPSON, L. M. (1933). *The calculus of finite differences*. MacMillan.

1940–49

KOLMOGOROV, A. N. (1941). Interpolation and extrapolation of stationary time series. *Bulletin Academy Science USSR Math. Ser.*, **5**.

GARDNER, M. F. and BARNES, J. L. (1942). *Transients in linear systems*, vol. 1. Wiley.

BODE, H. W. (1954). *Network analysis and feedback amplifier design*. Van Nostrand.

JAMES, H. M., NICHOLS, N. B., and PHILLIPS, R. S. (1947). *Theory of servomechanisms*. McGraw-Hill. Reprinted by Dover, 1965.

MINORSKY, N. (1947). *Introduction to non-linear mechanics*, Edwards.

BROWN, G. C. and CAMPBELL, D. P. (1948). *Principles of Servomechanisms*. Wiley.

ANDRONOV, A. A. and CHAIKIN, C. E. (1949). *Theory of oscillations*. Princeton University Press.

JAEGER, J. C. (1949). *Introduction to the Laplace transformation*. Methuen.

SHANNON, C. E. and WEAVER, W. (1949). *The mathematical theory of communication*. University of Illinois Press.

WIENER, N. (1949). *Extrapolation, interpolation and smoothing of stationary time series*. Wiley.

1950–54

COURANT, R. and HILBERT, D. (1953). *Methods of mathematical physics*, vol. 1. Interscience.

TSIEN, H. S. (1954). *Engineering cybernetics*. McGraw-Hill.

1955

BRUNS, R. A. and SAUNDERS, R. M. (1955). *Analysis of feedback control systems.* McGraw-Hill.

CRAMER, H. (1955). *The elements of probability theory.* Wiley.

TRUXAL, J. G. (1955). *Automatic feedback control system synthesis.* McGraw-Hill.

1956

LANING, J. H. and BATTIN, H. B. (1956). *Random processes in automatic control.* McGraw-Hill.

1957

BELLMAN, R. E. (1957). *Dynamic programming.* Princeton University Press.

FELLER, W. (1957). *An introduction to probability theory and its applications,* vol. 1 (2nd edn). Wiley.

MURPHY, G. J. (1957). *Basic automatic control theory.* Van Nostrand.

NEWTON, G. C., GOULD, L. A., and KAISER, J. F. (1959). *Analytical design of linear feedback controls.* Wiley.

1958

ASELTINE, J. A. (1958). *Transform method in linear system analysis.* McGraw-Hill.

BENDAT, J. S. (1958). *Principles and application of random noise theory.* Wiley.

DAVENPORT, W. B. and ROOT, W. L. (1958). *Random signals and noise.* McGraw-Hill.

JURY, E. I. (1958). *Sampled-data control systems.* Wiley.

RAGAZZINI, J. R. and FRANKLIN, G. F. (1958). *Sampled-data control systems.* McGraw-Hill.

SAVANT, C. J. (1958). *Control system design.* McGraw-Hill.

1959

CHESTNUT, H. and MAYER, R. W. (1959). *Servomechanisms and regulating system design,* vol. 1. Wiley.

GILLE, J. C., PELEGRIN, M. J., and DECAULNE, P. (1959). *Feedback control systems.* McGraw-Hill.

MURPHY, G. J. (1959). *Control engineering.* Van Nostrand. Reprinted by Boston Technical Publishers, 1967.

TOU, J. T. (1959). *Digital and sampled-data control systems.* McGraw-Hill.

1960

DEL TORO, V. and PARKER, S. (1960). *Principles of control systems engineering.* McGraw-Hill.

KALMAN, R. E. (1960a). Contributions to the theory of optimal control. *Bol. Soc. Mat. Mexicana.* **5**, 102–119.

—— (1960b). The theory of optimal control and the calculus of variations. In *Proceedings of Santa Monica Conference, 1960.* University of California Press, 1963.

—— (1960c). On the general theory of control systems. In *Proceedings of First IFAC Congress, Moscow, 1960.* Butterworths, 1961.

KALMAN, R. E. and BERTRAM, J. E. (1960). Control system analysis and design via the second method of Liapunov. *ASME J. Basic Eng.* **82**, 371–400.

LAUER, H., LESNICK, R. N., and MATSON, L. E. (1960). *Servomechanism fundamentals* (2nd edn). McGraw-Hill.

PARZEN, E. (1960). *Modern probability theory and its applications.* Wiley.

TAYLOR, P. L. (1960.) *Servomechanisms.* Longmans.

THALER, G. J. and BROWN, R. G. (1960). *Analysis and design of feedback control systems.* McGraw-Hill.

WEST, J. C. (1960). *Analytical techniques for nonlinear control systems.* Van Nostrand.

WILTS, C. H. (1960). *Principles of feedback control.* Addison Wesley.

1961

BELLMAN, R. E. (1961). *Adaptive control processes.* Princeton University Press.

BOGOLIUBOV, N. N. and MITROPOLSKY, Y. A. (1961). *Asymptotic methods in the theory of nonlinear oscillations.* Gordon and Breach.

BROWN, B. M. (1961). *The mathematical theory of linear systems.* Chapman and Hall (second edition, 1965).

CHANG, S. S. L. (1961). *Synthesis of optimum control systems.* McGraw-Hill.

FORRESTER, J. W. (1961). *Industrial dynamics.* MIT Press.

GOLDBERG, S. (1961). *Introduction to difference equations.* Wiley.

GRAHAM, D. and MCRUER, D. (1961). *Analysis of nonlinear control systems.* Wiley.

GNEDENKO, B. V. and KHINTCHINE, A. Y. (1961). *An elementary introduction to probability theory.* Freeman.

LA SALLE, J. and LEFSCHETZ, S. (1961). *Stability by Liapunov's direct method.* Academic Press.

RAVEN, F. H. (1961). *Automatic control engineering.* McGRAW-Hill (2nd edn, 1968).

1962

BELLMAN, R. E. and DREYFUS, S. E. (1962). *Applied dynamic programming.* Princeton University Press.

CHOKSY, N. H. (1962). Time lag systems. In *Progress in control engineering,* vol. 1 (ed. R. H. Macmillan *et al.*). Heywood.

CLARK, R. N. (1962). *Introduction to automatic control systems.* Wiley.

KU, Y. H. (1962). *Analysis and control of linear systems.* International Textbook Co.

MONROE, A. J. (1962). *Digital processes for sampled data systems.* Wiley.

PARZEN, E. (1962). *Stochastic processes.* Holden Day.

PONTRIAGIN, L. S., BOLTYANSKII, V. G., GAMKRELIDZE, R. V., and MISHCHENKO, E. F. (1962). *The mathematical theory of optimal processes.* Interscience (Wiley).

THALER, G. J. and PASTEL, M. V. (1962). *Analysis and design of nonlinear feedback control systems.* McGraw-Hill.

1963

BARBE, E. C. (1963). *Linear control systems.* International Textbook Co.

DOUCE, J. L. (1963). *Introduction to the mathematics of servomechanisms.* English Universities Press.

GIBSON, J. E. (1963). *Nonlinear automatic control.* McGraw-Hill.

HOROWITZ, I. M. (1963). *Synthesis of feedback systems.* Academic Press.

KALMAN, R. E. (1963a). Mathematical description of linear dynamical systems. *Journal SIAM on Control*, **1**, No. 2, 152–192.

—— (1963b). New methods in Wiener filtering theory. In *Proceedings of first symposium on engineering applications of random function theory and probability theory* (ed. F. Kozin and J. L. Bogdanoff). Wiley.

KUO, B. C. (1963). *Analysis and synthesis of sampled-data control systems.* Prentice Hall.

TOU, J. T. (1963). *Optimum design of digital control systems.* Academic Press.

ZADEH, L. A. and DESOER, C. A. (1963). *Linear system theory: the state space approach.* McGraw-Hill.

1964

ARIS, R. (1964). *Discrete dynamic programming.* Blaisdell.

HAYASHI, C. (1964). *Nonlinear oscillations in physical systems.* McGraw-Hill.

LEE, R. C. K. (1964). *Optimal estimation, identification and control.* MIT Press.

LUENBERGER, D. G. (1964). Observing the state of a linear system. *IEEE Trans. on military electronics*, **MIL-8** 74–80.

MERRIAM, C. W. (1964). *Optimization theory and the design of feedback control systems.* McGraw-Hill.

ROBERTS, S. M. (1964). *Dynamic programming in chemical engineering and process control.* Academic Press.

SHINNERS, S. M. (1964). *Control system design.* Wiley.

TOU, J. T. (1964). *Modern control theory.* McGraw-Hill.

WEBB, C. R. (1964). *Automatic control.* McGraw-Hill.

1965

ÅSTRÖM, K. J., BOHLIN, T., and WENSMARK, S. (1965). Automatic construction of linear stochastic dynamic models for stationary industrial processes with random disturbance using operating records. *Tech. Paper* TP18.150, *IBM Nordic Lab.*

BELLMAN, R. E. and KALABA, R. (1965). *Dynamic programming and modern control theory.* Academic Press.

COX, D. R. and MILLER, H. D. (1965). *The theory of stochastic processes.* Methuen.

DE RUSSO, P. M., ROY, R. J., and CLOSE, C. M. (1965). *State variables for engineers.* Wiley.

DEUTSCH, R. (1965). *Estimation theory.* Prentice Hall.

DORF, R. C. (1965). *Time-domain analysis and design of control systems.* Addison-Wesley.

DREYFUS, S. E. (1965). *Dynamic programming and the calculus of variations.* Academic Press.

FEL'DBAUM, A. A. (1965). *Optimal control systems.* Academic Press.

FREEMAN, H. (1965). *Discrete-time systems.* Wiley.

JOHNSON, C. D. (1965). Singular solutions in problems of optimal control. In *Advances in control systems.* Vol. 2.

LANGILL, A. W. (1965). *Automatic control systems engineering*, vols 1 and 2. Prentice Hall.

LINDORF, D. P. (1965). *Theory of sampled-data control systems.* Wiley.

NASLIN, P. (1965). *The dynamics of linear and non-linear systems.* Gordon and Breach.

PAPOULIS, A. (1965). *Probability, random variables, and stochastic processes,* McGraw-Hill.

SCHULTZ, D. G. (1965). The generation of Liapunov functions. In *Advances in control systems,* vol. 2.

SCHWARZ, R. J. and FRIEDLAND, B. (1965). *Linear systems.* McGraw-Hill.

STERN, T. E. (1965). *Theory of nonlinear networks and systems.* Addison-Wesley.

1966

ATHANS, M. and FALB, P. L. (1966). *Optimal control.* McGraw-Hill.

BAYLISS, L. E. (1966). *Living control systems.* English Universities Press.

D'AZZO, J. J. and HOUPIS, C. H. (1966). *Feedback control system analysis and synthesis.* McGraw-Hill.

GUPTA, S. C. (1966). *Transform and state variable methods in linear systems.* Wiley.

JOHNSON, C. D. and WONHAM, W. M. (1966). Another note on the transformation to canonical (phase variable) form. *IEEE Transactions on Automatic Control,* **11**, 609–610.

KALMAN, R. E. (1966). On structural properties of linear constant multivariable systems, *Proceedings of Third IFAC Congress, London, 1966,* Inst. Mech. Engrs.

MILHORN, H. T. (1966). *The application of control theory to physiological systems.* Saunders Press.

MILSUM, J. H. (1966). *Biological control systems analysis.* McGraw-Hill.

NEMHAUSER, G. L. (1966). *Introduction to dynamic programming.* Wiley.

SORENSON, H. W. (1966). Kalman filtering techniques. In *Advances in control systems,* vol. 3.

1967

AOKI, M. (1967). *Optimization of stochastic systems.* Academic Press.

BARRIERE, PALLU DE LA (1967). *Optimal control theory.* Saunders.

BROMBERG, P. V. (1967). *Matrix analysis of discontinuous control systems.* Elsevier, 1969. Russian original 1967.

DISTEFANO, J. J., STUBBERUD, A. R., and WILLIAMS, I. J. (1967). *Feedback and control systems.* Schaum, McGraw-Hill.

DORF, R. C. (1967). *Modern control systems.* Addison-Wesley.

ELGERD, O. I. (1967). *Control systems theory.* McGraw-Hill.

EVELEIGH, V. W. (1967). *Adaptive control and optimization techniques.* McGraw-Hill.

JACOBS, O. L. R. (1967). *Introduction to dynamic programming.* Chapman and Hall.

KUO, B. C. (1967). *Automatic control systems* (2nd edn). Prentice Hall.

LAPIDUS, L. and LUUS, R. (1967). *Optimal control of engineering processes.* Blaisdell.

LEE, E. B. and MARKUS, L. (1967). *Foundations of optimal control theory.* Wiley.

LIEBELT, P. B. (1967). *An introduction to optimal estimation.* Addison-Wesley.

OGATA, K. (1967). *State space analysis of control systems*. Prentice Hall.

SCHULTZ, D. G. and MELSA, J. L. (1967). *State functions and linear control systems*. McGraw-Hill.

VERNON, J. B. (1967). *Linear vibrations and control system theory*. Wiley.

WONHAM, W. M. (1967). On pole assignment in multi-input controllable linear systems, *IEEE trans on Automatic Control*, **12**, 660–665.

1968

ALLEN. R. G. D. (1968). *Macroeconomic theory*. Macmillan.

BUCY, R. S. and JOSEPH, P. D. (1968). *Filtering for stochastic processes with applications to guidance*. Interscience.

CHEN, C. F. and HAAS, I. J. (1968). *Elements of control systems analysis*. Prentice Hall.

GELB, A. and VANDER VELDE, W. E. (1968). *Multiple-input describing functions and nonlinear system design*. McGraw-Hill.

HSU, J. S. and MEYER, A. U. (1968). *Modern control principles and applications*. McGraw-Hill.

DRANSFIELD, P. (1968). *Engineering systems and automatic control*. Prentice Hall.

FLÜGGE-LOTZ, I. (1968). *Discontinuous and optimal control*. McGraw-Hill.

KOPPEL, L. B. (1968). *Introduction to control theory with applications to process control*. Prentice Hall.

LARSON, R. E. (1968). *State increment dynamic programming*. Elsevier.

PLANT, J. B. (1968). *Some iterative solutions in optimal control*. MIT.

SAGE, A. P. (1968). *Optimum systems control*. Prentice Hall.

SAUCEDO, R. and SCHIRING, E. E. (1968). *Introduction to continuous and digital control systems*. Macmillan.

SENSICLE, A. (1968). *Introduction to control theory for engineers*. Blackie.

STARK, L. (1968). *Neurological control systems*. Plenum Press.

TIMOTHY, L. K. and BONA, B. E. (1968). *State space analysis*. McGraw-Hill.

WONHAM, W. M. (1968). On a matrix Riccati equation of stochastic control. *SIAM J. Control.* **6**, 681–697.

1969

BRONSON, R. (1969). *Matrix methods*. Academic Press.

BRYSON, A. E. and HO, Y. (1969). *Applied optimal control*. Ginn.

CITRON, S. J. (1969). *Elements of optimal control*. Holt, Rinehart and Winston.

FORRESTER, J. W. (1969). *Urban dynamics*. MIT Press.

HARRISON, H. L. and BOLLINGER, J. G. (1969). *Introduction to automatic controls* (2nd edn). International Textbook Co.

JACOBS, O. L. R. (1969). Introduction to random processes. Chap. 5 of *Modern control theory and computing* (ed. D. Bell and A. Griffin). McGraw-Hill.

KALMAN, R. E., FALB, P. L., and ARBIB, M. A. (1969). *Topics in mathematical system theory*. McGraw-Hill.

MCCAUSLAND, I. (1969). *Introduction to optimal control*. Wiley.

MEDITCH, J. S. (1969). *Stochastic optimal linear estimation and control*. McGraw-Hill.

MELSA, J. L. and SCHULTZ, D. G. (1969). *Linear control systems*. McGraw-Hill.

NAHI, N. E. (1969). *Estimation theory and applications.* Wiley.

NASLIN, P. (1969). *Essentials of optimal control.* Boston Technical Publishers.

PRIME, H. A. (1969). *Modern concepts in control theory.* McGraw-Hill.

TAYLOR, P. L. (1969). *Servomechanisms* (2nd edn). Longman.

VIDAL, P. (1969). *Non-linear sampled-data systems.* Gordon and Breach.

WATKINS, B. O. (1969). *Introduction to control systems.* Macmillan.

1970

ÅSTRÖM, K. J. (1970). *Introduction to stochastic control theory.* Academic Press.

BARNETT, S. and STOREY, C. (1970). *Matrix methods in stability theory.* Nelson.

BOX, G. E. P. and JENKINS, G. M. (1970). *Time series analysis.* Holden Day.

BREIPOHL, A. M. (1970). *Probailistic system analysis.* Wiley.

BROCKETT, R. W. (1970). *Finite dimensional linear systems.* Wiley.

CADZOW, J. A. and MARTENS, H. R. (1970). *Discrete-time and computer control systems.* Prentice Hall.

CHEN, C. T. (1970). *Introduction to linear system theory.* Holt, Rinehart and Winston.

D'ANGELO, H. (1970). *Linear time-varying systems.* Allyn and Bacon.

DAVENPORT, W. (1970). *Probability and random processes.* McGraw-Hill.

DESOER, C. A. (1970). *Notes for a second course on linear systems.* Van Nostrand-Reinhold.

GUPTA, S. C. and HASDORFF, L. (1970). *Fundamentals of automatic control.* Wiley.

HEALEY, M. (1970). *Principles of automatic control* (2nd edn). English Universities Press.

JACOBSON, D. H. and MAYNE, D. Q. (1970). *Differential dynamic programming.* Elsevier.

JAZWINSKI, A. H. (1970). *Stochastic processes and filtering theory.* Academic Press.

KIRK, D. E. (1970). *Optimal control theory.* Prentice Hall.

KUO, B. C. (1970). *Discrete-data control systems.* Prentice Hall.

MAYR, O. (1970). *The origins of feedback control.* MIT Press.

OGATA, K. (1970). *Modern control engineering.* Prentice Hall.

ROSENBROCK, H. H. (1970). *State-space and multivariable theory.* Nelson.

SPEEDY, C. B., BROWN, R. F., and GOODWIN, G. C. (1970). *Control theory.* Oliver and Boyd.

TAKAHASHI, Y., RABINS, M. J., and AUSLANDER, D. M. (1970). *Control and dynamic systems.* Addison-Wesley.

TOWILL, D. R. (1970). *Transfer function techniques for control engineers.* Iliffe.

WILLEMS, J. L. (1970). *Stability theory of dynamical systems.* Nelson.

1971

ANDERSON, B. D. O. and MOORE, J. B. (1971). *Linear optimal control.* Prentice Hall.

ÅSTRÖM, K. J. and EYKHOFF, P. (1971). System identification—a survey. *Automatica*, **7**, 123–162.

BARNETT, S. (1971). *Matrices in control theory.* Van Nostrand-Reinhold.

BOUDAREL, R., DELMOS, J., and GUICHET, P. (1971). *Dynamic programming and its application to optimal control.* Academic Press.

FITZGERALD, R. J. (1971). Divergence of the Kalman filter. *IEEE Trans. on Automatic Control*, **AC–16**, No. 6, 736–47, Dec.

FORRESTER, J. W (1971). *World dynamics.* Wright Allen Press.

McFARLAND, D. J. (1971). *Feedback mechanisms in animal behaviour.* Academic Press.

KUSHNER, H. (1971). *Introduction to stochastic control.* Holt, Rinehart and Winston.

PATCHELL, J. W. (1971). *On the structure of stochastic control laws.* D. Phil. dissertation, Oxford.

PATCHELL, J. W. and JACOBS, O. L. R. (1971). Separability, neutrality, and certainty equivalence. *International J. of Control*, **13**, No. 2, 337–342.

RUBIO, J. E. (1971). *The theory of linear systems.* Academic Press.

SAGE, A. P. and MELSA, J. L. (1971a) *Estimation theory.* McGraw-Hill.

—— (1971b) *System identification.* Academic Press.

1972

AGGARWAL, J. K. (1972). *Notes on nonlinear systems.* Van Nostrand-Reinhold.

BHATTACHARYYA, S. P. and PEARSON, J. B. (1972). On error systems and the servomechanism problem, *Int. J. Control*, **15**, 1041–1062.

EVELEIGH, V. W. (1972). *Introduction to control systems design.* McGraw-Hill.

GRAUPE, D. (1972). *Identification of systems.* Van Nostrand-Reinhold.

JACOBS, O. L. R. and PATCHELL, J. W. (1972). Caution and probing in stochastic control, *International J. Control*, **16**, No. 1, 189–199.

KREYSIG, E. (1972). *Advanced engineering mathematics* (3rd edn). Wiley.

KWAKERNAAK, H. and SIVAN, R. (1972). *Linear optimal control systems.* Wiley-Interscience.

MAYNE, D. Q. (1972). A canonical model for identification of multivariable linear systems. *IEEE Trans. on Automatic Control*, **17**, 728–729, Oct.

NOTON, M. (1972). *Modern control engineering.* Pergamon.

SHINNERS, S. M. (1972). *Modern control system theory and application.* Addison-Wesley.

SMYTH, M. P. (1972). *Linear engineering systems.* Pergamon.

1973

IFAC/IFORS (1973). *International conference on dynamic modelling and control of national economies, Warwick.* IEE conference publication no. 101.

MACFARLANE, A. G. J. (1973). Trends in linear multivariable theory, *Automatica*, **9**, No. 2, 273–277, March.

MEDITCH, J. S. (1973) A survey of data smoothing for linear and nonlinear dynamic systems, *Automatica*, **9** No. 2, 151–162, March.

MELSA, J. L. and SAGE, A. P. (1973). *An introduction to probability and stochastic processes.* Prentice Hall.

1974

HUGHES, D. J. and JACOBS, O. L. R. (1974). Turn-off, escape, and probing in nonlinear stochastic control. *IFAC symposium on stochastic control, Budapest.*

Index

adjoint system, 156
  variables, 156, 176
analytic design, 90, 103 ff., 265, 268
Argand diagram, 5, 9
autocorrelation function,
  continuous-time, 249
  discrete-time, 246 ff.
    at output of continuous-time linear
      system, 257–8
    at output of discrete-time linear system,
      248

bandwidth, 98
bang-bang control, 180 ff.
Bayes's rule, vii, 239, 275, 297 ff., 299, 310,
  314
Bernoulli distribution, 239
bias of estimates, 298
bilinear transformation, 54–5
binomial distribution, 239
block diagram, 20
  of sampled-data system, 24
Bode diagram, 9, 64
  used to choose gain, 64–5
  stability criterion, 64
Bode's theorem, 25
Brownian motion, 250

calculus of variations, 171 ff., 185
canonical form, 324–5
  for single-variable systems, 131, 325
Cauchy distribution, 262
caution, 335, 337 ff.
central limit theorem, 242, 246, 249
certainty-equivalence, 275, 283, 308, 335, 336
certainty-equivalent control law, 281, 288,
  337
characteristic equation, 5, 72
  roots of, 5, 47, 70, 120, 204
chattering, 196
companion matrix, 119, 131, 324–5
complementary function, 4, 36, 48 ff., 70,
  120, 122, 159
conditional mean, 278 ff., 301
conformal mapping, 59–60

control effort, 105, 114
control theory, v
  courses in, vii–viii
  history of, vi
  optimal, 137 ff., 169 ff.
  stochastic, 265 ff.
controllability, 123 ff., 146, 310
  matrix, 129
controlled process, 35
controlled variable, 35
controller, 35
  calculation of gains and time constants,
    93 ff.
  dual functions of stochastic, 276, 297,
    334 ff.
  transfer functions approximated by elec-
    trical circuits, 95–6
controlling variable, 35
convergence of estimates, 298
cost of control, 114, 147, 270, 287
Coulomb friction, 199
covariance matrix, 243

damping ratio, 7, 49, 71, 121, 189–90, 212–13
  optimal value, 151
  relationship with phase margin, 62–3
describing function, 210, 217, 218 ff. 220–21
  dual input, 222
  of backlash, 226
  of deal zone, 226
  of ideal relay, 226
  of relay with dead zone, 226
  of relay with hysteresis, 226
    and dead zone, 226
  of saturation, 226
  of sum of non-linearities, 225
  relay control system, 218–20
desired value, 35, 268
  known in advance, 154
  in pre-specified class of functions, 158
desired-value functions, 36, 159
detectability, 161
disturbances, 42, 162, 268
dual functions of stochastic controller,
  276, 297, 334 ff.